Artificial Intelligence Applications in Battery Management Systems and Routing Problems in Electric Vehicles

S. Angalaeswari
Vellore Institute of Technology, India

T. Deepa
Vellore Institute of Technology, India

L. Ashok Kumar
PSG College of Technology, India

IGI Global
PUBLISHER of TIMELY KNOWLEDGE

A volume in the Advances in Computer and Electrical Engineering (ACEE) Book Series

Published in the United States of America by
IGI Global
Engineering Science Reference (an imprint of IGI Global)
701 E. Chocolate Avenue
Hershey PA, USA 17033
Tel: 717-533-8845
Fax: 717-533-8661
E-mail: cust@igi-global.com
Web site: http://www.igi-global.com

Library of Congress Cataloging-in-Publication Data

Names: Angalaeswari, S., 1981- editor. | Deepa, T., 1978- editor. | Kumar, L. Ashok, editor.
Title: Artificial intelligence applications in battery management systems and routing problems in electric vehicles / editors, S. Angalaeswari, T. Deepa, and L. Ashok Kumar.
Description: Hershey, PA : Business Science Reference, [2023] | Includes bibliographical references and index. | Summary: "The Handbook of Research on Battery Management Systems and Routing Problems in Electric Vehicles focuses on the integration of renewable energy sources with the existing grid, introduces a power exchange scenario in the prevailing power market, considers the use of the electric vehicle market for creating cleaner and transformative energy, and optimizes the control variables with artificial intelligence techniques. Covering key topics such as artificial intelligence, smart grids, and sustainable development, this major reference work is ideal for government officials, industry professionals, policymakers, researchers, scholars, practitioners, academicians, instructors, and students"-- Provided by publisher.
Identifiers: LCCN 2022043215 (print) | LCCN 2022043216 (ebook) | ISBN 9781668466315 (hardcover) | ISBN 9781668466322 (paperback) | ISBN 9781668466339 (ebook)
Subjects: LCSH: Electric vehicles--Batteries. | Battery management systems. | Lithium ion batteries--Mathematical models.
Classification: LCC TL220 .A776 2023 (print) | LCC TL220 (ebook) | DDC 629.22/93--dc23/eng/20221004
LC record available at https://lccn.loc.gov/2022043215
LC ebook record available at https://lccn.loc.gov/2022043216

This book is published in the IGI Global book series Advances in Computer and Electrical Engineering (ACEE) (ISSN: 2327-039X; eISSN: 2327-0403)

British Cataloguing in Publication Data
A Cataloguing in Publication record for this book is available from the British Library.

Advances in Computer and Electrical Engineering (ACEE) Book Series

Srikanta Patnaik
SOA University, India

ISSN:2327-039X
EISSN:2327-0403

Mission

The fields of computer engineering and electrical engineering encompass a broad range of interdisciplinary topics allowing for expansive research developments across multiple fields. Research in these areas continues to develop and become increasingly important as computer and electrical systems have become an integral part of everyday life.

The **Advances in Computer and Electrical Engineering (ACEE) Book Series** aims to publish research on diverse topics pertaining to computer engineering and electrical engineering. **ACEE** encourages scholarly discourse on the latest applications, tools, and methodologies being implemented in the field for the design and development of computer and electrical systems.

Coverage

- Microprocessor Design
- Chip Design
- VLSI Design
- Analog Electronics
- Digital Electronics
- Qualitative Methods
- Sensor Technologies
- Optical Electronics
- Programming
- VLSI Fabrication

IGI Global is currently accepting manuscripts for publication within this series. To submit a proposal for a volume in this series, please contact our Acquisition Editors at Acquisitions@igi-global.com or visit: http://www.igi-global.com/publish/.

The Advances in Computer and Electrical Engineering (ACEE) Book Series (ISSN 2327-039X) is published by IGI Global, 701 E. Chocolate Avenue, Hershey, PA 17033-1240, USA, www.igi-global.com. This series is composed of titles available for purchase individually; each title is edited to be contextually exclusive from any other title within the series. For pricing and ordering information please visit http://www.igi-global.com/book-series/advances-computer-electrical-engineering/73675. Postmaster: Send all address changes to above address.

Titles in this Series

For a list of additional titles in this series, please visit: www.igi-global.com/book-series/advances-computer-electrical-engineering/73675

Structural and Functional Aspects of Biocomputing Systems for Data Processing
U. Vignesh (Manipal Institute of Technology, Manipal, Karnataka, India) R. Parvathi (Vellore Institute of Technology, India) and Ricardo Goncalves (Department of Electrical and Computer Engineering (DEEC), NOVA School of Science and Technology, University in Almada, Portugal)
Engineering Science Reference • © 2023 • 315pp • H/C (ISBN: 9781668465233) • US $270.00

Energy Systems Design for Low-Power Computing
Rathishchandra Ramachandra Gatti (Sahyadri College of Engineering and Management, India) Chandra Singh (Sahyadri College of Engineering and Management, India) P. Srividya (RV College of Engineering, India) and Sandeep Bhat (Sahyadri College of Engineering and Management, India)
Engineering Science Reference • © 2023 • 310pp • H/C (ISBN: 9781668449745) • US $270.00

5G Internet of Things and Changing Standards for Computing and Electronic Systems
Augustine O. Nwajana (University of Greenwich, UK)
Engineering Science Reference • © 2022 • 308pp • H/C (ISBN: 9781668438558) • US $250.00

Theory and Applications of NeutroAlgebras as Generalizations of Classical Algebras
Florentin Smarandache (University of New Mexico, USA) and Madeline Al-Tahan (Lebanese International University, Lebanon)
Engineering Science Reference • © 2022 • 333pp • H/C (ISBN: 9781668434956) • US $270.00

Antenna Design for Narrowband IoT Design, Analysis, and Applications
Balachandra Pattanaik (Wollega University, Ethiopia) M. Saravanan (Vel Tech Rangarajan Dr. Sagunthala R&D Institute of Science and Technology, India) U. Saravanakumar (Muthayammal Engineering College, India) and Ganesh Babu T R (Muthayammal Engineering College, India)
Engineering Science Reference • © 2022 • 261pp • H/C (ISBN: 9781799893158) • US $250.00

Handbook of Research on Advances and Applications of Fuzzy Sets and Logic
Said Broumi (Laboratory of Information Processing, Faculty of Science Ben M'Sik, University Hassan II, Casablanca, Morocco & Regional Center for the Professions of Education and Training (CRMEF), Casablanca-Settat, Morocco)
Engineering Science Reference • © 2022 • 944pp • H/C (ISBN: 9781799879794) • US $435.00

IGI Global
PUBLISHER of TIMELY KNOWLEDGE

701 East Chocolate Avenue, Hershey, PA 17033, USA
Tel: 717-533-8845 x100 • Fax: 717-533-8661
E-Mail: cust@igi-global.com • www.igi-global.com

Editorial Advisory Board

Table of Contents

Detailed Table of Contents

Section 2
Electric Vehicles: Modelling and Control

Preface

Electric vehicles have started to occupy the majority share in the public roads. When compared with the IC engine counterparts they are more efficient, zero emission and convenient. Even with the better advantages, they suffer from glitches like battery management and charging facilities. Vehicle commutator must know about remaining charge in the battery and nearest charging station to plan the journey. This is possible only by the proper battery management system (BMS). BMS helps to retrieve the maximum power from battery.

Battery Management System (BMS) monitors performance of the battery packs present in the electric vehicle. BMS evaluates the state of charge (SoC) of the battery. It estimates SoC by measuring different parameters like current, voltage and temperature of the battery. It predicts the battery level, which the critical parameter in deciding remaining charge state, especially in electric vehicles. It helps in maximizing the efficiency of the battery and also improves the charging/discharging performance. BMS also estimate state of health and depth of discharge by measuring various parameters of the battery. It is integrated with vehicle management system. BMS system consists of both hardware and software components. BMS hardware components include automobile grade sensors and controllers. BMS performs additional functionalities like protection, diagnosis and data management.

Electric Vehicle Routing Problem is an optimization problem that deals with the framing of effective route plans for electric vehicle while fulfilling a set of batter power related constraints. This problem aims to find the optimized schedule for electric vehicles (EV) while considering the set of operational constraints like geography of charging stations, vicinity of charging station, number of charging points available in particular station, charging speed and time, power drawn from grid, and vehicle's battery capacity and state of charge. Routing Problem in electric vehicles can be considered as a special case of traditional traveling salesman problem with different operational constraints. Operational constraints include recharging policy, charging technology used, and availability of battery swapping stations. EV's characteristic parameters like travel distance, speed, vehicle tonnage, and gradient of road are also considered for optimized route plan. Routing problem is well utilized in EV based logistic delivery, public transportation system and EV based shuttle services.

Section 1: Battery Management System

Lithium-ion batteries are popular due to their low weight, high energy density, and long lifespan. These batteries must be safe, reliable, and long-lasting from consumer electronics to transportation particularly in industries such as transportation that require durability and safety. The design, installation, and management of battery cells (BMS) have a significant impact on battery safety and reliability by protecting the pack against damaging events such as over-charges and under-charges. They can also identify potentially dangerous conditions within the battery pack, which are vital in ensuring the reliability and longevity of systems utilizing lithium-ion batteries.

A Battery Energy Storage System (BESS) is generally used for storing the excess power produced from the solar PV panel and it supplies the stored electrical energy whenever it is needed. The accurate estimation of the State of Charge (SOC) of the battery is considered to be very essential for proper power management and reliable operation of the battery. Probabilistic Neural Network (PNN) approach is proposed for BESS based on SOC monitoring system. PNN controller is used for many DC applications. The performance of the proposed battery monitoring model is evaluated by using MATLAB simulation.

Chapter 3 focuses on how the voltage difference discomposes the battery pack's performance by using an optimized control strategy. It helps in battery operated vehicles to charge and work efficiently. The control strategy involves the proper switching of the resistors connected to the charging circuit. The thermal balance of the resistors is balanced using the thermal power dissipation modelling in simulation. The models are tested under variable cell voltage conditions in MATLAB Simulink. The technique for voltage error balancing is termed passive cell balancing in BMS for the arrangement of a multiplicity of cells in stacks. The major issue of the passive cell balancing arises due to the State of Charge (SoC) imbalance in the batteries. The charging conditions of the battery are taken into the parameter estimation for the control technique of the balancing technique.

Chapter 4 as per the author proposes Harris Hawk Optimized Algorithm-HHOA Based on Fuzzy Logic Controller with battery management system in electric vehicles. Extended Kalman Filter (EKF) is used to monitor and control BMS operation.

Section 2: Electric Vehicles – Modelling and Control

To reducing greenhouse toxic gases. Electric vehicles, including Battery vehicles, Hybrid vehicles, Plug-In Hybrid vehicles, Fuel Cell Electrical Vehicles are the most commonly used transports recently. Main important systems in Electric vehicles battery technologies and charging methods. More specific analysis of the worldwide market situation of EVs moving forwards to the future of transport. EV development impacts among significant of environment, power system, and other related sectors.

Chapter 6 presents e-Bike model developed in Simulink/Simscape. The performance of the vehicle at different conditions is observed and studied through simulation. The variations in parameters like speed of vehicle, current drawn by motor, modulated current and voltage waveforms at different conditions of e-Bike is discussed in this work. Enhancement of suspension performance of the vehicle model using sliding mode controllers Performance is presented in Chapter 7 and 8.

Section 3: Analytics in Electric Vehicles

intelligent demand and supply of electrical vehicles, data analytics in electric vehicle and autonomous electric vehicle and edge computing are discussed in Section 3. This intelligent system involves identifying the predicting trends in the market analysis based on the expenses and revenues patterns of a particular group of people in a city or location. Beside this the system needs to offer the insight into the direction for decision making of the market on demand of vehicle as per the supply with respect to the type of roads, season, and gender and age group patterns. Apart from this the tax rules makes a huge difference in this context. In some zones many governments encouraging by giving lots of subsidies or give tax relaxations to certain cc vehicles.

Data analytics landscape assessment addresses the incorporation of electric vehicles with green, smart urban environments. It serves as a blueprint for the objectives and alternatives for the incorporation of electric vehicles into smart future cities. Chapter 11 discusses a software that can be installed on a computer to control how an autonomous controller operates in a vehicle. It is a stand-alone device that can be configured to function with a broad range of distant sensors and control mechanisms. As a direct consequence of the rapid spread of new technologies, edge computing has developed into a more widespread practice. Users may be able to have a more individualized and interactive experience in their day-to-day lives thanks to edge computing and edge devices.

Section 4: Latest Trends in Electric Vehicles

Section 4 proposes the latest trends in electric vehicles include vehicle-to-grid and electric vehicle integrated demand response management and hybrid optimization based on genetic routing and scheduling scheme with secure transmission in wireless sensor networks.

The safe and efficient functioning of micro grids (MGs) has always depended on the best scheduling of MGs with renewable uncertainty. Electric vehicles (EVs) can back-feed power to the electrical grid to provide a variety of ancillary services, including demand response (DR), peak-load management, voltage support, and frequency regulation. This technology also enables EV holders to reduce their energy costs and make money through price arbitrage. Vehicle-to-grid (V2G) technology is developing quickly, and research is being done to improve its usability and use. The benefits of V2G include ancillary services, active power support, reactive power compensation, and promoting the use of renewable energy sources. In contrast, the challenges of V2G include battery degradation, effects on distribution equipment, elevated investment cost, and social obstacles.

Node mobility and congestion generate frequent connection failures and packet losses, affecting the QoS performance of the protocol. We used a Hybrid Optimization System (HOS) in this research to improve wireless network routing and transmission. To maintain network lifetime and consistent communication, wireless sensor networks need an energy-efficient protocol. This may be accomplished by using a high-quality routing approach.

S. Angalaeswari
Vellore Institute of Technology, India

T. Deepa
Vellore Institute of Technology, India

L. Ashok Kumar
PSG College of Technology, India

Acknowledgment

We are always thankful to the Almighty for their perseverance and achievements. We would like to express our sincere thanks to Dr. G. Viswanathan, Chancellor, Vellore Institute of Technology for providing us this opportunity. We would also like to thank the Vice Presidents of Vellore Institute of Technology for their continuous support and encouragement to complete this book successfully. The editors owe their gratitude to Shri L. Gopalakrishnan, Managing Trustee, PSG Institutions, and PSG Management for their wholehearted cooperation and great encouragement in this successful endeavor.

Dr. S. Angalaeswari would like to thank her husband Mr. R. Sendraya Perumal for his motivation and support during this book compilation. She is thankful to her son Mr. S. Dhianesh and daughter Ms. S. Rithika Varsha for their constant encouragement and moral support along with patience and understanding. She would like to extend her gratitude to her parents, in-laws for their persistent support throughout the path of success.

Dr. T. Deepa would like to express her special thanks to her son, her husband, her parents, in laws and her friends for their motivation during this book writing. She would like to dedicate this work to her lovable son.

Dr. L. Ashok Kumar would like to take this opportunity to acknowledge those people who helped me in completing this book. I am thankful to all my research scholars and students who are doing their project and research work with me. But the writing of this book is possible mainly, because of the support of my family members, parents, and sisters. Most importantly, I am very grateful to my wife, Y. Uma Maheswari, for her constant support during writing. Without her, all these things would not be possible. I would like to express my special gratitude to my daughter, A.K. Sangamithra, for her smiling face and support. I would like to dedicate this work to her.

We like to acknowledge the support rendered by our colleagues and friends in several ways throughout this book compilation. We would like to thank Mr. Selvasundar K, Ms. Maladhi D., Ms. Kripalakshmi T., Mr. Marabathina Maheedhar, and Ms. Senthamilselvi V. for helping us in several ways to complete this book.

We would like to extend our gratitude to each of the authors who have contributed their chapters in this book, without them, this would not be possible. We acknowledge and appreciate the effort and time dedication by all the authors to complete their book submission on time with quality and efficient work.

We like to recognize and thank the reviewers of this book who have done wonderful job by reviewing each chapter carefully and gave their comments for the enhancement of the chapters.

We finally thank the editorial board members in bringing the book in most successful manner in this span of time. The editors would like to extend our heartfelt thanks to all the editorial staff of IGI global team for their immediate response and assistance throughout this journey.

Section 1

Battery Management Systems

Chapter 1
Importance of Battery Pack Design and Battery Management Systems in Electric Vehicles

Hemavathi S.
https://orcid.org/0000-0002-7672-9530
CSIR-Central Electrochemical Research Institute, Chennai, India & Academy of Scientific and Innovative Research (AcSIR), Ghaziabad, India

Srirama Srinivas
Indian Institute of Technology, Madras, India

A. S. Prakash
Central Electrochemical Research Institute, India

ABSTRACT

From consumer electronics to transportation, lithium-ion batteries are popular due to their low weight, high energy density, and long lifespan. These batteries must be safe, reliable, and long-lasting, particularly in industries such as transportation that require durability and safety. The design, installation, and management of battery cells (BMS) have a significant impact on battery safety and reliability by protecting the pack against damaging events such as over-charges and under-charges. They can also identify potentially dangerous conditions within the battery pack, which are vital in ensuring the reliability and longevity of systems utilizing lithium-ion batteries. This Lecture presents information on the role of battery pack design and battery management systems in Electric vehicles and discusses the methodologies, features, and functions of the battery pack with BMS in protecting and preserving the lifespan of lithium-ion batteries.

DOI: 10.4018/978-1-6684-6631-5.ch001

1. INTRODUCTION

In this improving technology day by day with the smart grid and energy storage applications, the battery (Nitta et al., 2015) has become the most prominent energy storage device that is attracting attention. There is huge importance in developing higher power and energy density cells, which is as equal to improving the Battery Management Systems (BMS) (Lelie et al., 2018), (Andrea, 2010) performance for making the battery a safe, reliable, and cost-efficient solution. For several years, nickel-cadmium was the only battery that could be used in lightweight devices such as wireless networks and digital computing. and in the mid-nineties, nickel-metal-hydride and lithium-ion developed, challenging to gain consumer acceptance. Battery technology today is dominated by lithium-ion batteries, which are the fastest-growing and most exciting. Modern vehicles and applications for energy storage (Becker et al., 2012; Richard, 2018) are extremely popular because lithium-ion batteries are light in weight, have a high energy density, have a long lifespan, have large capacities, are quick to charge, and have low self-discharge levels. Continuous monitoring and control are essential to ensuring the safety and reliability of these batteries. BMS has many core functions which estimate the battery behavior so protecting it from damages is very challenging (Chen et al., 2020). BMS monitors the temperature, voltage, and current, detects any failure, and prevents that, collects the data through a communication channel for further parameter analysis (Hoque et al., 2017; Kumar, 2020).

It is only applicable to gasoline-powered vehicles or those powered by internal combustion engines (ICEs) that a BMS indicator estimates the driving range. Batteries and chargers are connected by an electronic component known as a battery management system (BMS) (Zhia et al., 2017). In an EV based on batteries, the BMS can be seen as a brain or as one of the main components. BMS provides energy management and safety protection, which are its two most important functions. Battery management systems may do this to control output, guide charging and discharging, and alert users when battery packs are low on charge or discharge (Wang et al., 2018). Furthermore, they prevent battery damage by providing vital precautions. An energy management system (EMS) tracks the battery's condition and performs evaluations such as State of Charge (SOC) and State of Health (SOH). Using model-based estimation methods, it is necessary to accurately predict battery characteristics, such as voltage, current, and state of charge, which are all significant battery operating states, but not directly measurable, using state of charge, state of health, and state of health estimation (SOH). To achieve this, a battery model that is accurate must be selected. These modeling techniques can be used to design a control strategy (Lua et al., 2013; Zhang et al., 2014). When multiple cells are connected in series, cell balancing (Kıvrak et al., 2020; Nikolian et al., 2014; Wang et al., 1998) is essential to maximizing battery life and capacity.

2. NECESSITY FOR BMS

It is essential to have accurate values for these parameters to determine the efficiency of the entire system. Measurement of these parameters is difficult due to the lengthy response times, the high cost, and the need to stop the vehicle, so they cannot be applied in real-time (Rivera-Barrera et al., 2017). A Li-Ion cell's SOA determines its temperature, voltage, and current. A battery that is overcharged or working outside of a safe temperature range may fire if these limits are exceeded, and a battery that is over-discharged may be permanently damaged. In addition, a cell might lose its lifespan if the current is discharged too fast or the cell is charged too fast. BMSs are essential for balancing cells with dif-

Figure 1. Peripheral coordination point for a battery pack

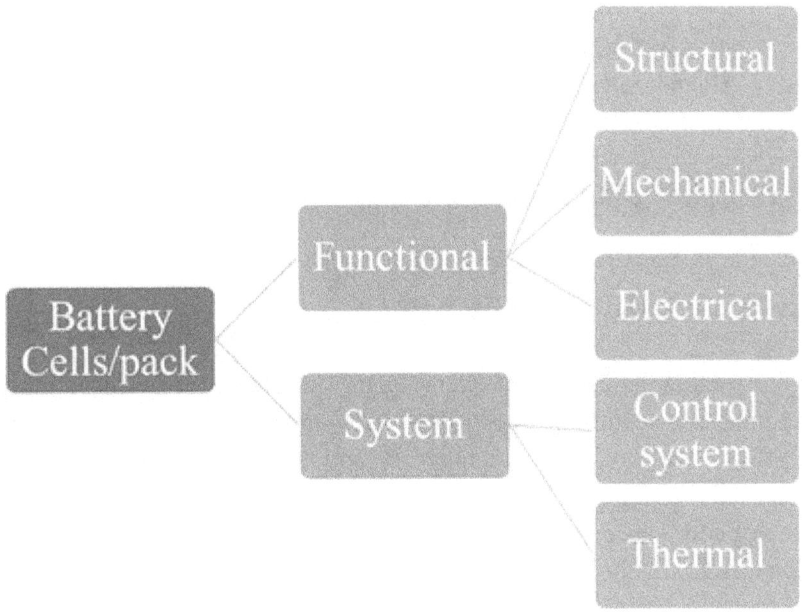

ferent SOC levels, capacities, and internal resistance since they solve the problem of uneven capacity and internal resistance between cells. Battery packs are then less able to charge and discharge because of this decrease in power and a decrease in capacity of the li-ion batteries (Chang, 2013). Below is the figure.1 Peripheral coordination point for a battery pack for the idea of the requirements of the battery.

Mechanical - For safety reasons, all safety-related mechanical components have been incorporated into this interface, such as cell spacers, dampers, pressure relief valves, seals, and gaskets.

Structural - To protect a battery pack against humidity, dirt, and other environmental factors, it must be enclosed in a case with a cover. It is also necessary for the vehicle to be vibration-isolated and crash-worthy. Battery packs are thus equipped with structural elements such as endplates, tie-rods, and cross-members for protection.

Thermal - A Li-ion battery pack with a temperature between 25 and 30 degrees Celsius will maximize its energy capacity. The thermal distribution must also be uniform throughout the pack. Thermal management systems are generally integrated with battery packs to ensure this stability. Fans, pumps, heat exchangers, cooling and heating fluids, insulation coating, and fan transfer ducts are typical components of such systems.

Electrical - Battery packs generate electricity at a specific voltage in response to the drive cycle requirements of electric vehicles. To route power to the EV driveline, a series of bus bars and cables are used, along with fuses, circuit breakers, contactors, and relays.

Control Systems - Measurements of voltage, current, pressure, temperature, and humidity are used to monitor the battery pack's condition.

Safe and Support - To achieve the necessary level of vibration isolation for reliable operation, EV batteries are normally mounted in a vehicle using mounting brackets and axles. Providing support to the chassis and body of the vehicle improves the vehicle's crashworthiness overall

2.1 Why Need BMS?

- Protects the operation of the host application and detects and responds to unusual conditions.
- Protects battery cells not to get damaged in any failure conditions
- Prolongs the battery life when operating in normal conditions.
- Maintains the battery state where it can satisfy the functional requirements.
- Tells the Host computer to make the best use of the battery pack, it also controls charging.

3. ESSENTIAL IN BMS AND BATTERY PACK

BMS is necessary to protect the cells from overvoltage, Undervoltage, overcurrent, overtemperature, and under-temperature based on the battery profile's programmed minimum and maximum values. BMS offers intelligent, Effective cell-by-cell balancing to expand the battery's functional range. The health of both individual cells and the entire pack is monitored by BMS and if either the pack or individual cells are faulty, error trouble codes are generated. The standard BMS must provide a wide list of features that protect the individual cells and the battery pack. These control strategies include:

3.1 The Functionality of BMS: (Figure 2)

- Detecting and controlling High voltage: Controls and detects the Voltage (Liu & Chien, 2017), current, temperature, Thermal management, and Fault detections.
- Protection: Provides overcharge, short circuit, and over-temperature Protection (Wang et al., 2015).
- Computation: Recording data, estimations, communicating, and reporting.
- Performance management: Cell balancing, Power computation.
- Estimation: SOH (Zhang & Lee, 2011), SOL, SOC.

3.2 Why Safety is Critical for Lithium-Ion Battery Applications

The 2-Wheeler and 3-Wheeler will lead the adoption, followed by the Bus and eventually the Car. Buses and cars won't likely make a big transition to electric vehicles (EV) until at least 10 years from now, while 2-wheelers and 3-wheelers will in the next 5 years. developed batteries for use in a range of applications, including 2- and 3-wheelers, vehicles, buses, tractors, and more. These batteries came in all cell form factors, including cylindrical, prismatic, and pouch. Are starting to manufacture batteries in big quantities after addressing the underlying technologies. Are setting up our first assembly line, which will generate a sizable number of batteries for 2- and 3-wheelers, to do that. To reduce fuel consumption and pollution, drivetrain electrification is a viable alternative to conventional vehicle architecture emissions. The performance of electric and hybrid vehicles depends strongly on the battery pack.

3.3 The Factor in the Battery Pack

- Choosing a battery for application (Depending on the voltage, current, chemistry, usage, performance, Thermal condition weight, size, and reusable)

Figure 2. BMS Features Block diagram

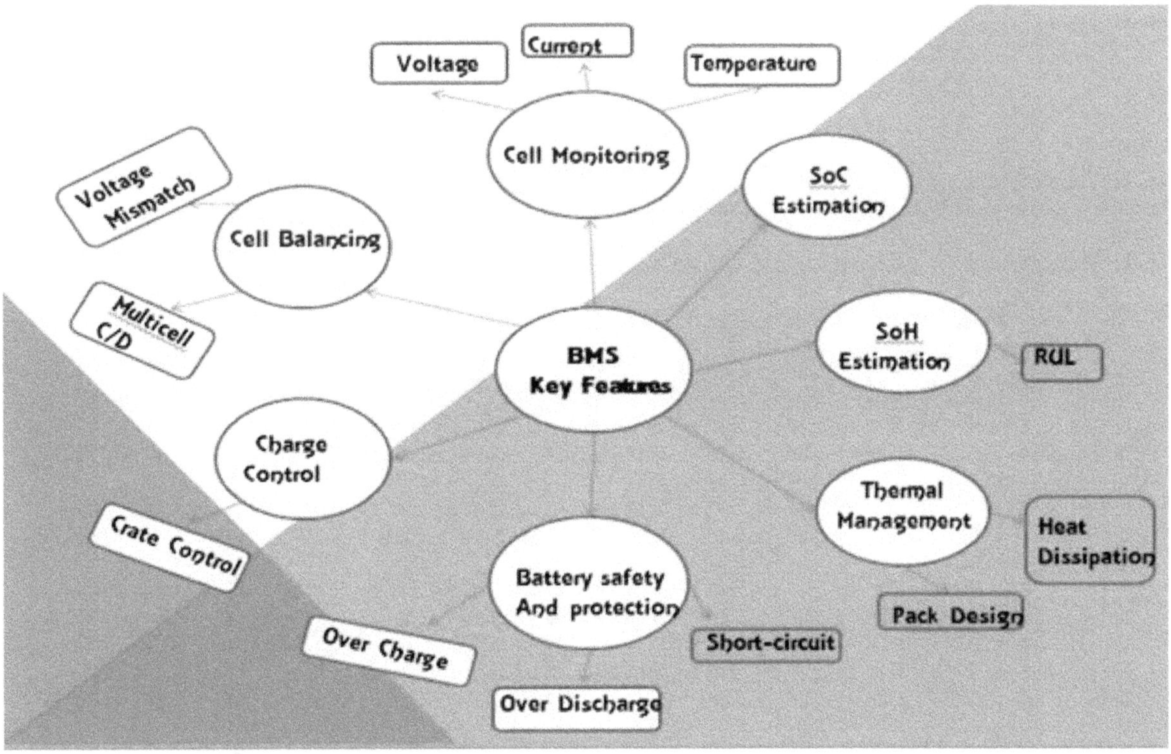

- Choosing the right battery pack size (Design a battery pack based on the space available and the effectiveness of location)
- Thermal management systems can be chosen for the pack based on requirements.
- Performance of systems over their lifetimes and as they age
- Reusability of resources and the availability of resources

Powering electric vehicles requires battery packs that respond to pedal forces, which control throttle position, and that is lightweight, compactly packaged, ergonomic, structurally rigid, swappable, configurable, scalable, thermally stable, impact-proof, fireproof, simple to manufacture, simple to service, and low in cost. Consequently, the BMS must include a mechanism for balancing cells that enables the following outcomes: To allow the battery to achieve a higher state of charge (SOC).

- Enhancing battery pack performance
- Reducing the degradation of cells
- In order to prevent thermal runaway
- Increasing the driving range and the life of the cells
- Enhancing the safety of batteries

Figure 3. Scope of BMS

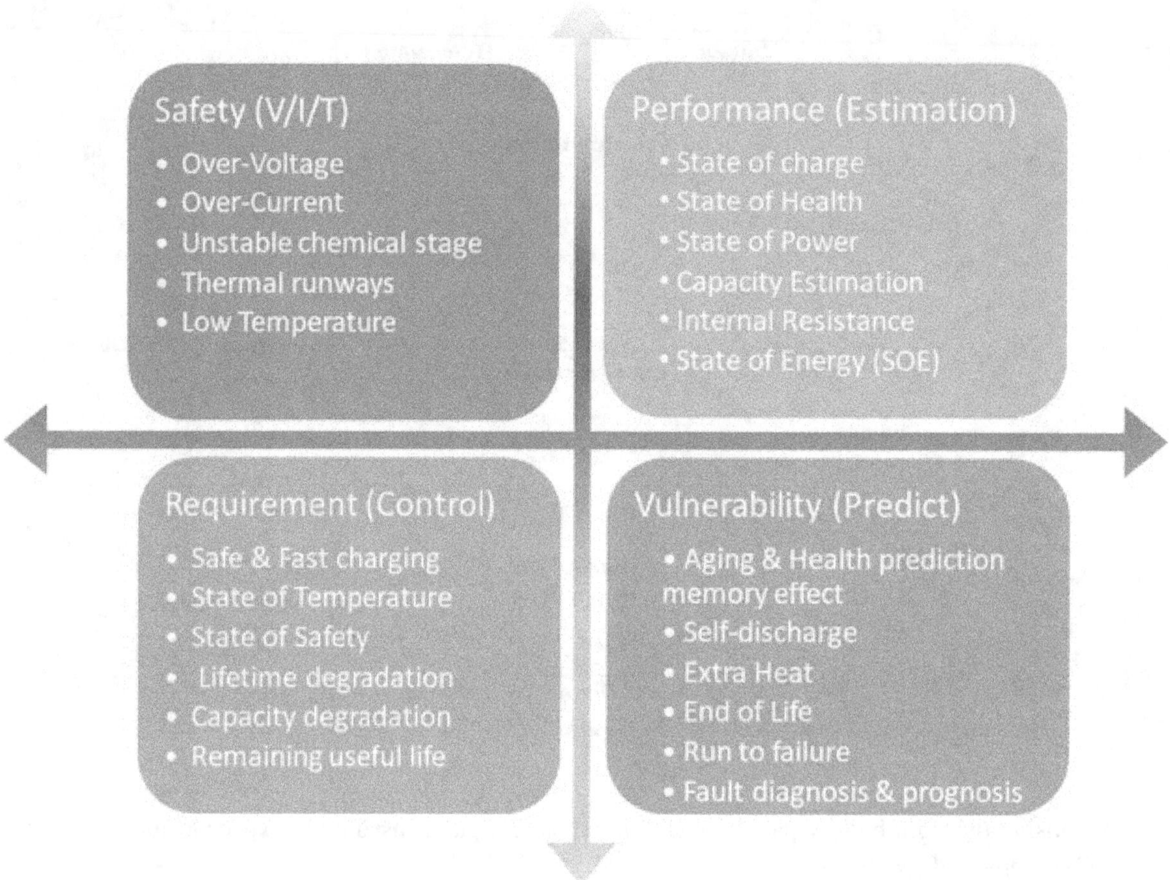

4. BATTERY MANAGEMENT SYSTEM AND ESTIMATE SYSTEM

Implementing advanced BMS monitoring and software solutions to accurately predict end-of-life, identify faulty cells, provide prognostics, improve battery longevity, support fast charging, etc. Below is the figure.3 gives a BMS high-level information for the scope understanding.

4.1 Battery Monitoring / Supervisor's: (Voltage-Current-Temperature):

The capacity and effectiveness of the Battery Pack determine the performance of a system. In order for the battery pack to operate properly, the BMS must maintain it effectively (Lelie et al., 2018). In energy storage applications, BMSs monitor cells, balance them, and protect them from temperature changes. They are sophisticated devices. Before the battery management system could take action, it would be necessary to assess the Lithium pack's conditions. Cells in a pack with varying voltage and current (Rahimi-Eichi et al., 2013), along with their temperature, are calculated for this purpose. In addition to measuring SOC and SOH, the BMS performs cell balancing and other tasks only based on these values.

Figure 4. In a series battery stack, the voltage of each cell is measured

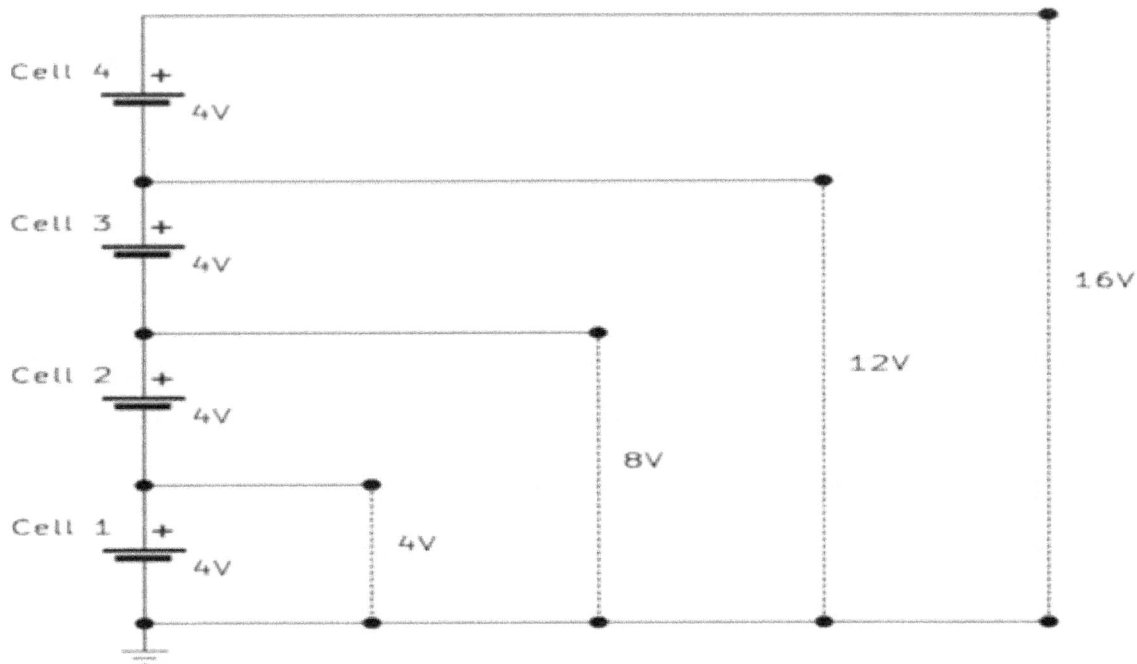

No matter the type of battery, whether it is a simple power bank or laptop battery, or a more complex solar or electric vehicle battery, monitoring its voltage and current is essential.

4.1.1 Voltage Monitoring

A lithium battery pack will be tested in this section by measuring the individual cell voltage. It is necessary to measure voltage differently depending on the type of battery technology (Jossen et al., 1999). Compared to other battery technologies, Li-FePo4 chemistry has the highest degree of voltage precision (Qaisar et al., 2013). A series connection of four lithium 18650 cells will be used to calculate the individual cell voltages.

A series-connected battery pack presents the challenge of not having a standard reference point for measuring individual cell voltage. Suppose that all four cells have a voltage of 4V in order to maximize efficiency. The voltage of the first cell can be determined easily with a microcontroller using the voltage of the second cell. As a result, the other end is connected to the ground. Nevertheless, we would measure the voltage of the following cell in conjunction with the voltage of the previous cell (Affanni et al., 2005). For instance, when measuring the fourth cell, all four cells can be measured simultaneously. Due to the inability to adjust the reference point from the ground, this is the case. Consequently, a circuit will be needed to estimate individual voltages. With a potential divider, voltage levels can be mapped down and calculated, but this method reduces the read value's resolution to >0.1V. As seen in the figure.4, we will measure the individual voltage by measuring the difference between the terminals of each cell using the Op-Amp Differential Circuit as shown in figure.5.

Figure 5. Different circuits to measure individual cell voltage

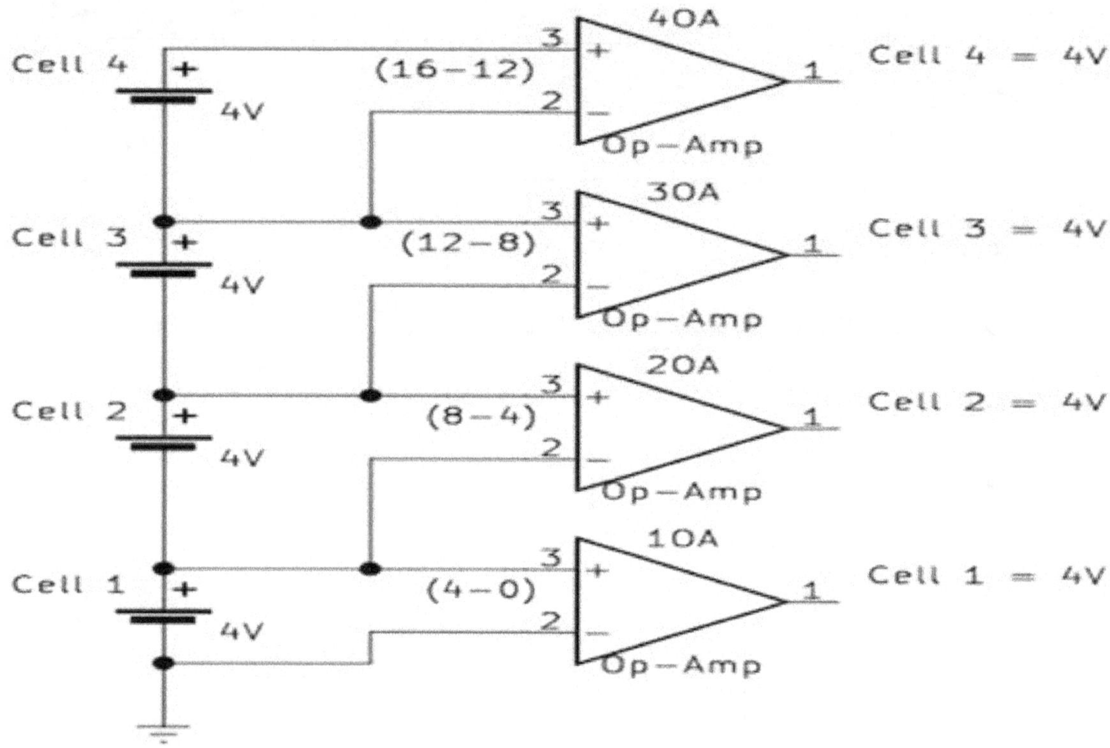

4.1.2 Current Monitoring

To determine the state variables of the battery system, current measurements require extreme precision. SOC can be calculated by integrating the total current using a precise calculation of the total current. In the following sections, we discuss and compare various current sensing methods that have been developed. Based on their sensor technology, current acquisition devices are classified into two types: galvanically connected and isolated (Isakova et al., 2011).

Shunt Resistor Method

High-precision voltage measurement devices are used in conjunction with low-resistance high-precision resistors to calculate current. The resistor is isolated from the current conducting direction, allowing the voltage drop to be measured proportionally to the current. As a result of the current flowing through the shunt, the shunt resistor heats up and causes a power loss (i2R). A resistor value that is sufficient to produce a voltage drop suitable for the measuring device is a compromise between keeping losses to a minimum and choosing a resistor value that minimizes losses. In addition, it is essential that a shunt resistor has minimal thermal drift. Precision resistors should have high long-term stability and a low-temperature coefficient if intense measurement accuracy is required. The shunt system can also measure direct and alternating currents. In addition to its simplicity and linearity, the measurement system can reach a wide bandwidth. Depending on the accuracy of the voltage measurement, the measurement range is limited. Figure.6 illustrates a standard schematic for shunt-resistant methods.

Figure 6. Schematic of Shunt Resistant Method

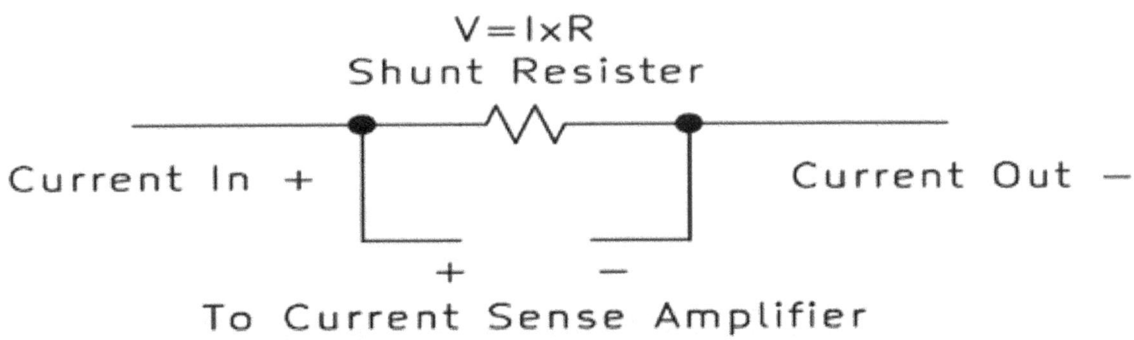

4.1.2.1 Hall Sensor Method

Additionally, there are contactless current sensors available that are available in the market alongside shunt resistors. Known as magneto resistive sensors (Jander & Schneider, 2005), UXgate sensors, and Hall sensors, these sensors are used to detect a magnetic field generated by currents. Based on the Hall Effect, Hall sensors can detect both AC and DC signals. When current flows through the Hall sensor, a voltage is generated proportional to the magnetic UX density and the current flowing through the sensor. Due to the Hall sensor's lack of resistance, the current direction will not be shunted, resulting in no additional conductivity loss. A further advantage of this approach is that it provides galvanic isolation. Signal conditioning does not require additional optocouplers or digital insulators when Hall sensors are used. It is common for Hall Effect sensors to have a limited bandwidth, which is typically only a few tens of kHz. In general, a shunt resistor is less expensive than a Hall sensor if the battery system designer needs greater bandwidth since Hall sensors are more expensive. Moreover, Hall Effect sensors must compensate for the temperature drift in their output signal when using this type of measurement.

4.1.3 Temperature Monitoring

A negative temperature coefficient (NTC) (metal oxide) or a positive temperature coefficient (PTC) (semiconductors) is commonly used to measure temperature in battery management applications. In response to rising temperatures, the resistance of the NTC decreases. Voltage drop is calculated by recording a steady flow of current through the resistor. Normal operating temperatures range from 50 to 150 degrees Celsius, and they're cheap, simple to use, linear over a narrow temperature range, but nonlinear over very low and very high-temperature ranges. BMS front-end ICs commonly use NTC-style thermistors and ADC channels for temperature measurement. To calculate temperatures accurately on the intelligent cell controller board, a nonlinear lookup table is required. It is more convenient to use sensors that have a digital interface and can measure temperature quickly. Figure.7 illustrates a specification for the 1-Wire digital interface (Maxim Integrated) integrated into this sensor (DS18B20). Texas Instruments (TMP100) and LM75 are among the companies that manufacture I2C-based sensors (Haq et al., 2014). It is easy to communicate with these sensors. In a battery pack, however, digital sensors must be placed far from high-power paths to avoid EMI-induced failures, even though they are known to be durable. A thermocouple is also capable of measuring temperature, but it requires more complex electronics compared to metal-based PTCs, such as PT100s.

Figure 7. Typical Application Schematic of BQ297000

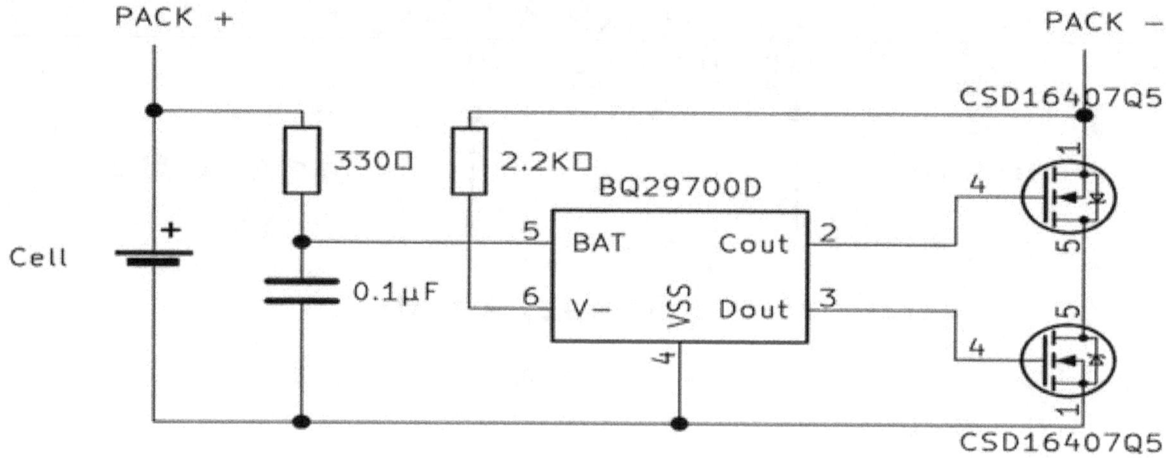

4.2 Battery Safety and Protection

Lithium-ion batteries are found in most electric two-wheelers because they are lightweight and relatively cheap, and fires often break out due to the rechargeable battery packs. Despite being an emerging industry, failures are bound to happen, even when these batteries are not dangerous. Fires may occur when the scooter is charging, discharging, while in use, or even when it stands idle.

- To determine the quality of a battery, three parameters need to be considered:
- Battery life and safety are the top priorities, followed by a performance
- There is no fire caused by summer heat or poor thermal management, so let's clear up this misconception
- Many battery fires are caused by short circuits, which lead to unregulated current
- There is a problem with the manufacturing quality of the battery being poor.
- An inadequate battery architecture is evident in the way the cells are arranged serially and parallelly as well as the way they are connected.

Failures in sensors and software intelligence lead to overcharging of cells under BMS control. Overcharging occurs because of inadequate cell management by the BMS. Battery safety depends on a fire-resistant design and a design that prevents fire from spreading. The battery will be able to survive a fire caused by the C-capacity cells, thereby ensuring its performance. It is possible for a battery to fail in many ways, including design or manufacturing flaws, damage to the battery, and inadequate cooling. for high-quality lithium-ion batteries to achieve this optimal level of charging, it is also the manufacturer's responsibility to specify specific internal controls. If the controls are overcharged or charged too quickly, they can be damaged. When temperatures exceed their optimum limit, lithium-ion batteries can no longer maintain a healthy cooling system. When operated beyond their optimum temperature, lithium-ion batteries can experience thermal runaway. Batteries don't catch fire in our pockets or backpacks because they are always at the right temperature. It is not uncommon for fire hazards to be caused by not only poor cell quality or grade, but also by defects in Battery Management Systems (BMS), as

Table 1. Important factors in protect of BMS and BTMS

Battery over-voltage (OV)	Battery overheating, smoke emissions, inflaming or explosion are very likely secondary chemical reactions. OV is typically close to 4 V.
Thermal runaway (OT)	Typically, an OV OT close to 60 °C will cause a positive temperature feedback mechanism
Battery under-voltage (UV)	A progressive breakdown of electrode materials occurs over a few cycles with LFP cells when UV is close to 2 volts
Battery over-current (OC)	This can lead to the melting of the battery contacts. Major safety issue: contactors are not able to be opened and the system cannot be put into its disabled safe state.
Battery under-temperature (UT)	Contactors lose their robustness, resulting in a reduction in performance

they can also lead to the failure of the battery as shown in the table.1. Despite the quality, poor cells can cause internal short circuits. The normal channel of current is short-circuited when anode and cathode are misinterpreted as being connected internally due to manufacturing faults, resulting in an unregulated current and a fire. When cells are packed together without any space between each other, they cannot survive without oxygen. It is essential to consider every nut and bolt in the absence of insulation.

4.2.1 Overcharge Detection

It is generally not recommended to charge lithium batteries above 4.2 V/cell, as this is considered a safe limit. Lithium-ion batteries can be damaged if they are charged beyond this limit (Liu et al., 2018). This can cause safety hazards like fires. With the help of a protection circuit, we will be able to overcome this kind of problem.

4.2.2 Over-Discharge Detection

In a Lithium-ion battery pack when the cell voltage reaches 2.5V means, the battery pack is completely discharged (Viera et al., 2006). Also discharging parameter of the lithium-ion cell if low is stressful and reduces cell lifetime characteristics. In general, the featured battery protection circuit has protection from over-discharge (Arasaratnam et al., 2013; Schwarz et al., 2015).

4.2.3 Sensors to Detect Short Circuits

Sensors to detect short circuits is an effective feature of a Battery Management System, which detects short circuits when a load is operated at a maximum voltage of 12V. Generally, this circuit has PNP transistors and 555 timers. Many protection ICs (Mutyala et al., 2014) switch the battery input and output of the circuit based on the above battery safety considerations. These battery protection ICs and MOSFETs are manufactured by Texas Instruments (BQ29700) (Farag, 2013), Seiko's S-8200A Series ICs, and DW01-P. The schematics for all Battery protection ICs are the same for BQ29700, on will be connected to the two MOSFETs denoted as charging (CHG) and discharging (DSG), and terminals. When the voltage at the cell terminals reaches a certain level, the maximum limit, the operation of the CHG MOSFET is turned off. Nevertheless, the cell can discharge through the CHG's body diode and through the DSG when the DSG is turned on. A rapid discharge will indicate a low cell voltage, causing

Figure 8. Typical Application Schematic of S-8200A Series

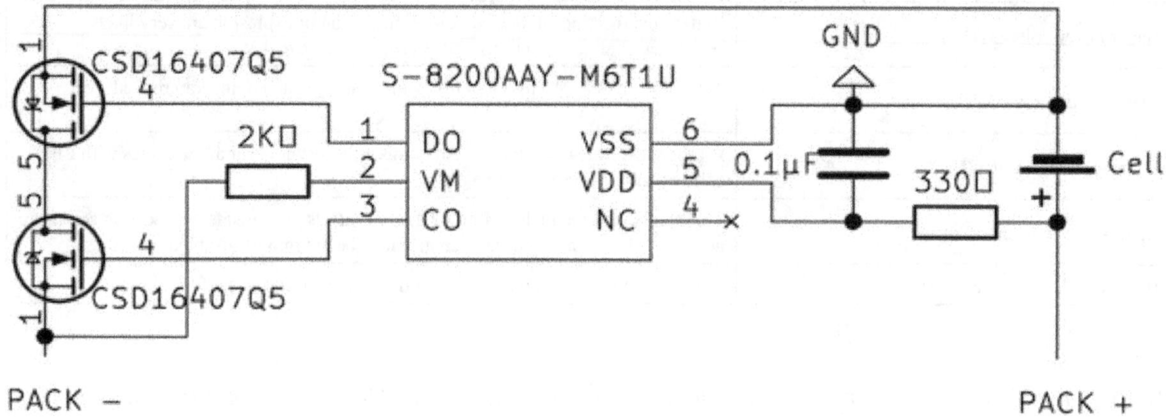

DSG to turn off and prevent further discharging. Using the CHG MOSFET and the DSG's body diode for additional cell charging is possible if the CHG MOSFET is on.

In Figure.8, Seiko's S-8200A (Plett, 2004) Series lithium cell protection IC's typical application schematic can be seen. In Figure.9, DW01-A protection IC's typical application schematic can be seen.

4.2.4 Batteries Protection Integrated Circuits: Considerations

ICs and MOSFETs are critical components of BMS design that determine overall efficiency. A battery protection IC's two most important parameters are its overvoltage and Undervoltage thresholds. The 4.2V charge during each charge period improves the safety of Li-ion batteries during overcharging. The batteries, however, have a longer lifespan when charged to 4.1V. Selecting an overcharge voltage threshold must hit a balance among charge per battery and lifespan. As well as cell life, the over-discharge safety threshold can have an impact on capacity/charge. The power per charge of a battery is greater if it is fully discharged. Battery life will be shortened, however, by this method. The battery safety circuit also requires MOSFETs.

Figure 9. Typical Function of DW01-A

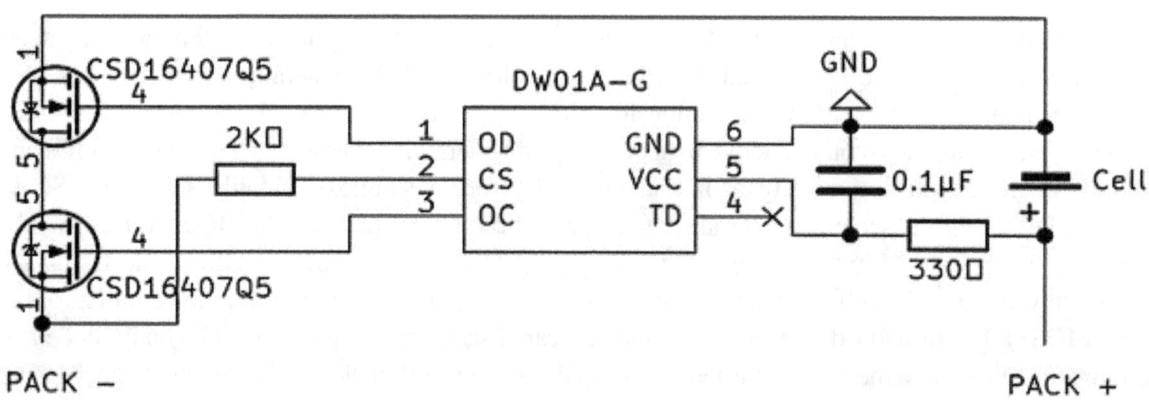

5. CHARGE CONTROL

Charge control is the act of controlling the charge current which may be added to or drawn from a Lithium-ion battery. To prevent overcharging, a charge controller (Hoque et al., 2017) offers protection against overvoltage. Similarly, it is also preventing the deep discharging of a battery. Therefore, it will reduce battery performance and increase cycle life, as well as posing a safety risk. For a battery cycle life to be protected, it depends heavily on battery chemistry and technology. The charge controller may be a standalone device that is integrated with a battery pack, or battery power devices (Kovalcik, 2015).

Charging Time: The batteries are balanced by low balancing currents when they are slowly charged. Whenever the battery is charged at a high C rate, high balancing currents are needed (Auger et al., 2013).

Charging Temperature: When designing a balancing resistor, the ambient temperature of the battery needs to be considered since battery pack temperatures rise more rapidly in hot climates than in cold climates (Meng et al., 2018).

Cost for Thermal Management: To support fast-changing requirements, thermal management systems dissipate much more energy in their balancing circuits. Choosing the right balancing resistor and using the right control algorithms will ensure thermal safety. Different cell chemistry results in different balancing currents and imbalance rates.

An Imbalance is Likely to Occur when: the life of the battery pack is reflected by this parameter, as well as the amount of imbalance present in it. A difference in the manufacturing batch can cause imbalances in the series. It is possible to decrease the number of lots produced to control this, but aged cells are more likely to become imbalanced (Kovalcik, 2015).

5.1 Charging Rates

Depending on the application or requirement, a battery can be charged at various charge rates. The following are typical charge rates (Hemavathi, 2020).

- Charging slowly at the rate of 0.1C over a period of 14-16 hours
- Charges quickly at 0.3C for three to six hours
- In less than an hour, can charge at 1.0C with Fast Charge.

5.2 Charge Control Methods

For various chemistries and applications, different types of charging control schemes have been developed. Below is figure.10 a list of the most popular ones.

5.2.1 Slow Charging Control Method

Semi-constant current is a common and cost-effective system. It uses a low current, so it doesn't produce heat, but it's slow; charging takes around 5 to 15 hours on average. This method's charge rate is normally 0.1C. This approach works with batteries and the charge method with a timer (Sironval, 2020) is easy and effective. In this methodology, the battery is charged at 0.2C for a fixed period of time, and then trickled at 0.05C for a short period of time (Meissner & Richter, 2003; Yao et al., 2007).

Figure 10. Charging Techniques

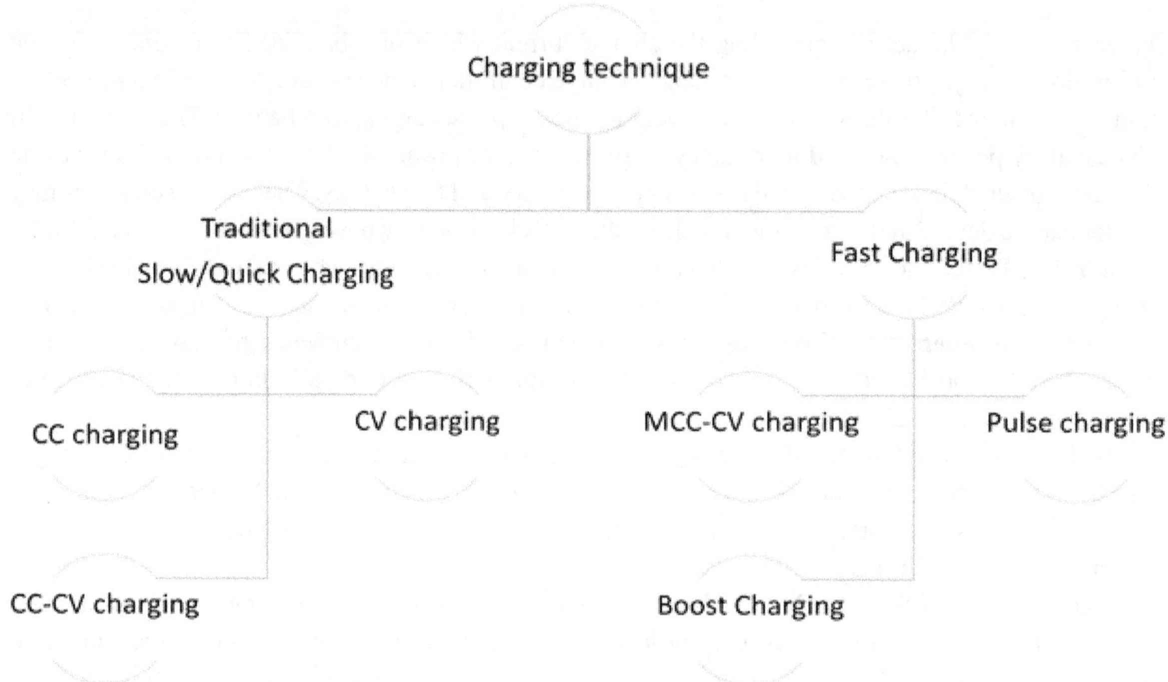

5.2.2 Quick/Fast Charging Control Methods

For various applications, there are numerous quick charging control methods: NDV (Negative Delta Voltage) (Freeman & Heacock, 1993), Voltage control, constant-current control, and constant-voltage control (CC/CV). Generally, the constant current mode is used when fast charging rates are mentioned. Depending on the cell chemistry, this cycle may take between 60% and 80%. Because the rate drops rapidly during the constant voltage cycle, it cannot be used to estimate when the battery will be fully recharged. Calculating the time needed to charge a lithium battery from empty by using the manufacturer's C rate would be impractical because the C rate changes as the battery's capacity changes during the charging process (Farag, 2013).

6. TYPES OF BATTERY MODELING

6.1 Battery Modeling

This section provides a brief overview of how the Battery model is conceptually in figure.11 to implement for protection. Two phases are involved in the implementation of the Battery model: static model and dynamic model

- Using a static model, it is possible to evaluate the constant current discharge characteristics of batteries undergoing constant current discharge tests.

Figure 11. Battery modeling and estimation needs

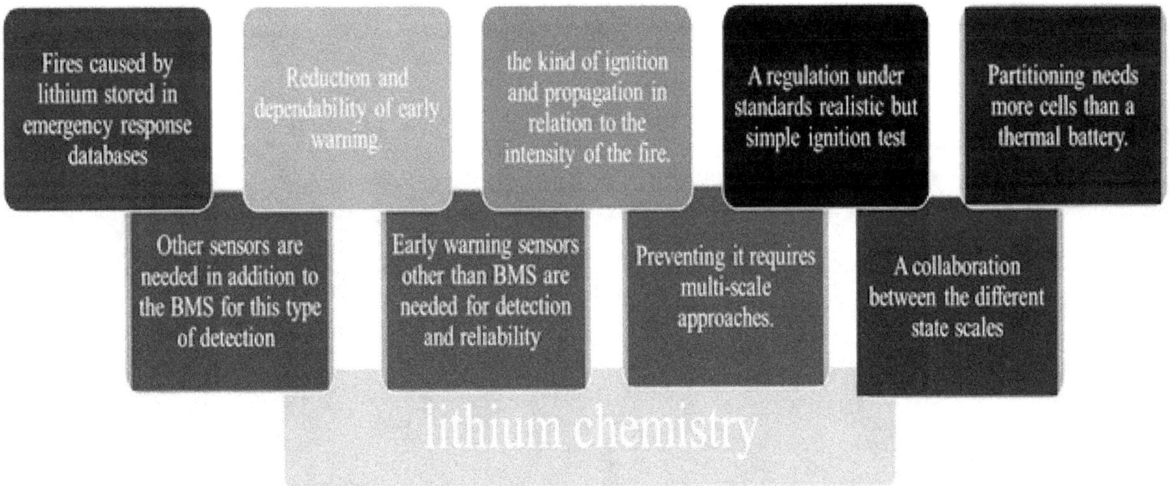

- Various data-driven approaches are used, such as neural networks, support vector regression, and polynomial regression.
- In increase to capturing the thermal behavior of the cell, the static battery model also captures its stability.
- To realize the static model, the necessary measurement data must be transformed.
- This model determines the electric parameters of a battery using the battery equivalent circuit approach.
- The developed static model is fed into the dynamic model to produce the output transient response.

Conclusively, accurate measurement data is required for the Battery model to be implemented successfully. It is necessary to follow developed test rules in order to obtain measurement data (Hemavathi, 2021).

Over the past decade, lithium-ion battery fault diagnosis has become increasingly popular in industry and academia. Lithium-ion batteries are widely diagnosed accurately using model-based fault diagnosis techniques because of their inherent advantages of low costs and high flexibility. Since ECM algorithms are simple formulations and require little computational power, they are accurate representations of battery dynamics in real-time applications. A Thevenin model extension containing an additional RC parallel circuit is frequently used as a substitute for Thevenin models. Equivalent circuit parameters such as state of charge, temperature, and current must be modelled differently depending on the operating conditions to accurately represent cell behaviour across a wide operating range, involving faults. There are several of these faults that can be removed in the event of an emergency to detect a set of conditions. The authors identify some battery operations as possible faults using offline electrochemical impedance spectroscopy (EIS).

A wide range of applications has been developed through the development of battery models with varying complexity as shown in the figure.12. Depending on their level of physical understanding, battery models fall into three categories: white box electrochemical models, gray box circuit-oriented models, and black box artificial neural network ANN models. In the battery modeling process, battery parameters

Figure 12. Battery modeling methods

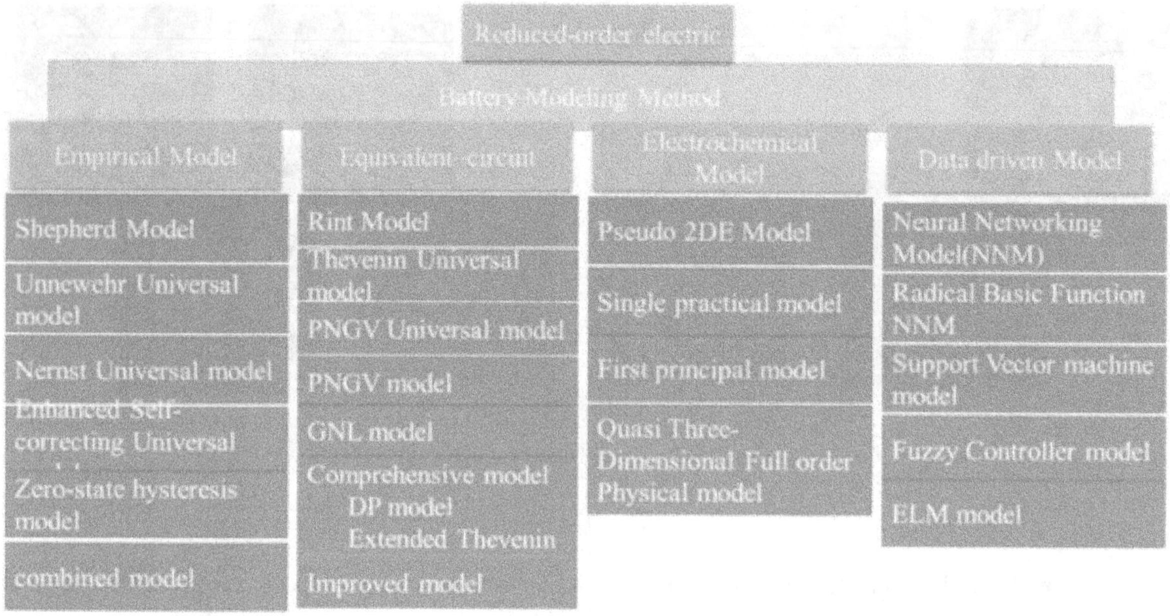

are used to model batteries. Modeling strategies include electrochemical, mathematical, circuit-oriented, and data-driven approaches for extracting battery parameters. An average battery's lifespan depends on its degradation capacity from its initial value. A battery's lifetime is reduced by acceleration when using an electric vehicle. It is necessary to apply ten times more power in order to overcome this disadvantage. In addition to increasing its specific energy requirements, a battery's efficiency and reliability should be improved through optimization techniques. Li-ion batteries have an average efficiency of 92% and LiFePO4 batteries have an average efficiency of 95%. There is an 80% discharge intensity in both types of cells. Design shows a considerable responsibility in the efficiency of any battery (Saji et al., 2019).

6.2 Mathematical Models and Electrochemical Models

The study of electrochemical reactions (PDE) can be undertaken by using partial differential equations (PDE), electrochemical models (ECMs), or physics-based models. Using this model, trade-offs can be analyzed and high accuracy is achieved by connecting physical parameters with the cell's internal electrochemical dynamics. There is a rapid spread of electromobility throughout the world. To ensure safe and reliable operation, mobile energy storage systems must perform well and be safe. Voltage and temperature must be monitored continuously for each battery cell. The Cell management controllers (CMCs) are specialized electronic monitoring systems that accomplish this task.

The Shepherd model, developed in the 1970s, was an early mathematically based battery model. Battery terminal voltage and current, current, internal resistance, and resistance all have a direct relationship. It was created with some approximations and assumptions while considering the electrochemical reaction between electrodes and electrolyte (Sajfar & Bullough, 1990).

The cell management controller (CMC) is an electronic element that monitors battery cells and contains one or more microcontrollers. To reliably measure the voltage, the CMC is linked to the +ve and -ve ends of the batteries via measuring leads. It is also connected to temperature sensors, which aid in the optimization of battery loads and efficiency. The exceptional to incorporate an ageing model into a BMS for degradation prediction is primarily influenced by the model's accuracy, regardless of the modelling approach used. It's difficult to believe that a model solely based on life-cycle data will perform well under all possible operating conditions over the course of a lifetime. As a result, software now has access to all the test's real-world battery operation data. Data from whole-battery durability tests, as well as data from ageing behaviour modelling, should be considered.

The sections that follow explain the fundamentals of this approach. Cell modules are formed by combining individual battery cells, each with its own CMC. battery and vehicle or endanger the occupant's safety The energy stored in electric vehicle batteries is comparable to that of a small explosive. Thermal runaways can occur as a result of overvoltage or undervoltage, resulting in battery failure. The voltage of each battery cell in a module, as well as the temperature of various points within the module and other conditions, are typically monitored by a battery monitoring integrated circuit (BMIC) or cell-balancer device.

The precision of these measurements, as well as the frequency of communications from the BMIC to the CMC and BMC, are critical to detecting and intervening before a potentially hazardous condition becomes dangerous. To return individual cell temperatures to an acceptable range, the BMC may stop regenerative charging or reduce power draw from a pack, or the driver may be alerted to such a condition via a "check engine" light on the dashboard. In any case, for the BMC to take appropriate corrective action as soon as possible, the BMICs must be capable of extremely accurate measurements as well as robust communications with the CMCs. Due to the abundance of electrical noise in the environment, designing an effective communication network for an EV is difficult. The robustness of BMIC and CMC communications is frequently determined by the overall design and routing of the network connecting the various devices in the BMS (Hemavathi, 2019).

6.3 Life-Cycle Analysis of Cells and Batteries

A typical lithium-ion cell life-cycle study begins with a series of degradation tests at a constant duty cycle. Lithium-ion batteries with high power for automotive applications in hybrid vehicles were used in this study for cell-level experiments. Below are the main specs for the deployed cell, which are taken from the cell itself. The nickel-cobalt-aluminum cathode chemistry makes it a well-known cell for its excellent calendar life and cycling ability. Several types of automated testing equipment were used to conduct the experiment, including environmental chambers, programmable electronic loads, and climate chambers. Test scenarios start with two full cycles, one discharge and one charge at 1 C current. After a relaxation period that is dependent on the adjusted temperature of the climate chamber, the cells are discharged to a specific SoC value before cycling begins. This article discusses the ambient temperature differences, start-SoCs, and load profiles between the cell and battery testing scenarios as well as the two batteries tested. The cells' performance was monitored during each testing scenario with capacity tests. During the capacity test, the constant current constant voltage (CCCV) method is used for each charge and discharge.

6.4 Modeling the Aging Behavior of Lithium-Ion Cells

Lithium-ion battery ageing models are a crucial part of the automotive development cycle. In the early stages of product development, this model is crucial since it models battery degradation based on battery operations. By estimating degradation behavior of lithium-ion cells based on information about real-world battery operations, an accurate ageing model ensures the battery's life expectancy as an important part of the battery management system. As part of continuous health monitoring, it measures vital battery parameters such as capacity and performs prognoses for their deployment. Furthermore, when such a health monitoring process is available, it generates data that a fleet management system in dealer workshops or via telematics can use.

7. ESTIMATION TECHNIQUES

As shown in figure.13 to state estimate, the model can accurately describe the dynamics of batteries in general, with each sub-model corresponding to a specific timescale within the larger multi-physics model. A battery's SOC, SOP, or SOE can change rapidly due to changes in microscopic electrochemical parameters. The macroscopic temperature distribution updates on an intermediate timescale due to the physical structure of batteries and heat transfer characteristics. Battery SOH happens on the slowest timescale and is characterized by slow-changing parameters such as an increase in internal impedance and resistance, as well as a decrease in capacity. To determine the battery safety state, consider the device's safety requirements as well as the System designs and management for SOC/SOE/SOH/SOP/SOT/SOS. For the development of precise and dependable battery management, several critical enabling technologies are required, including the ability to estimate and monitor vital internal conditions. The states of charge (SOC), energy (SOE), health (SOH), power (SOP), and temperature (SOT) are essential for efficient battery charging, thermal control, and health management. Battery state estimation has already gained significant traction due to the number of methodologies that have been reported in the literature. It aims to provide concise, comprehensive insights into key topics, technological challenges, and future research prospects in this area rather than exhaustively cataloging all significant publications.

One of the primary goals of this article is to encourage investigators and practitioners in this showground to underwrite new thoughts and tools for estimating the SOC/SOE/SOH/SOP/SOT/SOS and SOOS of power batteries, which are important for developing next-generation battery management systems. It would be extremely beneficial to decide immediately in the direction of obtaining raw materials or concentrating on more countries to supplement India's future raw materials. Another potential stumbling block is the scarcity of raw materials. A number of hopeful battery technologies are prepared to speed up the implementation of electric vehicles. Which incorporate market-leading Li-ion batteries as well as new batteries in development like Solid-state batteries (SSBs), Li-S, Metal-air, and Na-ion. Nevertheless, reserve accessibility may become a significant restriction in the future. Furthermore, early implementation of a closed recycling loop in which every bit of raw material used in a battery tense to be reused at the end of its life helps improve resource availability while establishing a sustainable battery life cycle and various estimation techniques as highlighted in the table.2.

Figure 13. To Estimate the state of batteries and their function

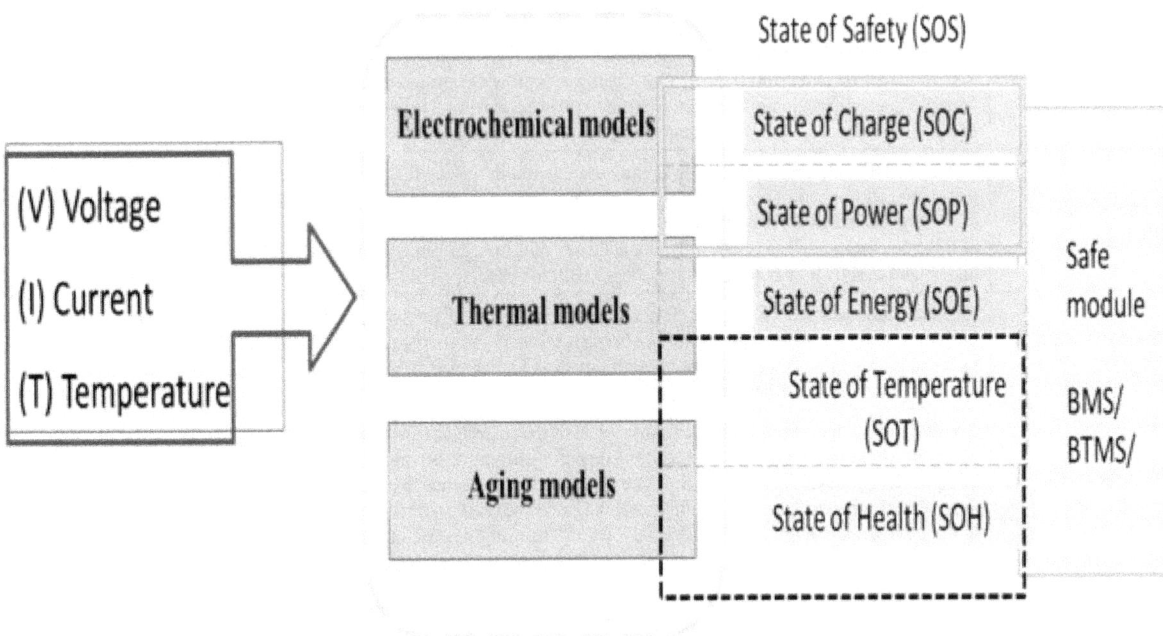

7.1 State of Charge - SOC

Batteries are measured by their SOC as shown in figure.14, which is the ratio between their available capacity and their maximum capacity. When a battery is fully charged, for instance, it has a SOC of 1 (or 100%). In the case of a battery with a SOC of 0 (or 0%), it indicates that it is completely discharged. Direct measurement, computational intelligence, model-based, and mixed methods are the four types of methods available for estimating the battery state of charge. Directive measurement methods include voltage translation and Coulomb counting. Voltage translation can be used to measure open-circuit voltage (OCV), while Coulomb counting can be used to calculate SOC from measured current (iB) over time, with information on initial SOC (SOCinitial) and maximum capacity (Cmax). Nevertheless, both methods have limitations. The former requires a battery to be resting and disconnected from an external circuit for a long period of time before the OCV can be measured. This results in an ineffective test. There may be a number of reasons why the latter experiences unrecoverable problems.

The most common errors are incorrect initial charging conditions, errors resulting from cumulative integration, and noise contamination. Additionally, Coulomb counting methods fail to account for nonlinear effects such as the rate capacity effect and recovery effect when estimating battery capacity variations. Current, capacity, coulomb efficiency, self-discharge rate, and initial SOC are all affected by the initial SOC, accurate self-discharge rate, and highly accurate current measurement. Also, an algorithm based on battery models was developed because of the study. For the estimation of battery impedance and state of charge (SOC), particle swarm optimization was used as a parameter identification algorithm. With a second algorithm, parameters can be identified using regression, SOC can be estimated using sliding-mode observers, and capacity can be estimated using two-point methods. The parameters of an online system were identified by a fast upper triangular and diagonal recursive least square algorithm.

Table 2. Estimate of Battery function and features

State of Charge (SOC):	BMS is effected by SOC in numerous ways. Accessible capacity is expressed as a percentage of nominal capacity, referred to as SOC. SOC functions similarly to a fuel gauge when used with a fuel tank. In addition, SOC refers to both the anode and cathode electrodes' charge in battery electrochemistry. Lithium concentration in electrode particles causes variation in SOC. Due to the highly dependent relationship between lithium stored in electrodes and available charge, SOC can be calculated directly using mean lithium concentration.
State of Energy (SOE):	It has become widespread practice to predict the remaining driving distance of electric vehicles by calculating their battery's amount of energy. The battery's potential level will decrease significantly under different SOC states during the discharge process, resulting in different energy changes based on the same charge change. Because of this, the commonly used SOC only characterises the battery's remaining power, and a nearby intuitive way to understand exactly how much available energy is in the battery.
State of Health (SOH):	As electrochemical batteries age, their performance will gradually deteriorate due to side reactions. Because of this, lithium inventory and active materials are being lost due to the ageing process. Popular relationships of capacity fade and internal resistance, SOH is a popular method for calculating battery ageing. Batteries can be measured for the End-of-Life (EOL) when their capacity fades by 20% and/or their internal resistance increases 100%.
State of Power (SOP):	Considering various operational constraints explicit to the design, a battery's SOP is thought-out for the purpose of the threshold current and voltage, and the power it can source or engage from the motor vehicle powertrain. Temperature, voltage, current, and SOC are some of the constraints that affect cell performance. Battery SOP is defined in this paper as the maximum power it can deliver or receive while still operating safely. The three modes of online SOP estimation are CC, CV, and CCCV. The result has been the development of several OLPMs for online SOP estimation.
State of Temperature (SOT):	Developing a rigorous definition of the State of Temperature (SOT) is difficult. Existing literature reports either the internal temperature or the temperature distribution. Battery thermal dynamics are revealed macroscopically by the temperature distribution. To facilitate control and effortlessness of implementation, battery essential, usual, and surface temperatures often represent SOT. Microscopically, SOT consists primarily of a heat source and structure-based heat conduction.
State of Safety (SOS):	Battery disaster levels are classified differently by countries and organizations, but there will be minor differences. It is also possible to measure danger levels numerically. Using hazard severity and probability as examples, hazard risk is calculated. The SOS can be calculated by using a probability function based on battery dynamics, such as V, T, charging/discharging currents, and IR. In this definition, SOS decreases with an increase in abuse function.

A smooth variable structure filter was used to estimate SOC, and a recursive total least square algorithm was used to estimate the maximum capacity and state of health. The second and third algorithms can accurately and robustly monitor lithium-ion batteries. Measure inactive current and detect unwanted leakage current caused by faulty BMS outputs, high-precision ammeters are placed between the cell inputs on the BMS and the cell outputs on the cell modules. Leakage current measurement is used to detect errors during BMS deactivation to avoid damage to battery cells. A hysteresis model also guides electrical circuit condition monitoring. Additionally, they are suitable for embedded BMS applications since they are low in complexity. A fast recursive least squares algorithm identified the parameters of an online system. SOC was estimated using a smooth variable structure filter, and maximum capacity and state of health were estimated using a recursive total least square algorithm. Using the second and third algorithms, lithium-ion batteries can be monitored accurately and robustly. Detect leakage currents from faulty BMS outputs caused by the inactive current. The BMS and cell modules are equipped with high-precision ammeters. To prevent damage to battery cells, leakage current measurement is used to detect errors during BMS deactivation. Hysteresis models guide electrical circuit condition monitoring, as well as being applicable to embedded BMS applications because they are low in complexity.

To improve battery performance, balancing cells is critical. In addition to the Kalman filter's precision, dynamic noise, and error of SOC values, consideration of the Kalman filter's precision is not necessary. A lack of interest will make it difficult to solve your problem. The initial SOC value should be

Figure 14. Categorization of state of charge estimation method

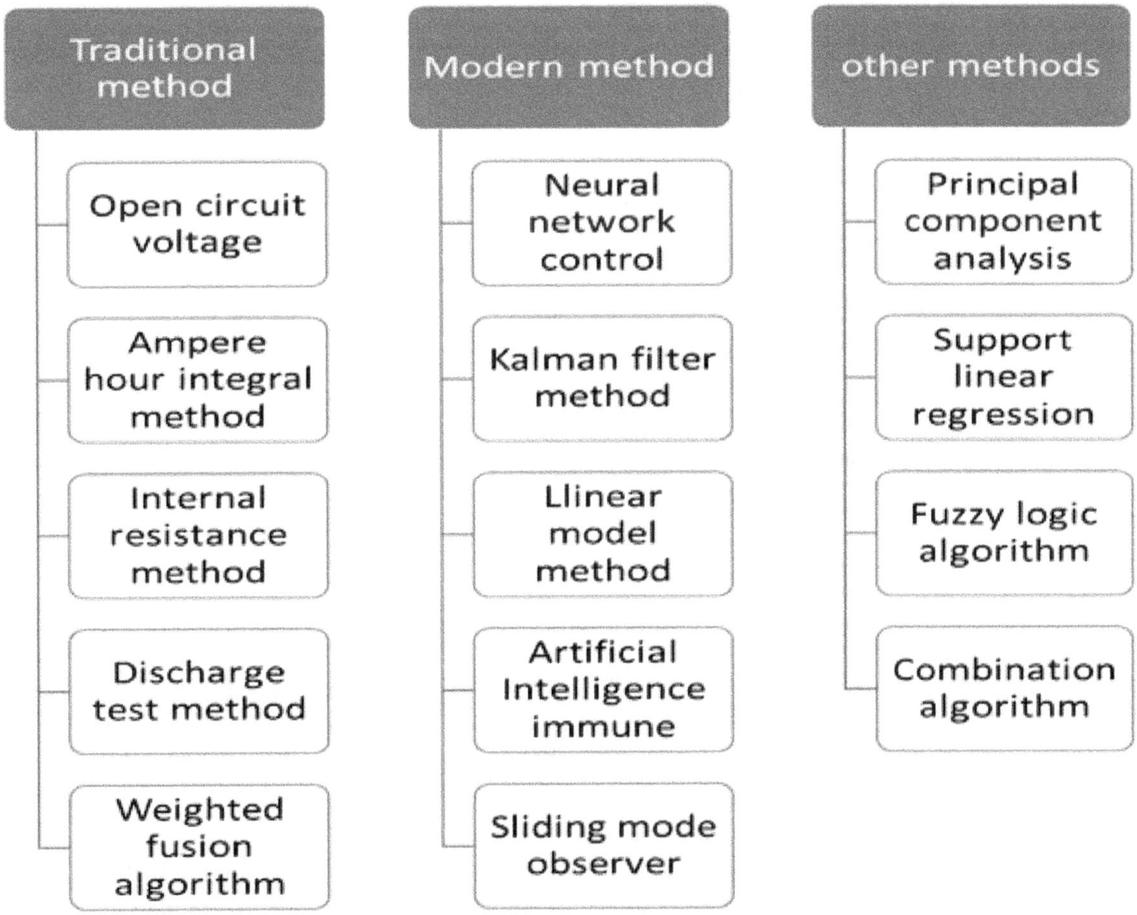

taken into account as well as various variables, such as current, voltage, capacity, Coulomb efficiency, and self-discharge rate. Performance depends heavily on the state of charge of the battery pack.y pack. In order to design a battery management system (BMS), it is essential to estimate the state of charge accurately. A study was conducted to test new methods of estimating and modeling SOC. When estimating the performance characteristics of the machinery, Kalman filters are utilized to estimate component characteristics according to operating conditions. System identification methods were used to model and control many mechanical systems. Control theory-driven DC motor systems were analysed to identify closed-loop DC motor system notifications that can be used to simulate nanocomposite machining and characterize complex materials. Artificial intelligence and machine learning models of manufacturing systems have shown excellent prediction consistency. Because batteries behave highly. In EV systems, SOC, SOE, and PL are difficult to model due to their highly nonlinear behavior under different operating settings. A microprocessor model that operates in real-time. Using newly developed methods, the state of charge of batteries was estimated and modeled. The SOC of lithium-ion battery packs was determined using a fractional order model and a fractional Kalman filter. The state of charge of an electric vehicle's lithium-ion battery pack is determined using Venin fractional order modelling. For estimating EV SOC, support vector machines (SVMs), single particle models (SPMs), and RC equivalent cell models (ECMs) were compared. Researchers have used several types of artificial neural networks to estimate SOC in E-mobility systems. Using a simplified genetic algorithm and artificial neural networks, we achieved a prediction accuracy of 95%. SOC predictions were reduced by 175% using recurrent neural networks.

7.2 State of Health

SOH, caused by battery aging and wear, is characterized by battery degradation and diminished capacity. Battery capacity and impedance components are commonly used to quantify power degradation and capacity fade. It is recommended that batteries be replaced when they reach 80% of their original capacity, in accordance with SOH standards. There are some arbitrary elements in this criterion, despite its simplicity. Comparing the relative impedance of the original and subsequent batteries is another way to detect SOH. There may be a difference in capacity based on other discharge profiles than the continuous current discharge profile, which is valid for 80% capacity only. Since internal cell resistance is one of the factors affecting SOH, synchronized measurements are better. In order to measure SOH, it is necessary to synchronize the current and voltage measurements.

7.3 SOP and SOF Estimation

As well as its remaining charge, the amount of power a battery can deliver/store limits its capability. Keeping the frequency of the power system within a certain range is the primary goal of power system planning and operation. A battery's grid power capability constantly changes over its discharge cycle as DER increases in power distribution networks, according to several studies. A change in power capability is caused by an increase in internal impedance and a decrease in internal OCV. While the battery is operating, the BMS must ensure the manufacturer's upper and lower operating voltages are met. Therefore, much research is aimed at estimating how much power can be generated. The SOP and SOF of a battery determine its power capacity. An electrical circuit battery model is used in a SOP to predict how much power a battery can provide within the specified voltage limits.

7.4 State of Function

In a SOF, the power requirements of a battery are described, and the battery's capability is assessed. Logic yes/no functions can therefore be defined as SOFs. SoFs of 1 indicate that batteries are capable of meeting demand, while SOFs of 0 indicate that they cannot. Battery performance is typically determined by checking the voltage at the minimum terminal, Vmin, as the SOF criterion. There should be a significant difference between this voltage and the voltage at the threshold terminal. Generally, SOFs should be defined as Pp, Pd, and Pmax. The Pp corresponds to the battery's maximum power (Noura et al., 2020). The battery will barely meet the power demand if the SOF is zero, for instance. With the help of SOC, SOH, temperature, and battery fault state, it is possible to calculate the SOF comprehensively. It illustrates the interdependence between the SOF, SOC, and SOH The SOF is calculated by dividing the battery's remaining available energy by the maximum amount of energy that can be stored in the instability and various categories as per the specification showcased in figure.15.

Figure 15. Estimate state with full architecture

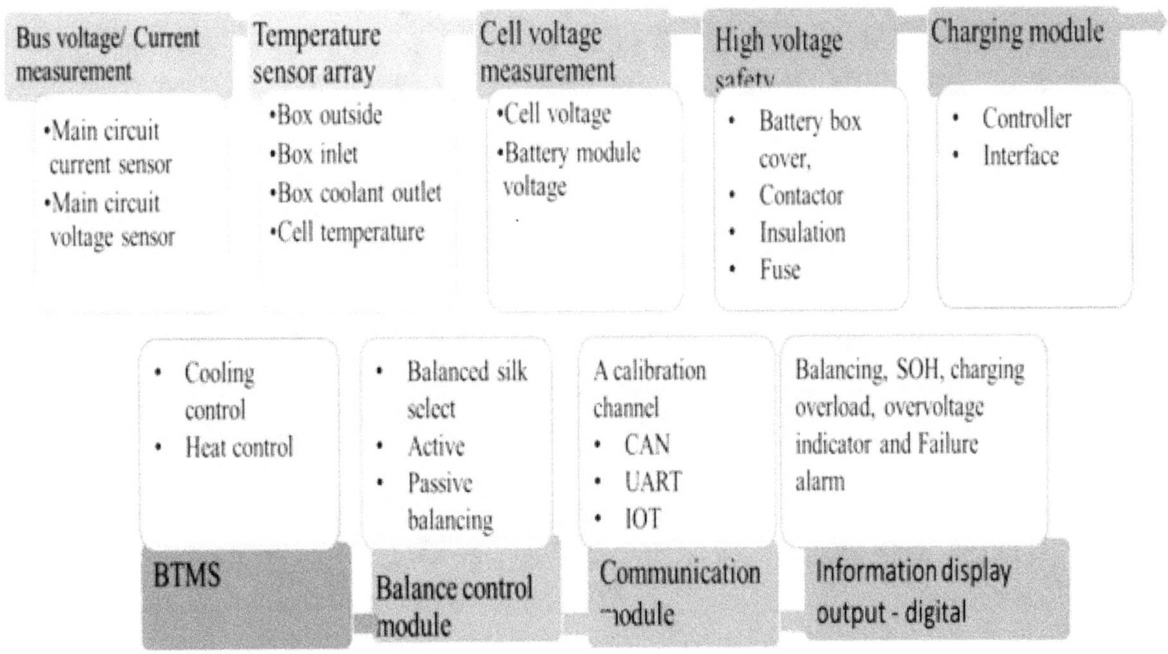

8. BATTERY THERMAL MANAGEMENT SYSTEM

A thermal simulation toolbox based on novel electro-thermal circuit models was developed to optimize the Lithium-Ion Battery Thermal Management System (BTMS) for automotive and energy storage applications. Anode, cathode, separator, electrolyte, and other physical and chemical properties simulation This platform performs in-depth thermal analyses of Li-ion cells under a variety of operating conditions and employs appropriate BTMS control strategies. Temperature sensors are also included, which are essential for optimizing battery loads. Battery cell voltages are used to determine current states of charge (SoC) and control communication between the battery and the vehicle (Hemavathi, 2020).

The command also instructs the battery cells to be balanced when necessary to prevent deep discharge or overcharging. If they are not completely discharged, these cells are then recharged at a high rate until they receive a full charging current. This is even if other partially charged cells receive a lower and aging current. Despite the simplicity and cost-effectiveness of this method, it converts energy into heat, which reduces efficiency as shown in the figure.16.

8.1 Active and Passive

In the case of non-equal voltages between cells, the full cell pack stops charging once some cells reach 100% SoC. It is imperative that the process of balancing is started at this point, not later, by dropping the fully charged cells to the level of the other cells in order to allow them to be powered simultaneously. It is possible to achieve balance in two ways: passively or actively. An algorithm applies resistance to the cells with the highest SoC in passive balancing.

Figure 16. Information in BTMS

- Materials that are critical to safety and cell design Key materials, including cathodes, anodes, electrolyte and separators.
- An important parameter of the BMS is the ratio of the SOC to the OCV.
- Resistance and reactance combine to form impedance.
- R1, R2, and Rint power ratings of the motor as a function of SOC
- Thermal behavior of different design parameters, such as geometrical effects and material properties, at the module level
- Battery pack level specifications for the structural design of the heat generation profile of the battery thermal management system at various temperatures and power demands
- The temperature is an important factor in the transfer of charge and discharge toward the cell chemistry and cell type.
- Thermal management specification at the cell level and standardised method of calculating the cooling coefficient of a cell towards thermal management analysis.
- While in action, the vehicle battery requires characteristics and specifications for thermal analysis.

Figure 17. Active and Passive Balance

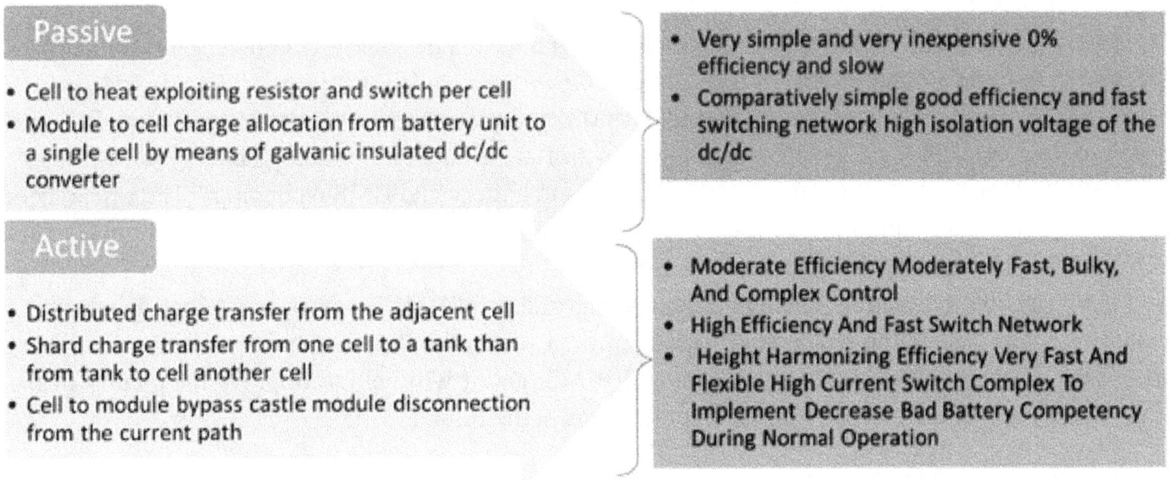

Cells that have not been discharged are then charged at much slower rates. It is not possible to charge the other partially charged batteries in the series circuit until the SoC is balanced. Simple implementation makes this method cost-effective. Energy, however, is converted into heat, which reduces the efficiency of the system. Active balancing, controlled by the BMU, involves the exchange of charge between the battery cells. A full battery cell transfers energy to a neighboring battery cell that has not yet reached its SOC during this process as shown in the figure.17,

- Matrix of transistors for semiconductor switching
- Charge transfer via capacitive devices, as well as
- A charge is transferred inductively

A passive balancing algorithm considers three important parameters to overcome the limitations of existing passive balancing systems: balancing time, power loss, and temperature rise. It is the balancing resistor that determines most of the three parameters. A balance resistor is determined in real time by considering the charging time, charging temperature, charging frequency, thermal effect, battery chemistry, imbalance likelihood, and charger usage. Lithium-ion batteries have a much higher energy density than nickel-cadmium batteries, which is one of the main characteristics of passive balancing. Energy densities of higher magnitude are theoretically possible. A lithium-ion battery's power per unit of energy is higher than other batteries. As a result of machine-learning algorithms, the resistor is selected to increase balancing speed without increasing system temperature or power loss. An evaluation of ML models takes into account power loss from the balancing resistor, balancing time, and mean system temperature.

CONCLUSION AND FUTURE SCOPE

A well-designed smart battery management system is very essential for energy storage (ESS) and electric vehicle (EV) applications. Nowadays, very rapid improvement in technology of lithium-ion batteries provides a higher power density, and energy density and reduces the maintenance cost of the system. This will happen only when we design proper BMS hardware circuits and software algorithms to monitor individual cells as well as battery packs and this guarantees the safety, accuracy, and efficiency of the battery in both EV and energy storage applications.

In most electronic applications, BMS and performance requirements are already provided. But still, plenty of Research and developments are going on in the field of battery monitoring, battery safety, battery production, charge control, etc., to meet the standards in battery and BMS technology. The researchers mainly focus on determining the best battery modeling and estimating the accurate algorithms to find SOC and SOH to predict the RUL and EOL exactly.

Identifying and estimating the highly accurate algorithms of BMS are also used to balance the individual cells and sustain the temperature profile of the Li-ion pack. So, the above work and discussion will assist to improve the basic understanding of BMS, and it will help to improve the quality of the BMS design in the scope of ESS and EV applications in the future.

ACKNOWLEDGMENT

The author would like to thank the Research and development in the battery division of Central Electrochemical Research Institute, Council of Scientific and Industrial Research in India, for the financial assistance. CSIR-CECRI manuscript communication number: CECRI/PESVC/Pubs/2023-010

Conflicts of Interest: The Author declares that there is no conflict of interest.

REFERENCES

Affanni, Bellini, Franceschini, Guglielmi, & Tassoni. (2005). Battery Choice and Management for New-Generation Electric Vehicles. IEEE Transactions on Industrial Electronics, 52(5), 1343–1349.

Andrea, D. (2010). *Battery Management Systems for Large Lithium-Ion Battery Packs*. Artech House. https://www.librarything.com/work/11739168/details/78008584

Arasaratnam, Tjong, Ahmed, El-Sayed, & Habibi. (2013). Adaptive Temperature Monitoring for Battery Thermal Management. In *2013 IEEE Transportation Electrification Conference and Expo (Itec)*. IEEE.

Auger, F., Hilairet, M., Guerrero, J. M., Monmasson, E., Orlowska-Kowalska, T., & Katsura, S. (2013). Industrial Applications of The Kalman Filter: A Review. *IEEE Transactions on Industrial Electronics*, *60*(12), 5458–5471. doi:10.1109/TIE.2012.2236994

Becker, J., Schaeper, C., & Sauer, D. U. (2012). Energy Management System for A Multi-Source Storage System Electric Vehicle. *2012 IEEE Vehicle Power and Propulsion Conference*, 407-412. 10.1109/VPPC.2012.6422664

Chang, W.-Y. (2013). The State of Charge Estimating Methods for Battery: A review. *International Scholarly Research Notices*, *2013*.

Chen, M., Lin, S., Song, W., Lv, J., & Feng, Z. (2020). Electrical and thermal interplay in lithium-ion battery internal short circuit and safety protection. *International Journal of Energy Research, 44*(8), 6745–6757. doi:10.1002/er.5411

Farag, M. (2013). *Lithium-Ion Batteries: Modelling and State of Charge estimation* [Ph.D. Dissertation].

Freeman, D., & Heacock, D. (1993). *Advanced Charger Designs Address Modernbattery Technologies. Proceedings of Wescon 39.*

Haq, N., Leksono, E., Iqbal, M., Sodami, F. N., Kurniadi, D., & Yuliartoet, B. (2014). "Development of Battery Management System for Cell Monitoring And protection. In *2014 International Conference on Electrical Engineering Andcomputer Science (Iceecs).* IEEE. 10.1109/ICEECS.2014.7045246

Hemavathi, S. (2019). Modeling and Estimation of Lithium Ion Battery State of Charge Using Intelligent Techniques. *IEEE International Conference on Computing, Power and Communication Technologies (Gucon 2019).*

Hemavathi, S. (2020). Modeling of Analog Battery Management System for Single Cell Lithium Ion Battery. *Energy Storage*, 1-9.

Hemavathi, S. (2020). Li-Ion Battery Health Estimation Based on Battery Internal Impedance Measurement. *1st National Conference on Innovations in Sustainable Energy and Technology (Iset 2020).*

Hemavathi, S. (2021). *Modeling and Energy Optimization of Hybrid Energy Storage System.* Hybrid Renewable Energy Systems. doi:10.1002/9781119555667.ch4

Hoque, M. M., Hannan, M. A., Mohamed, A., & Ayob, A. (2017). Battery charge equalization controller in electric vehicle applications: A review. *Renewable and Sustainable Energy Reviews, Elsevier, 75*(C), 1363–1385. doi:10.1016/j.rser.2016.11.126

Isakova, Y. I., Pushkarev, A. I., & Kholodnaya, G. E. (2011). A differential high-voltage divider. *Instruments and Experimental Techniques, 54*(2), 183–186. doi:10.1134/S0020441211020175

Jander, C. S., & Schneider, R. (2005). *Magnetoresistive Sensors for Non-Destructive Evaluation. In Advanced Sensor Technologies for Nondestruc-Tive Evaluation and Structural Health Monitoring* (Vol. 5770). International Society for Optics and Photonics.

Jossen, V., Späth, V., Döring, H., & Garche, J. (1999). Reliable Battery Operation—A Challenge for The Battery Management System. *Journal of Power Sources, 84*(2), 283–286. doi:10.1016/S0378-7753(99)00329-8

Kıvrak, S., Özer, T., Oğuz, Y., & Erken, E. B. (2020). Battery management system implementation with the passive control method using MOSFET as a load. *Measurement and Control, 53*(1-2), 205–213. doi:10.1177/0020294019883401

Kovalcik, M. (2015). Constructing A Lithium-Ion/Polymer Battery With An Imbed-Ded Protection Circuit Module (PCM). Michigan State University.

Kumar, M. (2020). Estimation of State of Charge and Terminal Voltage of Li-Ion Battery Using Extended Kalman Filter. *2020 6th IEEE International Energy Conference (Energycon).*

Lelie, M., Braun, T., Knips, M., Nordmann, H., Ringbeck, F., Zappen, H., & Sauer, D.U. (2018). Battery Management System Hardware Concepts: An Overview. *Applied Sciences, 8*(4). doi:10.3390/app8040534

Liu, G., Guo, L., Liu, C., & Wu, Q. (2018). Evaluation of Different Calibration equations for Ntc Thermistor Applied To High-Precision Temperature Measurement. *Measurement, 120*, 21–27. doi:10.1016/j.measurement.2018.02.007

Liu, P.-J., & Chien, L.-H. (2017). A High-Efficiency Integrated Multimode Batterycharger With An Adaptive Supply Voltage Control Scheme. *IEEE Transactions on Power Electronics, 33*(8), 6869–6876. doi:10.1109/TPEL.2017.2761816

Lua, L., Hana, X., Lia, J., Huab, J., & Ouyanga, M. (2013). A Review on The Key Issues for Lithium-Ion Battery Management In Electric Vehicles. *Journal of Power Sources, 226*, 272–288. doi:10.1016/j.jpowsour.2012.10.060

Meissner, E., & Richter, G. (2003). Battery Monitoring and Electrical Energy Man-Agement: Precondition for Future Vehicle Electric Power Systems. *Journal of Power Sources, 116*(1-2), 79–98. doi:10.1016/S0378-7753(02)00713-9

Meng, Luo, Ricco, Swierczynski, Stroe, & Teodor-Escu. (2018). Overview of Lithium-Ion Battery Modeling Methods for State-Of-Charge Estimation in Electrical Vehicles. Applied Sciences, 8(5), 659.

Mutyala, M. S. K., Zhao, J., Li, J., Pan, H., Yuan, C., & Li, X. (2014). In-Situ temperature Measurement In Lithium Ion Battery By Transferable Flexible thin Film Thermocouples. *Journal of Power Sources, 260*, 43–49. doi:10.1016/j.jpowsour.2014.03.004

Nikolian, A., de Hoog, J., Fleurbaey, K., Timmermans, J. M., Omar, N., van den Bossche, P., & van Mierlo, J. (2014). Classification of Electric modeling and Characterization methods of Lithium-ion Batteriesfor Vehicle Applications. *Proceedings of the European Electric Vehicle Congress*, 2–5.

Nitta, N., Wu, F., Lee, J. T., & Yushin, G. (2015). Li-Ion Battery Materials: Present and Future. *Materials Today, 18*(5), 252–264. doi:10.1016/j.mattod.2014.10.040

Noura, N., Boulon, L., & Jemeï, S. (2020). A Review of Battery State of Health Estimation Methods: Hybrid Electric Vehicle Challenges. *World Electric Vehicle Journal, 11*(4), 66. doi:10.3390/wevj11040066

Plett, G. L. (2004). Extended Kalman Filtering for Battery Management Systems of Lipb-Based Hev Battery Packs: Part 3. State and Parameter Estimation. *Journal of Power Sources, 134*(2), 277–292. doi:10.1016/j.jpowsour.2004.02.033

Qaisar, S. M., Dallet, D., Benjamin, S., Desprez, P., & Yahiaoui, R. (2013). Power efficient Analog To Digital Conversion for The Li-Ion Battery Voltage Monitoring and Measurement. In *2013 IEEE International Instrumentation and Measurement Technology Conference (I2Mtc)*. IEEE.

Rahimi-Eichi, H., Ojha, U., Baronti, F., & Chow, M.-Y. (2013). Battery Manage-Ment System: An Overview of Its Application In The Smart Grid and Electric vehicles. *IEEE Industrial Electronics Magazine, 7*(2), 4–16. doi:10.1109/MIE.2013.2250351

Richard, B. (2018). State of Charge and Parameter Estimation of Electric Vehicle Batteries. doi:10.25071/10315/35324

Rivera-Barrera, Munoz-Galeano, & Sarmiento-Maldonado. (2017). Soc Estimation for Lithium-Ion Batteries: Review and Future Challenges. Electronics, 6(4).

Sajfar, M., & Bullough, R. (1990). Sealed Batteries In Transient Limit-Ing Distribution Networks-Methods of Measuring Their Internal Resistance. In *12th International Conference on Telecommunications Energy*. IEEE.

Saji, D., Babu, P. S., & Ilango, K. (2019). Soc Estimation of Lithium Ion Battery using Combined Coulomb Counting and Fuzzy Logic Method. In *2019 4th international Conference on Recent Trends on Electronics, Information, Communication & Technology (Rteict)*. IEEE. 10.1109/RTEICT46194.2019.9016956

Schwarz, R., Semmler, K., Wenger, M., Lorentz, V. R., & Arz, M. (2015). Sensorless Battery Cell Temperature Estimation Circuit for Enhanced Safety in Battery Systems. In *IECON 2015-41St Annual Conference of The IEEE Industrial Electronics Society*. IEEE.

Sironval, V. (2020). *Evaluating and Predicting the Respiratory Hazard of Particles Used in Li-Ion Batteries, for A Safe and Sustainable Development* [Ph.D. Dissertation]. Catholique De Louvain.

Viera, Gonzalez, Anton, Campo, Ferrero, & Valle-Dor. (2006). Nimh Vs Nicd Batteries Under High Charging Rates. In *INTELEC06-Twenty-Eighth International Telecommunications Energy Conference*. IEEE.

Wang, C., Gu, W., & Liaw, B. (1998). Micro-Macroscopic Coupled Modeling Ofbatteries and Fuel Cells: I. Model Development. *Journal of the Electrochemical Society*, *145*(10), 3407–3417. doi:10.1149/1.1838820

Wang, K., Gao, F., Zhu, Y., Liu, H., Qi, C., Yang, K., & Jiao, Q. (2018). Internal resistance and heat generation of soft package Li4Ti5O12 battery during charge and discharge. *Energy*, *149*, 364–374. doi:10.1016/j.energy.2018.02.052

Wang, L., Cheng, Y., & Zhao, X. (2015). Influence of Connecting Plate Resistanceupon Lifepo4 Battery Performance. *Applied Energy*, *147*, 353–360. doi:10.1016/j.apenergy.2015.03.016

Yao, Y., Chen, H.-Y., Huang, J., & Yang, Y. (2007). Low Voltage and Fastspeed All-Polymeric Optocouplers. *Applied Physics Letters*, *90*(5), 053509. doi:10.1063/1.2437683

Zhang, C., Li, K., Mcloone, S., & Yang, Z. (2014). Battery modelling methods for electric vehicles - A review. *2014 European Control Conference (ECC)*, 2673-2678. 10.1109/ECC.2014.6862541

Zhang, J., & Lee, J. (2011, August). A Review on Prognostics and Health Monitoring Ofli-Ion Battery. *Journal of Power Sources*, *196*(15), 6007–6014. doi:10.1016/j.jpowsour.2011.03.101

Zhia, L., Pengb, Z., Zhifua, W., Song, Q., & Yinana, R. (2017). State of Charge Estimation for Li-Ion Battery Based on Extended Kalman Filter. *Energy Procedia*, *105*, 3515–3520. doi:10.1016/j.egypro.2017.03.806

Chapter 2
Battery Monitoring System and SOC Enhancement Analysis Using Artificial Intelligence Techniques

Mohana Sundaram K.
KPR Institute of Engineering and Technology, Coimbatore, India

Kavya Santhoshi B.
Godavari Institute of Engineering and Technology (Autonomous), India

Chandrika V. S.
https://orcid.org/0000-0001-8417-4742
KPR Institute of Engineering and Technology, Coimbatore, India

ABSTRACT

A battery energy storage system (BESS) is generally used for storing the excess power produced from the solar PV panel and supplying the stored electrical energy whenever it is needed. The accurate estimation of the state of charge (SOC) of the battery is considered to be very essential for proper power management and reliable operation of the battery. SOC monitoring system for BESS based on probabilistic neural network (PNN) approach is proposed. The PV panel output voltage is enhanced and made distortion free by single ended primary inductance converter (SEPIC) with the assistance of PNN controller. The output thus obtained from the converter is used for many DC applications. The performance of the proposed battery monitoring model is evaluated by using MATLAB simulation.

DOI: 10.4018/978-1-6684-6631-5.ch002

INTRODUCTION

(a) Rise of Renewable Energy

The need for power is rising day to day while; non-renewable energy sources are rapidly exhausting. To devastate issues, effective and efficient electric power generation from RE source is necessary. Renewable energy sources are regarded as clean energy sources, since they produce little secondary waste, and have little influence on environment. The growth of renewable energy (RE) is crucial for addressing the energy problem and environmental pollution. Solar, wind, small hydro, biomass, tidal, and geothermal energy are some of the renewable resources that are commercially available for producing power. High power quality is a challenge in photovoltaic generation, yet PV power generation is becoming more popular these days. Generally, PV generator setup to supply nonlinear loads is preferred when it can also function as an active power filter (APF).

In power systems, the major concern lies in power quality. Delivery of quality power to the grid still remains a challenge to the researchers. Due to increased use of nonlinear loads, the effect of harmonics can be felt as a degradation of power quality. Grid-connected systems, nowadays, use PV power or any other form of non-conventional source due to numerous benefits. Among all the forms of non-conventional sources, PV power is viewed with great prospects due to easy availability. Over the past ten years, solar photovoltaic (PV) energy has grown in popularity among sources of renewable energy (RE).Solar PV based energy generating systems are widely used renewable resources because of their static structure, compact size, and nearly maintenance-free nature. It is anticipated that the photovoltaic (PV) system, which accounts for a sizeable share of global renewable energy, will contribute more to change in the power systems. As a result, many frequency management strategies for PV systems have been created. Since, PV systems are intermittent and engage low DC power, a converter operation is essential.

(b) Power Electronic Converters in PV-Grid Systems

Power electronic converters for use in PV-grid systems are predominantly used in recent times. This is because of the variation in the levels of voltage output which can be achieved by using boost/buck converters and also the ability of the converters to utilize the available form of supply to modulate and provide the output in the desired manner. In PV applications, boost converters are frequently employed for DC-DC conversion. A typical Boost converter is operated at a ratio of high duty cycle in order to generate a significant step-up voltage. However, the diode's reverse recovery time and the presence of parasitic elements cause a drop in the step-up ratio and reliability of the device. A DC-DC converter known as a "buck-boost converter" generates an inverted output voltage that is larger or lower than the input voltage. Although, typical buck-boost converter convert voltage in both step-up and step-down directions, its discontinuous input current prevents it from doing MPPT at its best without using a lot of decoupling capacitors. Generally, controller operation is important for a converter to extract maximum output power from PV systems. For standalone and grid-correlated photovoltaic applications, a number of maximum power point tracking (MPPT) algorithms with varying degrees of complexity, efficiency, and expenses have been proposed. The mechanism of Maximum Power Point Tracking (MPPT) functioning becomes challenging since DC current used by the PV panels is shared. The MPPT may not always be reached, even though the centralized inverter systems lower power loss inside power converters, the captured energy is reduced.

(c) Significance of SEPIC Converters

For sake of continuous input current, SEPIC operates on the combined concept of buck and boost converter. The converter acts as an excellent choice for PV techniques since the SEPIC idea is applicable for current and voltage applications. SEPIC converters, with their ability to boost the voltage, are widely preferred. Though, Cuk and SEPIC converters have uninterrupted input currents and can step down or up voltage, their high input power ripples still restrict the MPPT's efficiency because PV system will vary greatly near its maximum power point (MPP). In this work, SEPIC converters and a modified form of SEPIC converter for a PV-grid system that produces lesser harmonics compared to the standard SEPIC converter, is proposed. Also, the use of multiple SEPIC choppers is employed due to its low complexity.

A BRIEF REVIEW ON PV-GRID SYSTEMS

(a) Need for Power Conversion Systems (PCS)

In recent times, the world has started to recognize the need to invest and expand renewable energy sources for power generation in order to combat climate change and advance towards a carbon free future. Among all renewable energy sources, solar energy is gaining more prominence due to the steep decline in the price of PV panels. Overall natural change and energy supply is declining have animated changes in vehicular development. For the applications in later vehicles the cutting-edge innovations are at present being researched. Among such applications, power module mixture electric vehicles (FCV/HEV) are productive and promising competitors. The PV system is a type of power generation system that harnesses solar energy to generate electric power by utilizing the principle of photovoltaic effect (C. M. & P., 2021). The PV output voltage tends to fluctuate from time to time owing to factors like temperature, insolation and climatic conditions.

There are many DC-DC converters that are generally designed to carry out voltage boosting operation. These converters are classified on the basis of the isolation property. In PV applications, boost converters are often employed to execute DC-DC power conversion. An inductor which is present at the input of the boost converter provides voltage boosting capacity and allows the converter to operate in both continuous and discontinuous conduction modes, however only step-up voltage operation is achievable. The conventional buck-boost converter is able to carry out the voltage step up as well as voltage step down operation, however, in the absence of big decoupling capacitors, discontinuous input current inhibits its capacity to conduct optimal MPPT. The Cuk converter does the same operation of a buck boost converter and has continuous output and input current characteristics. Although the cuk converter has a high static gain, it is not appropriate for low voltage solar panels. The use of a cuk converter with a high duty cycle value during steady state condition is to be prevented since it affects the efficiency and raises current and voltage stress. The SEPIC converters possess continuous input currents and execute both voltage step-down and step-up operations; however, the high ripple content in the input current inhibits the solar PV system's maximum power extraction potential. The zeta converter, which belongs to the buck-boost converter family, is employed to raise or lower the output voltage. The presence of inductor at the output keeps the current steady and prevent ripples. It suffers to maintain the voltage during multiple output condition.

Hence, a proper Power Conversion System (PCS) is needed to transfer power from solar PV panel to grid and various loads. In grid connected PV system, the PCS comprises of a DC-DC converter to perform both voltage step up as well as voltage step down operation and an inverter to perform DC to AC power conversion (Sreekanth et al., 2017).

(b) Need for AI Techniques in PV-Grid Systems

The PV system delivers low output voltage that is converted into a high DC voltage as per the requirements by implementing different types of DC-DC converters like boost converter (Errouissi et al., 2016), buck-boost converter (Zadeh & Fathi, 2017), Cuk converter (de Morais et al., 2019), etc. The performance of the converter is considerably enhanced with the usage of efficient control techniques, which includes conventional controllers and Artificial Intelligence (AI) controllers (Ali, Wang, Alquhayz et al, 2021). In (Seidi Khorramabadi & Bakhshai, 2015), conventional Proportional Integral (PI) controller is used for the purpose of regulating the DC variables but the performance of the controller is affected by sudden changes in the PV parameters. To deliver improved DC voltage with minimum loss, AI techniques such as Artificial Neural Network (ANN) (Kavya Santhoshi et al., 2019; Santhoshi, Mohanasundaram, & Kumar, 2021) and Fuzzy Logic Controller (FLC) (Santhoshi, Sundaram, Hussain et al, 2021; Yap et al., 2021) are adopted. The ANN approach generally requires more computational time to train and design the system. When it comes to FLC approach, formulating the membership function is extremely difficult. Hence, in our proposed system, the PNN controller is used to address the issues encountered by other controllers that are discussed earlier.

(c) Need for MPPT & Intelligent Algorithms

The large scale implementation of PV system is limited by certain issues like high installation cost, variation of output power with changing weather conditions and incoming irradiance. In order to overcome these issues and to extract maximum power in the available weather condition, a technique called MPPT (Maximum power point tracking) is adapted. Several MPPT methods are developed to figure out the MPP (maximum power point) which include traditional, artificial intelligence and hybrid methods. The most commonly used traditional MPPT algorithms are Perturb and observe (P&O), incremental conductance (INC) and Hill climbing (HC). The HC and P&O methods are easy to implement and contributes for power loss due to the presence of high oscillations around the maximum power point (MPP). Under changing atmospheric conditions, the INC technique is precise and adaptable. It does not have simulation and experimental complications. However, the algorithms cited above are ineffective when the solar radiation varies and when estimating the exact perturbation size. Hence, intelligence algorithms such as fuzzy logic control (FLC) and artificial neural network (ANN) techniques are chosen as MPPT trackers to tackle the shortcomings of traditional MPPT algorithms under changing weather circumstances. The fuzzy logic based MPPT technique has a simple and robust design, handles the solar power system's non linearity and uncertainty problems. In fuzzy logic technique, the detection of membership function is a tough and time consuming task. The NN based control algorithms has multiple layer of neurons that are broadly utilized for quick tracking of PV power in uncertain environmental conditions. The neural network technique requires a longer processing time and huge data for training and modelling the system in order to obtain excellent tracking of MPP

GENERATING SWITCHING SIGNALS THROUGH ANN APPROACH

ANN has the potential of providing an enhanced methodology for deriving nonlinear models. It possesses the self-adapting capability, which is suitable for handling the parameter variations, uncertainties and non-linearities. Because of the ability of generalization and learning, it is generally utilized for identification.

To obtain the required switching signals, the approach called ANN is introduced. The ANN controller is utilized for rapid estimation and generation of reference voltage. The input voltage vector $\left(V_{DC,LOW}, V_{DC,LOW}^{ref}\right)$ is fed to the state generator as given by,

$$V = \left[V_{DC,LOW}^{ref} V_{DC,LOW}\right]^T \tag{1}$$

The obtained states from the state generator are termed as x_1 and x_2,

$$x_1 = V_e(m), x_2 = \frac{\delta x_1}{\delta m} \tag{2}$$

Where, $V_e(m) = V_{DC,L}^{ref}$ -$V_{DC,L}(m)$. The output error Z(m) is given by,

$$Z(m) = V_o(m) - V_o(m-1) \tag{3}$$

For computing I_β, the generated output voltage $V_o(m)$ is applied to the output state. With the assistance of neuron cell, the controlling signals are generated across the interrelated gatherings as given by,

$$V(m) = V(m-1) + \sum_{k=i}^{ii} W_k(m) a_k(m) \tag{4}$$

Where, W_k is the weight of the system.

By utilizing covariance algorithm, the weight of the neuron is updated by Hebb's rule, which is given by,

$$\Delta W_k(m) = F_k(Y(m), a_{k(m)}) \tag{5}$$

The change in weight of mth moment is characterized as,

$$\Delta W_k(m) = -c \frac{\delta F_k(m)}{\delta W_k(m)} \tag{6}$$

STANDARD SEPIC CONVERTER TOPOLOGY

The circuit of a standard SEPIC converter is shown in figure 1. The circuit comprises of two converters. One is a dc-dc boost converter which has the ability to boost the voltage up to six times. Another converter is the voltage source converter, which acts as an inverter. The capacitor together with this VSC forms an APF. AC voltage which is injected to the grid has reduced harmonics. The use of MPPT controller gives the added advantage of extraction of maximum power from PV unit. In this converter soft switching is not possible and due to high switching losses and poor voltage gain of the boost converter, efficiency becomes lesser.

Figure 1. Circuit of a standard SEPIC converter

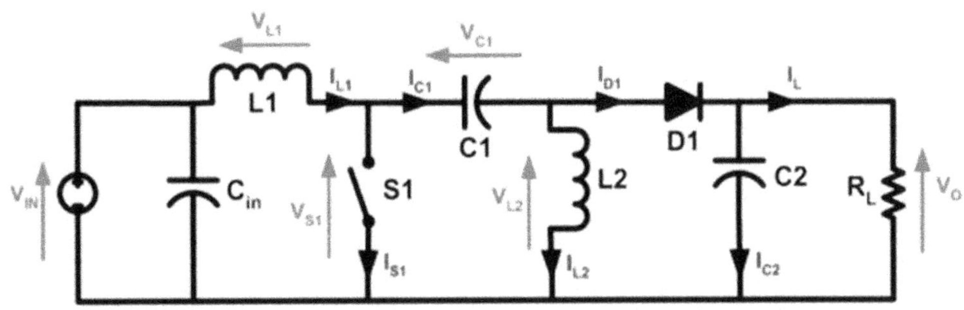

PROPOSED SOC MONITORING MODEL

The DC output voltage generated from the PV source is generally low and fluctuates because of sporadic nature of the solar energy. A DC-DC SEPIC converter is a high gain converter, used to enhance the PV source output voltage. The output of the converter varies and not remains constant with the changing weather conditions and solar intensity. The actual voltage V_{act} obtained from the converter is compared with the reference voltage V_{ref} and error is produced. This error when provided as input to the PNN controller, a control signal is generated. Based on this control signal, the PWM generator generates PWM pulses, which in turn decides the operation of switches in the converter and a controlled output without any distortion is produced. This controlled DC output is used for DC applications like EV battery charging. The block diagram of the proposed SOC monitoring model is illustrated in Figure 2.

Due to the intermittent nature of PV source, the Energy Storage System (ESS) like battery serves an important function in PV system based applications for providing reliable power supply and maintaining energy balance. The SOC of the battery is monitored in order to control the charging as well as the discharging operation of the battery with the assistance of PNN controller.

Figure 2. Block diagram of proposed SOC monitoring model

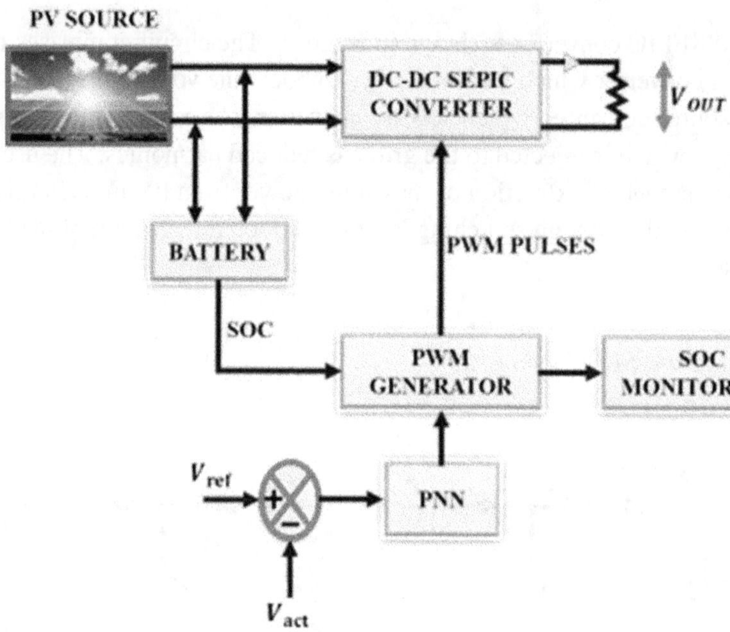

PROPOSED SYSTEM MODELLING

A) Solar PV System Modelling

A PV cell, which converts the energy emitted by the sun in the form of light into electricity, is a low voltage DC source. When number of solar cells is connected together, a solar panel is formed. The equivalent circuit of a PV device is represented in Figure 3.

The basic current equation of the PV device is given as,

Figure 3. Equivalent circuit of a PV device

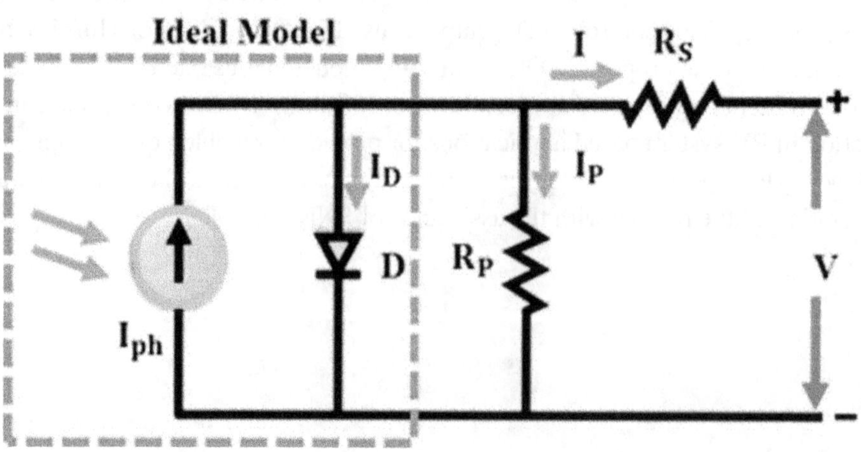

Figure 4. SEPIC converter circuit diagram

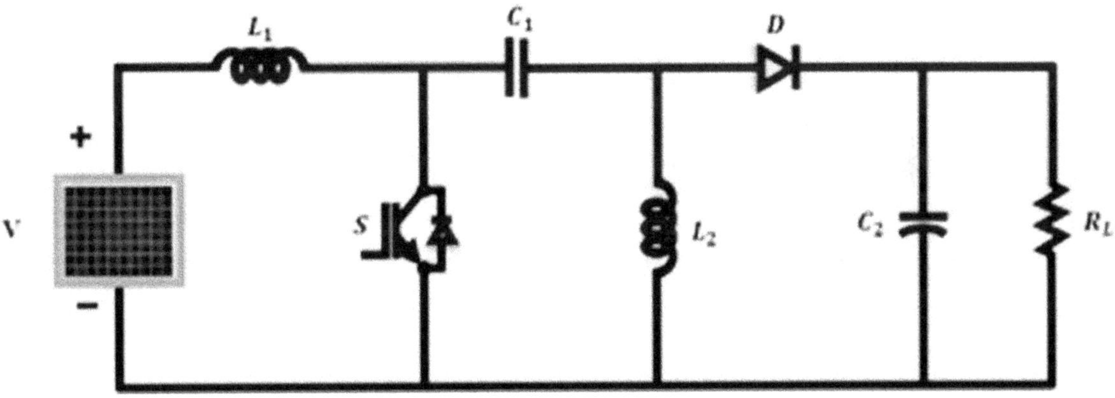

$$I = I_{ph} - I_D - I_P \tag{1}$$

$$I = I_{ph} - I_O\left[\exp\left(\frac{V + IR_S}{Nv}\right) - 1\right] - \frac{V + IR_S}{R_P} \tag{2}$$

Here, I_{ph} represents the current produced by incident light, the PV output current and voltage are specified by I and V respectively, the series and parallel resistance are denoted as R_S and R_P respectively and I_D specifies the diode saturation current.

B) SEPIC Converter

The Single Ended Primary Inductance Converter (SEPIC) is a DC-DC converter, which is similar to the Cuk converter. This converter is capable for operating in either buck or boost mode without reversing the polarity. The basic circuit diagram of the SEPIC converter is depicted in Figure 4.

Mode 1

During this mode of operation, the power switch S is closed i.e. ON condition and the diode D gets reverse biased. The capacitor C_1 discharges at this stage, while the inductors L_1 and L_2 start charging energy. The Figure 5 (a) illustrates the mode 1 operation of the SEPIC converter.

Mode 2

During this mode of operation, the power switch S is open i.e. OFF condition and the diode D gets forward biased. At this stage, the inductors L_1 and L_2 discharge the stored energy and the load resistance R_L get supply from the inductor L_2. The Figure 5 (b) illustrates the mode 2 operation of the SEPIC converter.

The output voltage of the SEPIC converter is,

Figure 5. (a) Mode 1 and (b) Mode 2

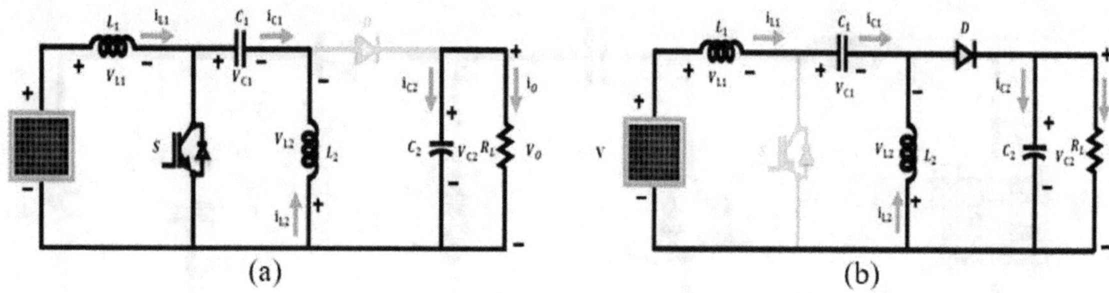

$$V_O = V\left(\frac{D}{1-D}\right) \tag{3}$$

From the above equation,

$$D = \frac{V_o}{V_o + V} \tag{4}$$

Here, D represents the duty cycle ratio of the power switch.
The value of the inductor L_1 is,

$$L_1 = \frac{VD}{\Delta i_{L_1} f} \tag{5}$$

The value of the inductor L_2 is,

$$L_2 = \frac{VD}{\Delta i_{L_2} f} \tag{6}$$

The value of the capacitor C_1 is,

$$C_1 = \frac{D}{R_L\left(\frac{\Delta V_{C_1}}{V_O}\right) f} \tag{7}$$

The value of the capacitor C_2 is,

$$C_2 = \frac{D}{R_L\left(\frac{\Delta V_O}{V_O}\right)f} \tag{8}$$

Here f is the switching frequency.

The output voltage of the converter is enhanced by using the PNN controller.

ENERGY MANAGEMENT SCHEME

The Battery Energy Storage System (BESS) is used with the PV system for storing energy when power generation is high and discharge it when there is an energy deficit. The State of charge (SOC) of a battery is an important parameter that specifies how much charge is available in a battery. The battery's durability, efficiency and reliability are improved with proper monitoring and estimation of SOC. Numerous immeasurable factors, such as battery chemistry, ambient environment, ageing factor and so on, influence SOC. A SOC estimation model for BESS based on a PNN approach is proposed in this paper. The PV system offers low DC output voltage, which is enhanced and controlled with the implementation of PNN assisted SEPIC converter. PNN is one kind of feed forward neural network that controls the charging and discharging process of the battery with proper SOC estimation. Continuous and stable power can be obtained from PV based system when there is no compromise made on the energy storage part.

PNN CONTROLLER FOR SEPIC CONVERTER AND SOC ESTIMATION:

The PNN is a type of artificial neural network that is capable of calculating the probability density function for a set of input data. It is used to solve classification problems and compute nonlinear decisions. The structure of PNN consists of several interconnected processing units called neurons, which are arranged in successive layers as illustrated in Figure 6. There are totally four layers in the PNN namely, input layer, pattern layer, summation layer and the output layer. The input layer receives the input and distributes it to the neurons present in the pattern layer, without performing any computations. The neurons of pattern layer perform computation after receiving the data from the input layer and it expresses the classification training sample. Each neuron of this layer is capable of expressing one classification. The final classification results are obtained from the output layer.

The output voltage obtained from the SEPIC converter varies from time to time and does not remain stable due to the variations in the solar intensity and temperature. Hence, a PNN controller is implemented for controlling the operation of switches of the converter for producing a more stable and controlled output with minimized ripple content. The actual voltage V_{act} obtained from the converter is compared with the reference voltage V_{ref} and an error is produced. This error when provided as input to the PNN controller, a control signal is generated that is fed to the PWM generator. The PWM generator produces PWM pulses, which control the switching operation of the converter for the generation of controlled output with no distortion. This controlled DC output is used for many DC applications. The SOC of the battery is frequently monitored in order to regulate the charging and discharging process. By using the PNN controller, the charging time of the battery is reduced and the battery life span is increased. If the

Figure 6. PNN structure

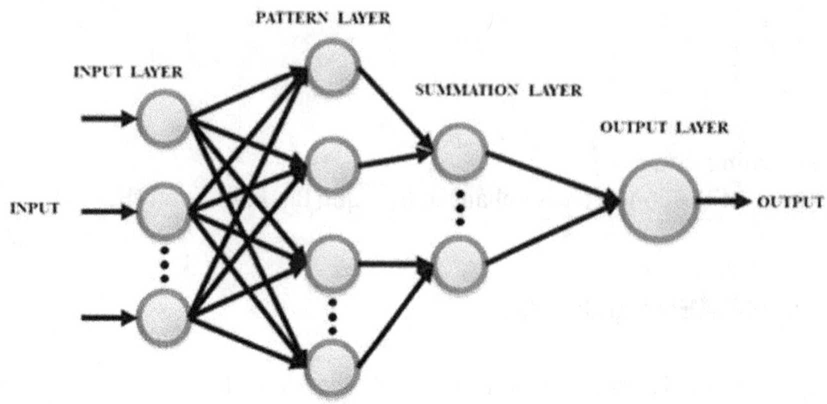

SOC of the battery is less than 60% then the charging process is initiated by the PNN controller. If the SOC of the battery is greater than 60% then the discharging process is initiated by the PNN controller.

SIMULATED RESULTS FOR IMPLEMENTATION OF PNN BASED SOC MONITORING SYSTEM

The PNN based SOC monitoring technique is the major focus of this section. In addition to monitoring of SOC, the PNN is also used for the control of SEPIC converter. The SEPIC is an effective converter, which has the ability to both step-up and step-down its input voltage. The operational performance of the proposed technique is analysed using MATLAB simulation. The parameters for PV and SEPIC converter are given in Table 1.

The transformation of PV output to a necessary level is accomplished using a SEPIC converter, which is a high gain DC-DC converter with non-inverted output in addition to minimum losses. The SEPIC converter provides a voltage of 350V and a current of 2500A as given in Figure 7.

The waveforms that represents the level of current and voltage in the battery is given in Figure 8. The battery current and voltage is 1A and 12V respectively. The surplus power generated from the PV is stored in battery. During the unavailability of primary source of power supply, the battery, acts as a secondary power source that discharges and transfers power to the load.

The waveforms for the SOC of the battery, which is 60% as depicted in Figure 9. The SOC of the battery is effectively monitored using PNN. When the SOC value goes below 60%, the PNN enables the charging of battery. The discharging of the battery is enabled when the SOC is above 60%.

MODIFIED SEPIC CONVERTER TOPOLOGY:

Block diagram of the PV system with improved SEPIC converter is shown in figure 10. The use of one way primary inductor in the topology of the improved SEPIC converter plays a key role in boosting the voltage and improving the overall power quality. The use of MPPT controller gives the added advantage of extraction of maximum power from PV unit. The capacitor together with the voltage source converter

Table 1. Specifications of parameters

Parameter	Values
Solar Panel	
Power	2000W
No. of panels	100W, 20 panels
No. of solar panels connected in series	36
V_{SC}	12V
V_{OC}	22.6V
I_{SC}	8.33A
SEPIC converter	
Inductors (L_1,L_2)	1mH
Capacitors (C_1,C_2)	22μF
Switching Frequency	10kHz

Figure 7. SEPIC converter (a) Output voltage and (b) Output current

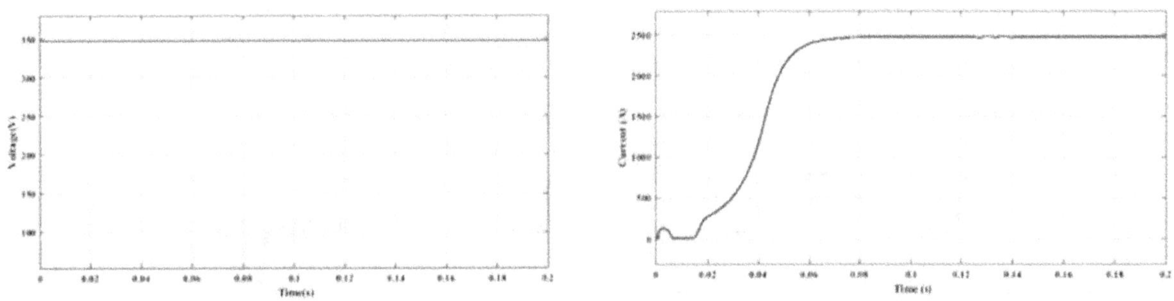

(a) (b)

Figure 8. Waveforms of battery (a) Current and (b) Voltage

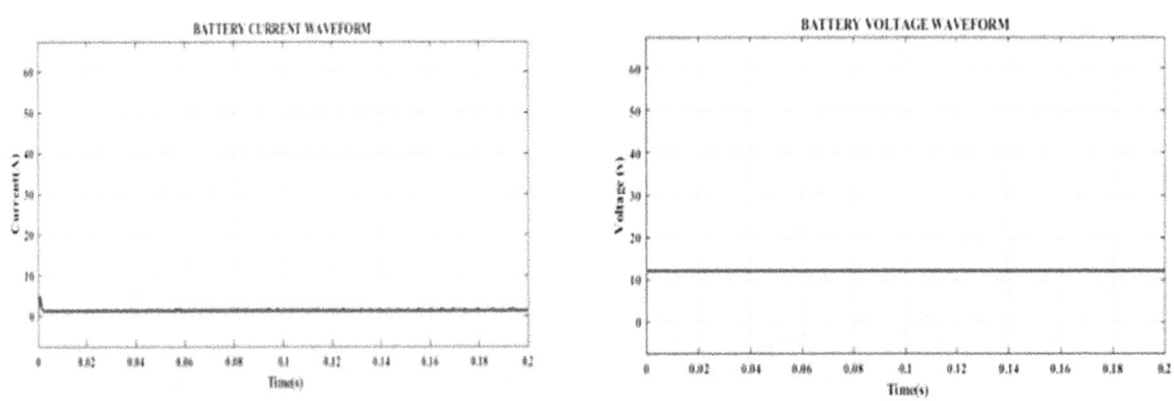

(a) (b)

Figure 9. SOC of Battery

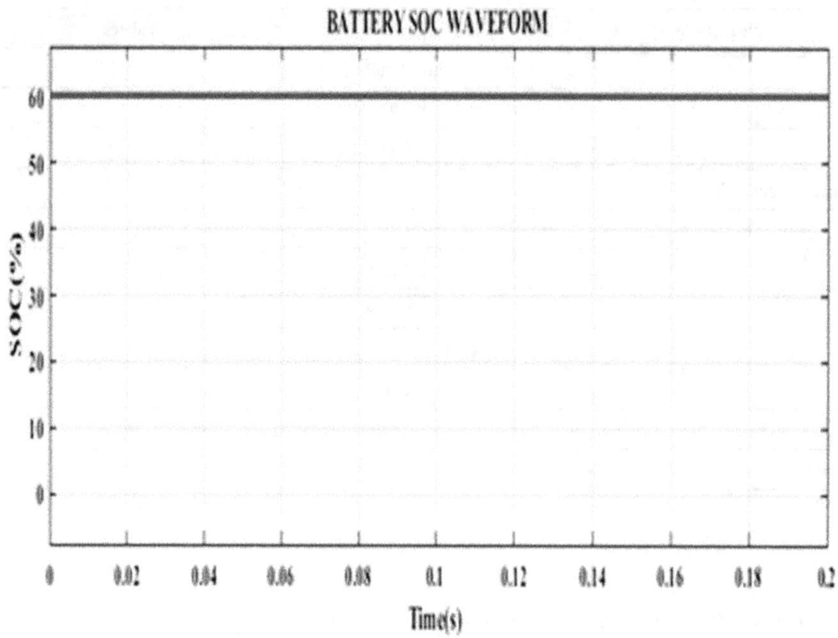

Figure 10. Block diagram of the proposed work with improved SEPIC converter

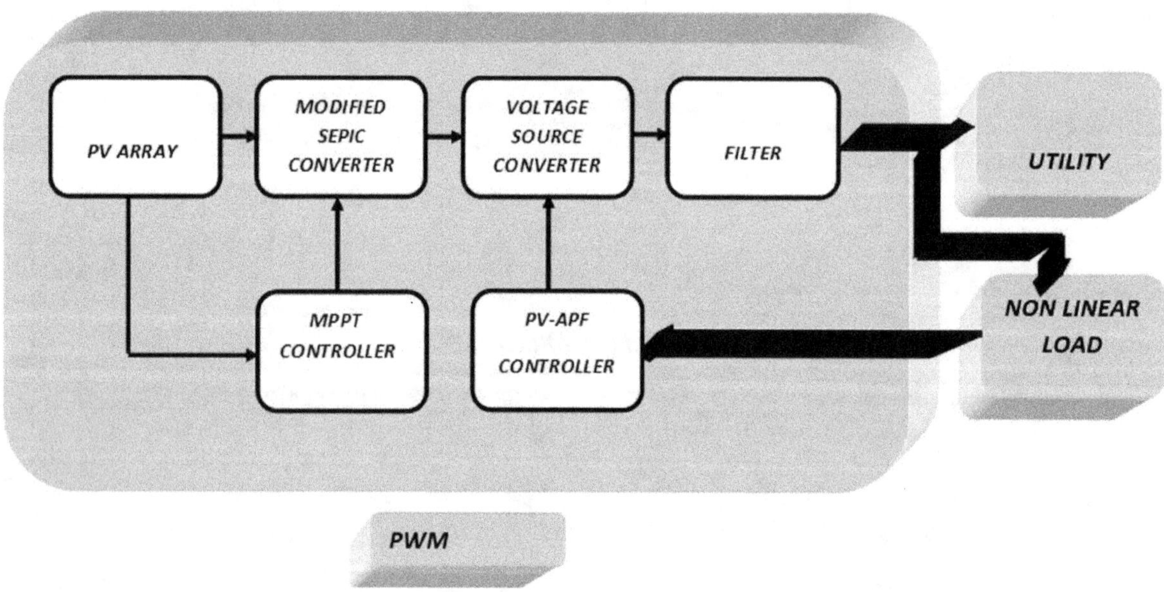

forms an APF. The improved one way primary inductor topology of the converter boosts the dc voltage up to ten times higher than the PV input voltage. Through the proposed converter, higher gain is obtained and soft switching becomes feasible while reducing the switching loss and increasing the efficiency.

The MATLAB simulink model of the proposed modified SEPIC converter circuit is presented in figure 11. The grid voltage and current with improved power quality as per standards are obtained and

Figure 11. Improved SEPIC converter with MPPT controller

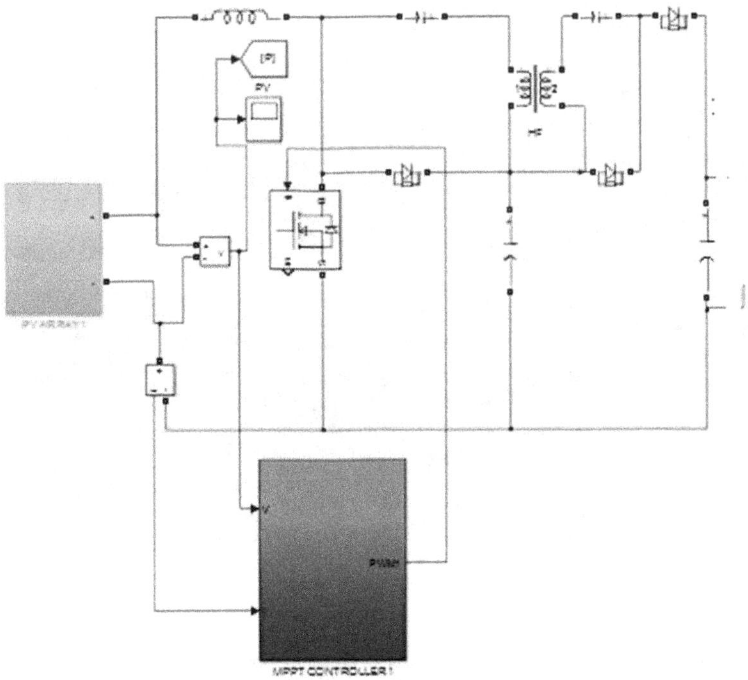

indicated in figure 12. Voltage is boosted up to ten times that of the input. It is observed that the proposed work ensures more reliability in terms of quality power delivered. PV voltage and SEPIC converter voltage are shown in figures 13 and 14 respectively. With the filtering action achieved, the output of voltage source converter shows significant improvement in the power quality as referred from figures 15 and 16. It can be observed that the voltage waveforms after filter are nearly sinusoidal compared to that before filter. From the simulation results of the improved SEPIC converter, it can be noted that the proposed topology yielded an output with reduced THD. The proposed improved SEPIC converter produced only 4.06% THD as depicted in the figure 17. The standard converter produced a THD of 28.64% as shown in figure 18.

PROPOSED BLOCK DIAGRAM OF PV SYSTEM WITH MULTIPLE SEPIC CHOPPERS AND DMPPT APPROACH

Solar energy is the most reliable renewable source, due to the effective use by recent technologies. In a PV system, electricity is generated by sunlight directly. The solar panel characteristic depends on two parameters, namely solar irradiation and temperature. The entire block diagram for standalone PV approach is demonstrated in Figure 19.

This work involves multiple PV panels, with two major components, a converter to boost up the system performance and a controller to stabilize the DC voltage. Here, multiple PV panels are connected which combines and works like a PV system. The DC power generated by PV panel is fluctuating in nature. In order to obtain enhanced voltage and to stabilize the power a converter along with controller

Figure 12. Grid voltage and current

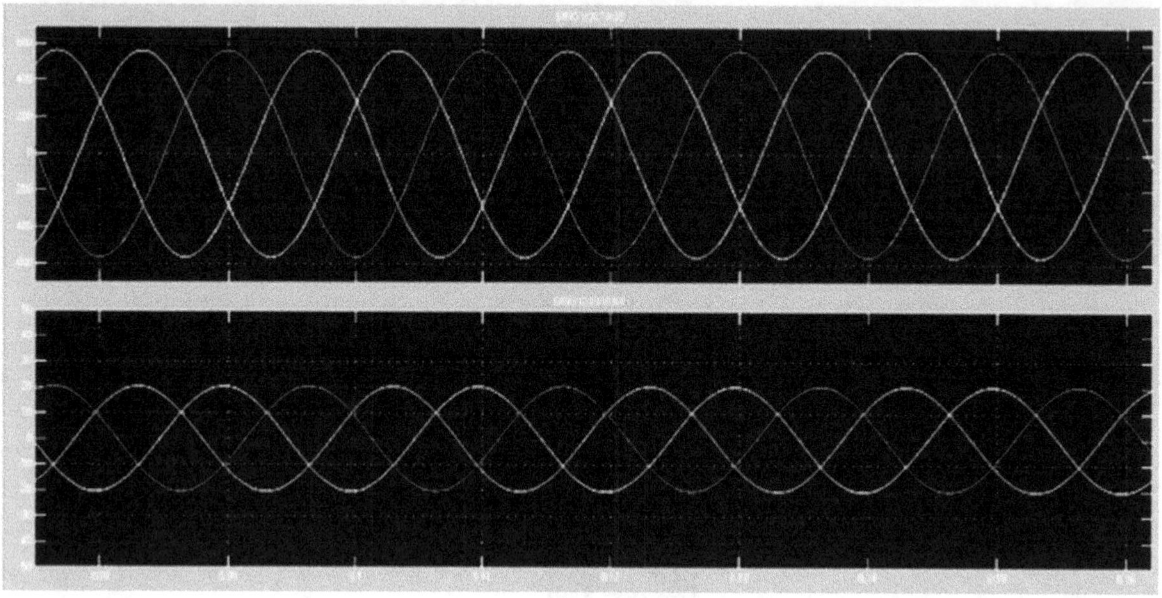

is implemented. Multiple SEPIC choppers are employed in this approach for power enhancement. To transfer maximum amount of power from panel to load, the duty cycle of multiple SEPIC chopper is controlled. This alters the load impedance by source, so as to match the point of maximum value of power from source. Hence, to attain a maximum output power from PV system DMPPT controller approach is introduced. Therefore, the obtained maximum power can be verified using MATLAB responses.

A) PV SYSTEM FORMULATION

Solar power generation relies heavily on PV cells. They create a solar PV effect with PN junctions and transform solar power into electrical energy. A single PV cell has a low output power. However, in practical applications, once they are arranged in a PV array, either in series or parallel, the needed power can be attained. Due to the nonlinear and time-varying output characteristics of PV arrays are

Figure 13. PV voltage after inverter operation

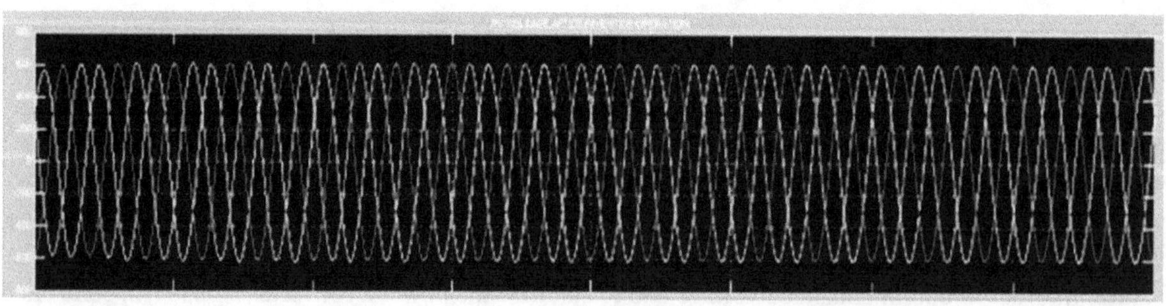

Figure 14. Sepic converter voltage

Figure 15. Voltage source converter voltage without filter

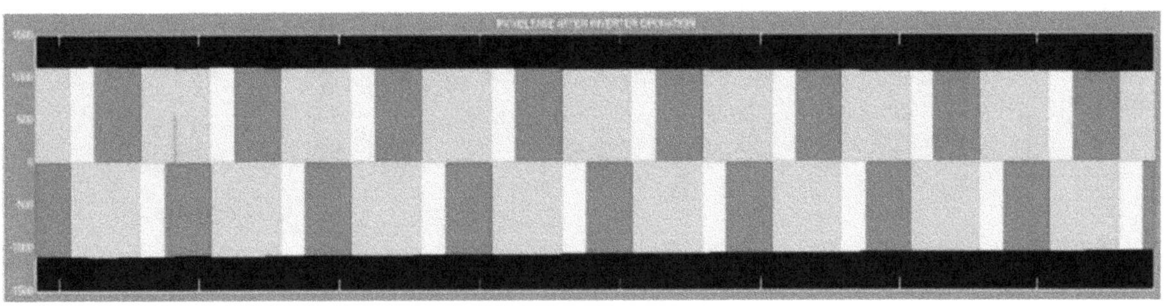

Figure 16. Voltage source converter with filter

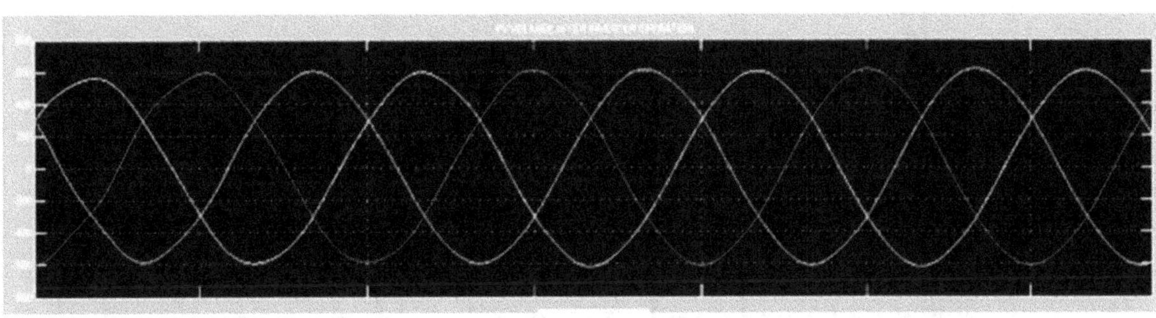

Figure 17. THD of the improved SEPIC converter based PV-grid system

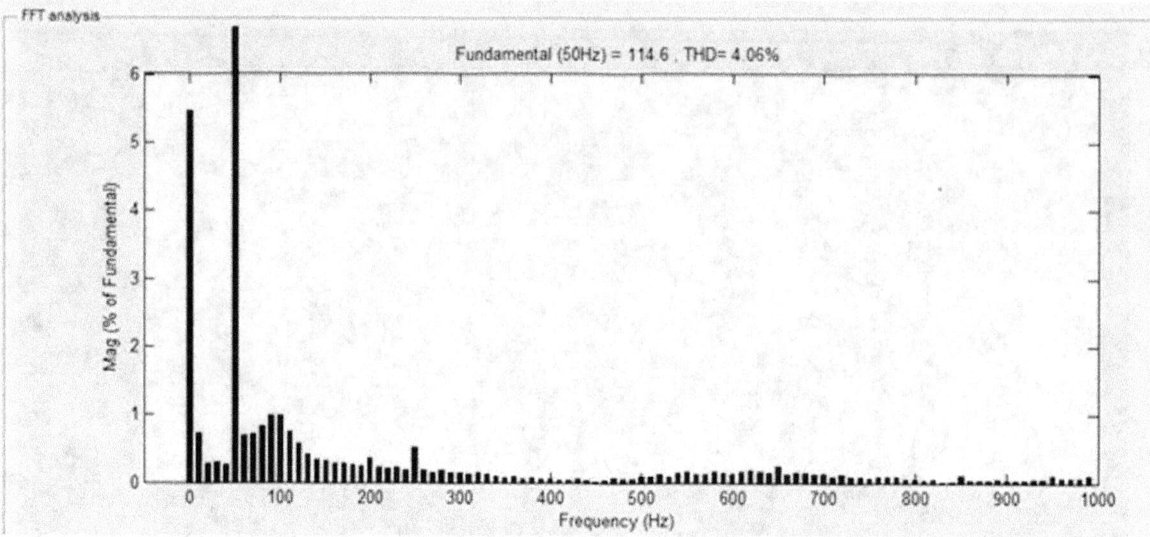

influenced by incident cell temperature, light intensity, load conditions, and intrinsic impedance can be altered, in which solar energy consumption is not always maximized. The equivalent Solar PV system is illustrated in Figure 20.

The simulation of PV cell is done using the mathematical equation of PV system equivalent circuit. Hence, to calculate output current I_{pv} Kirchhoff's Current Law (KCL) is used as shown in the following equation,

Figure 18. THD of the standard SEPIC converter based PV-grid system

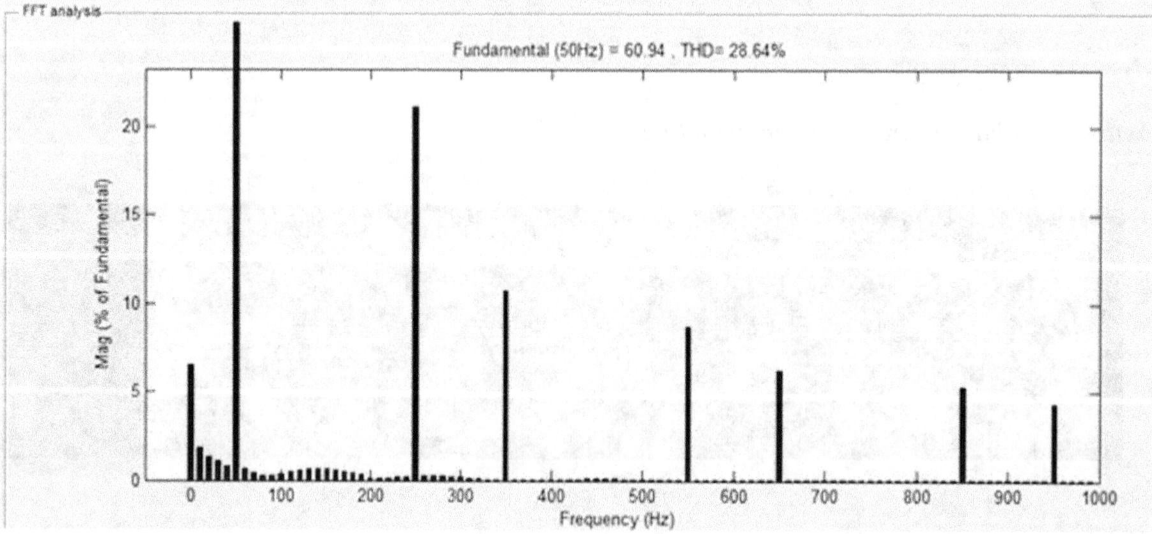

Figure 19. Proposed Block Diagram of PV System

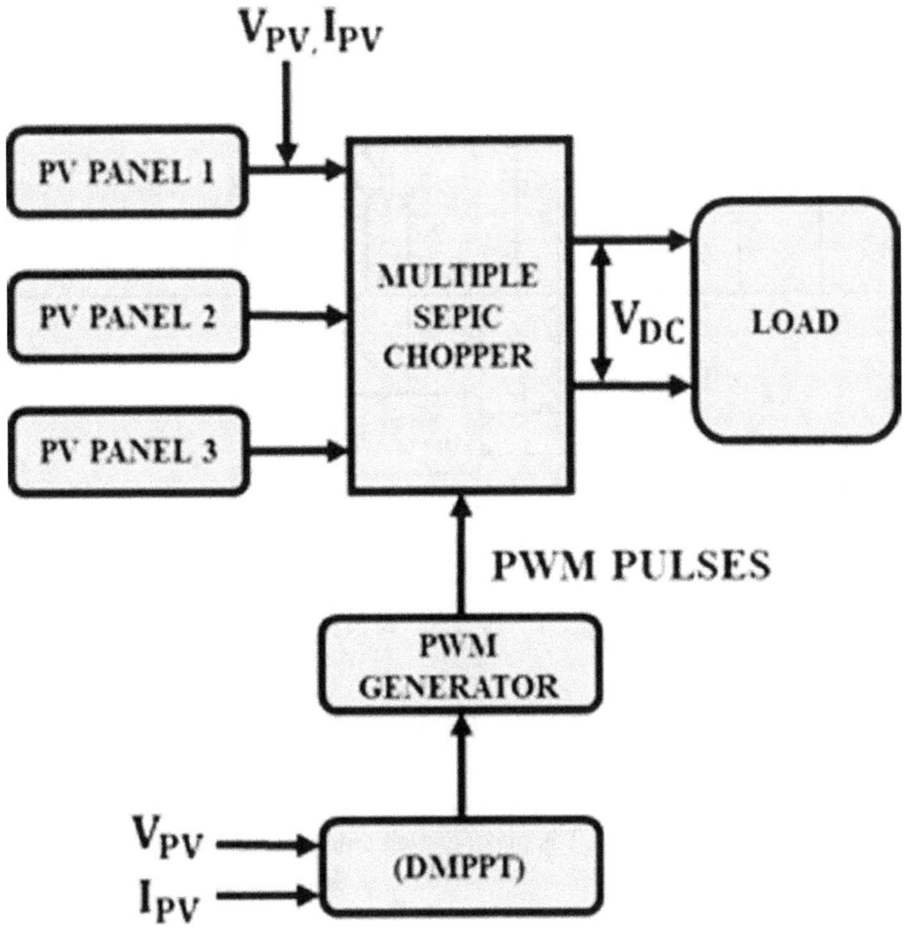

Figure 20. Equivalent PV System

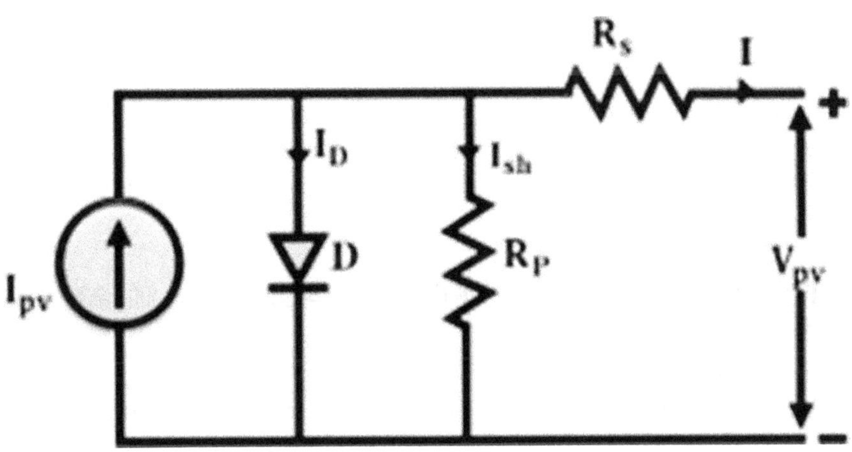

Figure 21. Circuit Model of SEPIC based PVU

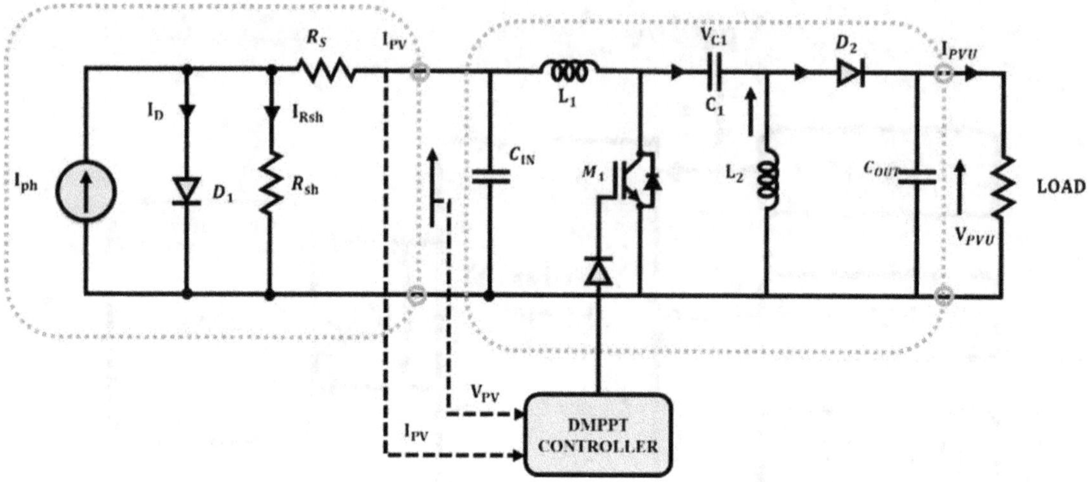

$$I_{pv} = N_p I_{sh} - N_p I_0 \left[\exp\left(\frac{q\left(V_{pv} + I_{pv} R_s\right)}{\left(N_s AKT\right)} \right) - 1 \right] - \left(\frac{V_{pv} + I_{pv} R_s}{R_{sh}} \right) \quad (9)$$

Where, I_{pv} is the current provided by Photon, whose value is influenced by the PV cell's light-receiving area, ambient temperature and brightness of the input light; the reverse current is denoted as I_D; the shunt current is represented as I_{sh}, which flows through the bypass resistor; R_S is the resistor in series connection; K denotes Boltzmann constant; T is the operating temperature and q as the charge of electron.

B) SEPIC CHOPPERS

A SEPIC chopper is alike a conventional buck-boost converter which essentially consists of boost converter accompanied by inverted buck-boost converter. The advantage of SEPIC converter is that output and input voltage have similar polarities. Figure 21 shows the circuit of SEPIC chopper with photovoltaic unit.

The operation of SEPIC converter involves two modes:

Mode 1: When the switch is in ON state, the inductor L_1 gets charged, and diode D becomes reversed biased with no conduction of current.

Mode 2: When the switch is in OFF state, the stored energy in inductor L_1 discharges and flows throughout the circuit. The diode D attains forward bias with conduction of current.

C) DISTRIBUTED MAXIMUM POWER POINT TRACKING (DMPPT)

For the solar panels, the general characteristics of VI curve represent a specific region where the maximum power is observed. The resulting power operates with a specific voltage and current rating. However finding the perfect combination of the voltage and current is the basic need of an MPPT algorithm. Pertub and observe is basically a search process where the voltage is being varied and in each step the voltage changes a particular level to obtain the best power. This can be termed as the hill climbing process as it is dependent on the gradient of the power level generally depicts like a hill. The Pertub and Observe algorithm P&O is the mostly used MPPT approach for its simplification. P&O method can only reach the global optimum if a proper strategy is being adopted. Hence, in this section, DMPPT approach is discussed.

In this approach the converter control operation is handled using DMPPT controller to harvest maximum output power for the load. This section of the proposed work, encloses multiple PV panels which is intermittent and induces low DC power. In order to boost up the voltage, multiple SEPIC choppers are employed with the adaption of DMPPT controller. The enhanced DC voltage is stabilized using the controller approach and corresponding maximum power is distributed to the load.

The function of an active power filter is fulfilled with the proposed circuit. The PV-APF controller is designed for improvement of power quality. The need for electrical power, which is now met by fossil fuels produce significant amount of carbon emissions into the environment. Environmental issues like greenhouse effects, climate changes, and global warming made engineers to develop eco-friendly technologies for generating electricity. One of the most crucial answers to the issue of energy production is development of solar energy systems. This proposed concept uses multiple PV systems which have been trending due to its high efficient applications. The PV panels are employed with SEPIC chopper for boosting the system. The non-inverting converter with its continuous output is fed with the DMPPT which avoids the transient and inrush in PV system. MATLAB software is implemented and the results of maximum power point are verified using the simulation responses. The most effective way to combat the negative impacts of mismatching phenomena, which have significant implications on the energetic effectiveness and dependability of the entire PV system, is through the DMPPT technique. However, a number of issues, such as climatic circumstances, the power stage's effectiveness, topological restrictions, the finite rating of silicon devices, and the suboptimal string voltage value, can still lower the DMPPT's energy efficiency.

DMPPT structure partition PV system into smaller arrays, called sub-arrays, to lower the number of MPPs in every sub-array. At that instant, a power converter with an MPPT approach that is much simpler than those used in Centralized Maximum Power Point Tracking (CMPPT) systems is attached to each sub-array. One of the most popular architectures of multiple DMPPT system, shown in Figure 22. Each panel is coupled with SEPIC converter to create DMPPT unit, which is then connected in series to sustain an inverter. The primary advantage of DMPPT is that every PV panel may still function at its maximum power point even in circumstances that are not ideal.

RESULTS FOR PV SYSTEM WITH MULTIPLE SEPIC CHOPPERS AND DMPPT

The proposed approach involves the working of multiple chopper to enhance the voltage with the use of DMPPT control approach. The entire work is implemented in MATLAB simulation and the following responses are obtained. The specifications considered for carrying out the simulation in MATLAB 2014a are tabulated in Table 2.

Figure 22. DMPPT Architecture Connected with Load

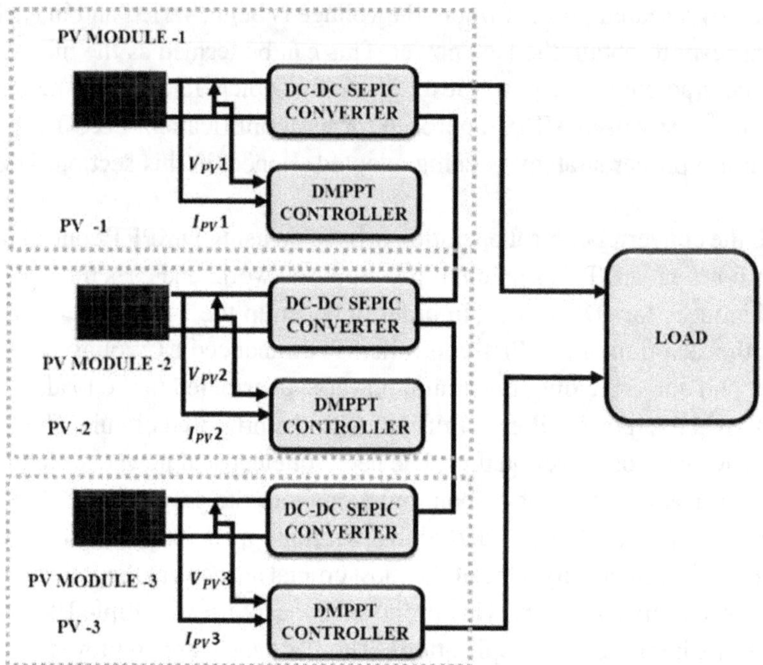

Figure 23 illustrates the input and output current waveform. It is shown that the input current fluctuates slightly and maintains a constant of 5A from 0.02s whereas the output current of constant 5A is maintained after a mild fluctuation.

Figure 24 demonstrates the input and output voltage waveform. It is shown that when a 35V constant input voltage is supplied the SEPIC chopper it generates an output voltage of 30V after mild fluctuation.

Table 2. Parameters specification

Specifications	Values
PV Panel	
Open Circuit Voltage (V_{oc})	22.54V
Short circuit current (I_{sc})	5.28A
Peak power voltage (V_{mp})	16.86V
Peak power current (I_{mp})	5.42A
Peak power (P_{mp})	1000W, 12 Panels
Number of series-linked cells (N_s)	24
SEPIC Converter	
Inductor, L_A,L_B	0.288mH
Capacitor, C_C,C_0	150µF
Switching frequency f_s	10KHz
Diode	MCD95

Figure 23. (a) Waveform of Input Current (b) Waveform of Output Current

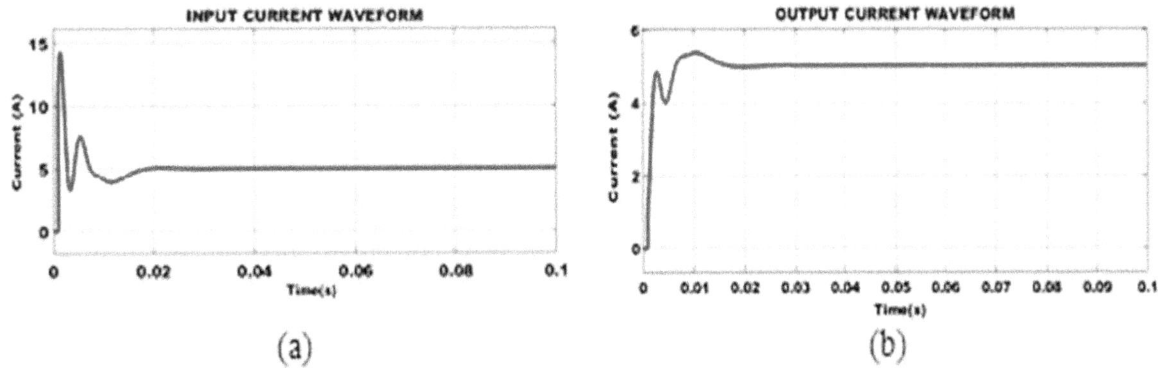

The representation of PWM waveform is depicted in Figure 25, when PWM pulse is applied to switches S_1,S_2 and S_3. For effective operation of the system a constant of 1V is applied.

Figure 26 demonstrates the comparison efficiency of Boost, Buck-Boost, Cuk and SEPIC converters where the efficiency lies on 87%, 89%, 91.5% and 93%. Similarly, voltage gain of these converters lies in the ratio of 1:1.5, 1:2, 1:3 and 1:8. By comparison, the proposed approach performs better with the efficiency of 93% and resulting in voltage gain of 1:8.The comparative results have been tabulated in table 3 below.

CONCLUSION

This research work focuses on a PNN based SOC monitoring approach. The ever-increasing need for electricity, along with the uncertainties surrounding the availability and price of fossil fuels, has prompted the installation of large-scale PV-based power generation systems in recent times. The PV system requires a suitable converter for effective conversion of the input voltage to a desired voltage level, so a SEPIC converter, which is a DC-DC converter with high voltage gain and non-inverted output is used. A BESS integrated with the PV system, acts as a secondary power source. The battery's State of Charge (SOC) is monitored in order to control the charging and discharging operations with the help of the PNN controller.

Figure 24. (a) Input Voltage Waveform (b) Output Voltage Waveform

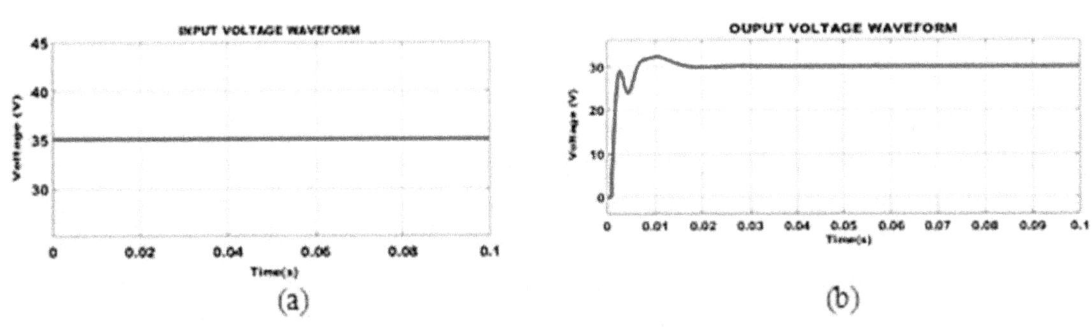

Figure 25. PWM Pulse Waveform to Switch $S_1S_2S_3$

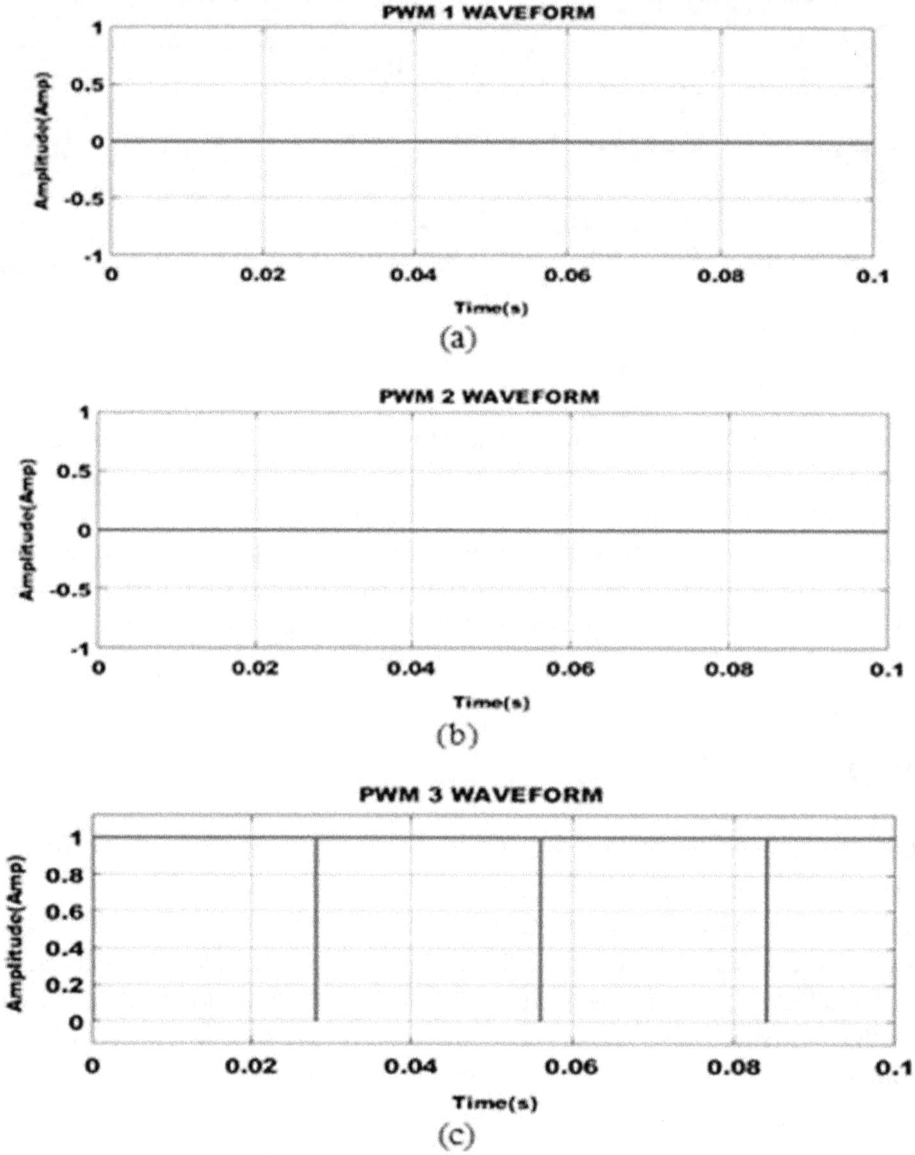

The improved topology of SEPIC converter is also proposed in this work. It is designed to achieve high power quality in the utility grid. It helps in attaining high voltage gain a well and works on soft switching principle. The overall system efficiency is improved in the proposed work. From the simulation results, it is evident that the overall system performance with respect to power quality improvement can be assessed to be better in case of improved SEPIC converter. Thus, the proposed work has proven to adhere to the Harmonics standards proposed by IEEE and IEC. In recent days renewable energy plays an important role in contemporary society, to produce energy and is a key input to all sectors. The commonly used renewable energy is solar energy due to its zero maintenance cost and low emission. The PV panels are used to generate electricity directly from sunlight. The proposed work uses multiple PV

Figure 26. (a) Comparison of Efficiency (b) Comparison of Voltage Gain

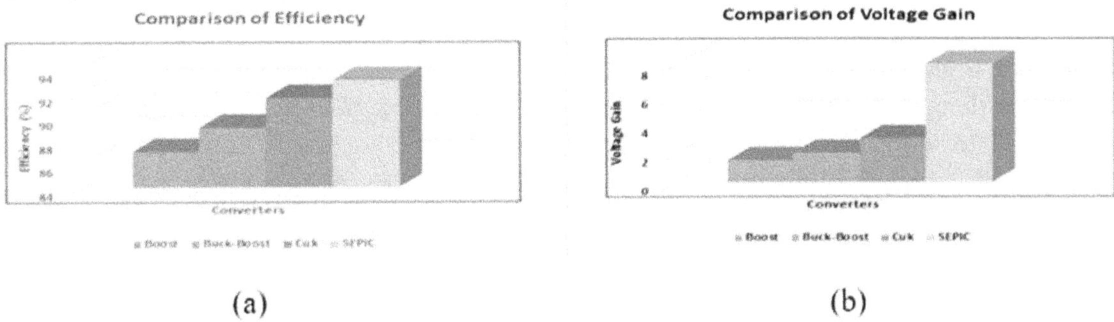

panels accompanied with multiple SEPIC chopper to extract enhanced power from the PV system. In order to attain maximum power, an intelligent MPPT controller approach is introduced. The DMPPT controller functions even in non-ideal conditions, therefore enabling the system to work effectively by attaining maximum output power.

A Design and implementation of intelligent MPPT with multiple choppers for PV System is presented in the last section of this work. The designed multiple SEPIC chopper achieves high voltage gain, with low duty cycle and reduced voltage stress, across the power semiconductor switches. The outcome of PV is not constant, DMPPT based MPPT is executed to control the SEPIC chopper. The obtained results indicate that the proposed approach maintains sustainable environment with enhanced efficiency and minimal harmonics. The tracking efficiency is 98.6% without considering the efficiency of the solar PV panel and converter.

Therefore, in this chapter, a detailed comparison of the results obtained by using a SEPIC converter and its improved topology for a PV-grid system is presented. The battery SOC analysis using PNN is also clearly presented. The authors intend to extend the analysis for the improved converter topologies used in PV-grid systems and hope that this research would pave a way for industrialists and researchers in increased use of renewable energy.

Table 3. Comparison of Efficiency and Voltage gain ratio of the studied converters

Type of converter	Efficiency in %	Voltage Gain Ratio
Boost	87	1:1.5
Buck-Boost	89	1:2
Cuk	91.5	1:3
SEPIC	93	1:8

REFERENCES

Ali, M. S., Wang, L., Alquhayz, H., Rehman, O. U., & Chen, G. (2021). Performance Improvement of Three-Phase Boost Power Factor Correction Rectifier Through Combined Parameters Optimization of Proportional-Integral and Repetitive Controller. *IEEE Access: Practical Innovations, Open Solutions*, *9*, 58893–58909. doi:10.1109/ACCESS.2021.3073004

C. M., NP., J. (2021). Realization of Cascaded H-Bridge Multilevel Inverter Based Grid Integrated Solar Energy System With Band Stop Generalized Integral Control. *IEEE Transactions on Industry Applications*, *57*(1), 764–773. doi:10.1109/TIA.2020.3031546

de Morais, J. C. S., de Morais, J. L. S., & Gules, R. (2019). Photovoltaic AC Module Based on a Cuk Converter With a Switched-Inductor Structure. *IEEE Transactions on Industrial Electronics*, *66*(5), 3881–3890. doi:10.1109/TIE.2018.2856202

Errouissi, R., Al-Durra, A., & Muyeen, S. M. (2016). A Robust Continuous-Time MPC of a DC–DC Boost Converter Interfaced With a Grid-Connected Photovoltaic System. In *IEEE Journal of Photovoltaics, 6*(6), pp. 1619-1629.

Kavya Santhoshi, B., Mohana Sundaram, K., Padmanaban, S., & Holm-Nielsen, J. B. (2019). K. K., P. Critical Review of PV Grid-Tied Inverters. *Energies*, *12*, 1921. doi:10.3390/en12101921

Santhoshi, B. K., Mohanasundaram, K., & Kumar, L. A. (2021). ANN-based dynamic control and energy management of inverter and battery in a grid-tied hybrid renewable power system fed through switched Z-source converter. *Electrical Engineering*, *103*(5), 2285–2301. doi:10.100700202-021-01231-7

Santhoshi, B. K., Sundaram, K. M., Hussain, A., Sanjeevikumar, P., Holm-Nielsen, J. B., & Kaliappan, V. K. (2021). Deep Learning for Fault Diagnostics in Bearings, Insulators, PV Panels, Power Lines, and Electric Vehicle Applications—The State-of-the-Art Approaches. *IEEE Access: Practical Innovations, Open Solutions*, *9*, 41246–41260. doi:10.1109/ACCESS.2021.3064360

Seidi Khorramabadi, S., & Bakhshai, A. (2015). Critic-Based Self-Tuning PI Structure for Active and Reactive Power Control of VSCs in Microgrid Systems. *IEEE Transactions on Smart Grid*, *6*(1), 92–103. doi:10.1109/TSG.2014.2354651

Sreekanth, T., Lakshminarasamma, N., & Mishra, M. K. (2017). A Single-Stage Grid-Connected High Gain Buck–Boost Inverter With Maximum Power Point Tracking. *IEEE Transactions on Energy Conversion*, *32*(1), 330–339. doi:10.1109/TEC.2016.2633365

Yap, K. Y., Beh, C. M., & Sarimuthu, C. R. (2021). Fuzzy logic controller-based synchronverter in grid-connected solar power system with adaptive damping factor. in *Chinese Journal of Electrical Engineering, 7*(2), pp. 37-49, .

Zadeh, M., & Fathi, S. H. (2017). A New Approach for Photovoltaic Arrays Modeling and Maximum Power Point Estimation in Real Operating Conditions. *IEEE Transactions on Industrial Electronics*, *64*(12), 9334–9343. doi:10.1109/TIE.2017.2711571

ADDITIONAL READING

Ali, M., Tariq, M., Lodi, K. A., Chakrabortty, R. K., Ryan, M. J., Alamri, B., & Bharatiraja, C. (2021). Robust ANN-Based Control of Modified PUC-5 Inverter for Solar PV Applications. *IEEE Transactions on Industry Applications*, *57*(4), 3863–3876. doi:10.1109/TIA.2021.3076032

Kavya Santhoshi, B., & Mohana Sundaram, K. (2020). A Cost-Effective PV-Based Single-Stage Conversion System for Power Backup. In H. Saini, T. Srinivas, D. Vinod Kumar, & K. Chandragupta Mauryan (Eds.), *Innovations in Electrical and Electronics Engineering. Lecture Notes in Electrical Engineering* (Vol. 626). Springer. doi:10.1007/978-981-15-2256-7_50

Singh, R., & Bansal, R. C. (2019). Optimization of an Autonomous Hybrid Renewable Energy System Using Reformed Electric System Cascade Analysis. *IEEE Transactions on Industrial Informatics*, *15*(1), 399–409. doi:10.1109/TII.2018.2867626

Chapter 3
Control of Cell Voltage Difference Balancing in Battery Management System Charging Circuits in Electric Vehicles

Kripalakshmi Thiagarajan
Vellore Institute of Technology, Chennai, India

Mahadeedhar Marabathina
Vellore Institute of Technology, Chennai, India

Maladhi D.
Vellore Institute of Technology, Chennai, India

Selvasundar K.
Vellore Institute of Technology, Chennai, India

Deepa T.
Vellore Institute of Technology, Chennai, India

Angalaeswari S.
https://orcid.org/0000-0001-9875-9768
Vellore Institute of Technology, Chennai, India

Subbulekshmi D.
Vellore Institute of Technology, Chennai, India

ABSTRACT

The growing development of battery technologies for the transposition of use of fuel in electric vehicles leads to the competent Battery Management System (BMS) design implementation. The chapter focuses on how the voltage difference perturbs the battery pack's performance by using an optimized control strategy. It helps in battery operated vehicles to charge and work efficiently. The control strategy involves the proper switching of the resistors connected to the charging circuit. The thermal balance of the resistors is balanced using the thermal power dissipation modelling in simulation. The models are tested under variable cell voltage conditions in MATLAB Simulink. The technique for voltage error balancing is termed passive cell balancing in BMS for the arrangement of a multiplicity of cells in stacks. The major issue of the passive cell balancing arises due to the State of Charge (SoC) imbalance in the batteries. The charging conditions of the battery are taken into the parameter estimation for the control technique of the balancing technique.

DOI: 10.4018/978-1-6684-6631-5.ch003

1. INTRODUCTION

Recently, Li-Ion batteries have risen to the top of the most popular battery categories. Li-Ion batteries with fast charge capabilities are crucial components of grid-connected and automotive systems because of their light weight, high capacity, and high energy density. A BMS is an electrical component that acts as the brain of a battery pack, manages output, and guards the battery against extreme harm (Hemavathi, 2020). Data collection is needed to study the battery's parameters, predict or prevent failures, and monitor temperature, voltage, and current via an interprocess connection.

The ratio of the energy now stored in the battery to its nominal capacity is known as the battery's state of charge (SOC). The cell balancing approach is one of the crucial elements of BMS (Vezzini, 2014; Yusof et al., 2016). The voltages of the individual cells will fluctuate over time without a suitable normalisation device, drastically lowering battery capacity. These problems will reduce the lifespan of the battery and the electric car range. A Li-Ion battery requires the use of a balancing device due to the imbalanced characteristics of the cells in the same series string that make overcharging possible. Passive and active balancing systems are frequently separated into these two types. However, because it is easier to operate and less expensive than an active balancing system, the latter is more widely used (Aizpuru et al., 2013; Cao et al., 2008; Daowd et al., 2011; Moore & Schneider, 2001; Quan, et al., 2018; Wei et al., 2017; Zheng et al., 2014).

Passive equalizing processes remove an excess charge from fully energized cells and bring about a condition in which all cells have a charge that is identical to the lowest cell charge by using resistor components (Loniza et al., 2016). Because it uses fewer components and is more reliable, this technique cuts the cost of the complete system. This method causes energy losses because of heat dissipation, which reduces the effectiveness of the system. As a result, this strategy is appropriate for low-power applications.

(Vulligaddala et al., 2020) presented an all-encompassing HEV and EV battery monitoring system. The seven Li-ion batteries in the design, which were implemented as an IC, may be tracked and balanced. The cell voltages of a stack of series-connected cells can be measured using a daisy chain of integrated circuits. Both the temperature and the cell voltages were measured using a 12-bit SAR ADC. In order to determine the reference voltage, this data can be transmitted to an external host utilizing a bidirectional synchronous voltage mode communication link.

For electric vehicles, Chen Duan (2018) proposes a solar energy harvesting and storage system-based battery balancing method (EVs). To prevent the energy loss that occurs in traditional active and passive battery balancing systems, the battery module with the lowest SOC/voltage during discharge can be charged using solar energy or energy from a storage cell. According to the simulation results, the suggested approach can save 2.1%–3.3% of the 50Ah battery pack's overall capacity per 13.2 miles. The battery modules can be balanced when the solar current is between 6-7A. Even when the solar current is restricted to 2-3 A or when using cell-storage mode with the same current restriction, the balance is still remarkably successful. The passive cell balancing technique will aid in preserving the same SOC and voltage across cells for low-powered electric vehicles. in order to prevent uneven voltage across cells and excessive cell heating.

For the energy balance of battery cell strings, Claudio (2016) offers an offline approach based on convex optimization that addresses a number of control issues. Using the provided method, seven balancing circuit topologies are compared. The outcomes revealed that applying the suggested technique can result in a variety of benefits. All topologies, except for the CB topology, show loss reduction compared to a no-control technique. But the most obvious benefits for all topologies in terms of energy losses are the variances in power loss between cells.

(Mestrallet et al., 2014) For cell active balancing, a multiphase converter is recommended. The working basis of the architecture is spontaneous cell balancing, which provides energy to the cells as required without the use of voltage or current sensors. The building was planned and built to balance eight cells. This structure's principal drawback is its reliance on converter variables like maximum current, switching frequency, and nominal dc bus voltage, which has an immediate impact on injector volumes.

Table 1. Causes and remedies for cell imbalance (Omariba et al., 2019)

Imbalance Types	Causes	Solution	Reference
Imbalance of state of charge	Ø Electrochemical characteristics difference	Ø Energy sharing technique Ø Capacity screening	(Diao et al., 2018; Qi et al., 2014; Zheng et al., 2017)
Differences of Impedances	Ø Overcharging or discharging Ø Current balancing in the equalization circuit Ø Battery aging	Ø A charge or discharge that is not necessary if there is no capacity difference Ø Maximum power transfer control Ø During charging or discharging change in voltage over current	(Huang et al., 2017; Lawson, n.d.; Qi et al., 2014)
Overall capacity difference	Ø Cells with different characteristics in the module Ø Initial charge capacity differences	Ø Impedance reduction technique Ø Effective balancing	(Arachchige et al., 2017; Lin, 2017; Qi et al., 2014)

Figure 1. Cell Balancing

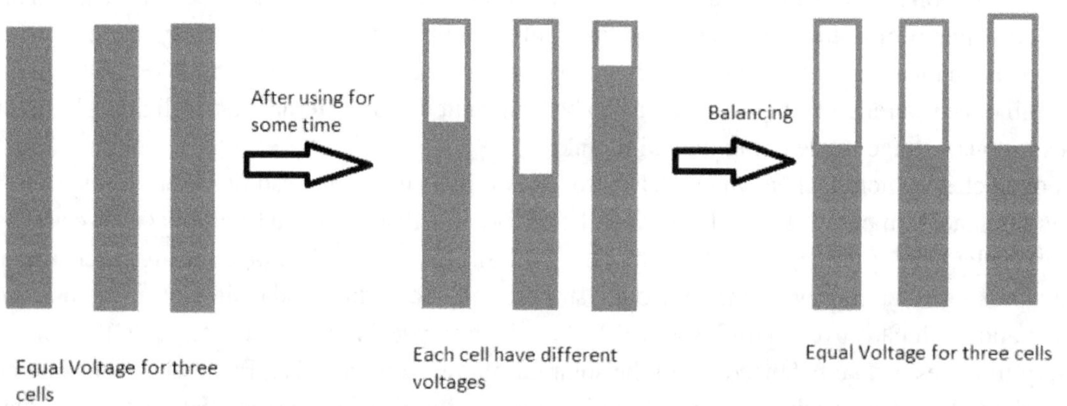

2. PURPOSE OF PASSIVE CELL BALANCING IN BATTERY MANAGEMENT SYSTEM

The main aim of the design of BMS is to monitor the wellness of the serially connected batteries which enables the maintenance of the Soc level even in fault conditions. The purpose of Li-batteries balancing is mainly to the overvoltage condition that shortens the battery life. Charging an overcharged battery damage the condition of the entire system. The balancing objective is equal to SoC in all batteries connected and cascaded together. Passive cell balancing discharges the overcharged battery and the unuseful power is discharged as heat. Hence the temperature monitoring control for BMS is mandatory. Active balancing transfers energy from battery to battery. Balancing during charging acquires two things when a battery reaches maximum voltage and from the initial beginning of the charging.

3. PROPOSED CELL BALANCING TECHNIQUE

The size of the resistor determines how much current or power is lost in a passive balancing circuit. The balancing current cannot be changed while the circuit is in use because the resistance in a traditional cell balancing circuit is maintained constant. The imbalance or discrepancy between Li-Ion battery cells in a battery pack gets worse with time if they have aged and experienced changes in their characteristics. The balancing circuit required a greater current loss over time to compensate for the larger unequal SoCs (State of Charge) between the cells brought on by ageing. For a battery pack that isn't too old, the balancing current produced with a fixed current loss might still be sufficient, but as the battery cells get older, the balancing period will grow longer. A higher current loss is required to keep the same balancing time for a larger unbalanced SoC. Consequently, a fixed resistor cannot satisfy the older cells' greater need for current loss (Barsukov & Instruments, n.d.).

In order to specifically meet the increased demand for balancing power from the older cells, PWM is utilized in passive balancing to alter the amount of balancing current. We can modify the amount of power lost during the passive balancing process. Equation (1) can be used to change the balancing power for the duty cycle.

$$P_m = (U \times \text{duty cycle})^2 / R \tag{1}$$

The current loss should be adjusted as high as practically practicable with a low enough resistance to cover the anticipated current loss level required for balancing the older battery cells in order to achieve the highest degree of flexibility in power control. When the MOSFET acts as the balancing switch, there are limitations. The battery cell voltage, which is only 3.5–4V, drives the MOSFET's gate voltage. Conventional MOSFETs may operate in the saturation region due to their low gate voltage and high drain current, which could reduce the drained power of passive balancing. When building a circuit, this behavior should be taken into consideration by choosing a MOSFET with the appropriate properties (Supanat et al., 2019). It is ideal to compare cell voltages when all of the cells are in the same state, regardless of whether a load is present or not. The cell voltage is frequently assessed after the balancing procedure. This operation prevents the balancing process from continuing. The off-period, which occurs throughout each cycle of the PWM signal when utilizing the PWM approach, can be assumed to represent the no-balancing period. This enables a continual balancing operation and the completion of the cell voltage monitoring throughout the off-period.

SoC of cell calculated by $SOC = SOC \pm \left(\frac{Q}{FCC}\right)100\%$ (2)

Figure 2. Passive Cell Balancing technique

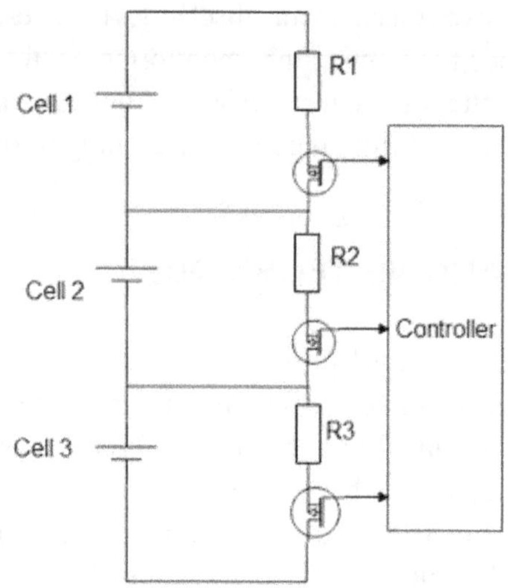

Where Q represents the current integration value, FCC stands for full charging capacity, and + and - signify charging and discharging, respectively.

The power loss across each resistor $P_{loss} = \frac{V^2}{R}$ watts (3)

Since V is the voltage across the resistor, R is a resistor connected across cells.

The passive cell balancing method has the advantages of a simple design, affordability, and durability.

Four serially connected battery cells were given a suggested and built Battery Management System (BMS). The suggested design improves accuracy, performance, and safety in cell control by utilizing a variety of specialized integrated circuits for battery management applications, such as gate drivers, power MOSFET transistors, and differential operational amplifiers (Guran et al., n.d.). A safe way to acquire voltage that enables the detection of a small differential voltage in the presence of high common mode voltages is through the use of high-common mode voltage amplifiers.

Using a power resistor and an internal resistance MOSFET as a balancing resistor, successful research into BMS with passive balancing can be found in (Negrohol et al., 2017). The proposed BMS can be utilized to drain the LiFePO4 battery cell of extra energy (overcharge). The maximum voltage of the cell is well restricted, according to experimental findings on 15 LiFePO4 series batteries, and the suggested BMS can keep the voltage of the cell in line with the reference balancing voltage. According to the results of this experiment which was conducted in (Negrohol et al., 2017), the considerable discrepancy between the cells with the highest voltage and the largest capacities causes the balancing time to take a very long time (lowest voltage).

The constant voltage charging method and the constant current charging approach are contrasted in (Shukla & Patel, 2022). It has been determined that the continuous current charging technique is the most effective way to recharge lithium-ion batteries. As can be seen from the results of a simulation of passive cell balancing, energy loss through the resistor during the MOSFET switching process causes both cells to be balanced at 3700 seconds.

In this proposed technique IGBT was used for switching operations due to the advantages. The benefits of IGBTs that motivate their implementation in this suggested work (ESCTUFF4U, n.d.) are their straightforward drive circuit, low on-resistance, quick switching speed, low switching loss, low gate drive requirement, and low on-stage power dissipation.

4. HoW DOES A BATTERY MANAGEMENT SYSTEM MAINTAIN A HEALTHY BATTERY

1) Discharge Control: The BMS's primary goal is to maintain the battery in a secure posture. BMS must therefore safeguard the battery from excessive charging and draining.
2) Determination of SoC: Battery charge is the main objective, which is consistently measured by coulomb counting or direct voltage measurement. Utilizing a voltmeter, one may measure the voltage across a battery. If the voltage falls, the soc likewise falls.
3) Determination of health status: Here, the internal resistance or conductance of the battery is used to gauge how much energy is lost during operation.
4) Logbook: The BMS continuously tracks the health of both the new and the old batteries and monitors their performance.
5) Communication: The BMS system saves data that humans can quickly access and use to make choices.

The battery's voltage, temperature, and level of charge are all constantly monitored by the battery management system. In a battery pack, if a few cells have lower voltages than the others, the BMS will use an active cell balancing mechanism to share the voltage from the high-voltage cell to the low-voltage cell in order to make all the cells' voltages equal. In the passive cell balancing method, the high-voltage cell will lose voltage through a resistor in the form of heat until the voltage approaches the low voltage. Similarly to this, all cells will have the same level of charge.

5. RESULTS AND DISCUSSION

This section displays the simulation outcomes for the passive cell balancing technique. Here we consider two cases one is a series connection and the second case is a series and parallel connection. Initially, three battery cells rating 2.05Ah with a nominal voltage of 3.35V are connected in series with different SoC. For cell1, cell2, cell3, and cell4 the SoC is 80%, 70%, 20%, and 90% respectively. Three battery cells were balanced, and they now have the same SoC and Voltage of 19.97% and 3.441V, respectively, and their temperature has dropped from 30°C to 25°C. Figure 3, Figure 4, and Figure 5 show the SoC response, voltage response, and cell temperature respectively.

For series and parallel connection balancing the cells here, eight battery cells are taken into account with a rating of 2.05Ah and a voltage of 3.35V. Together with another set of four battery cells that are joined together in series, these four battery cells are connected in parallel and series to form a battery pack. Initially, the SoC of cells is different. For cell1 and cell5 the Soc is 80% for cell2 and cell6 the

Figure 3. SoC response of cells

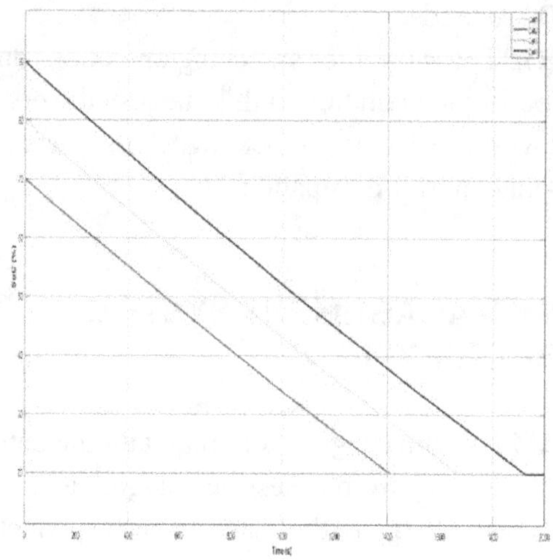

Figure 4. Voltage response of Cells

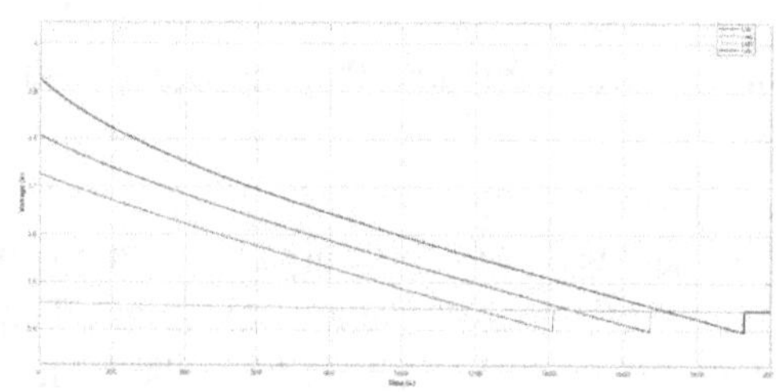

Figure 5. Temperature response of Cells

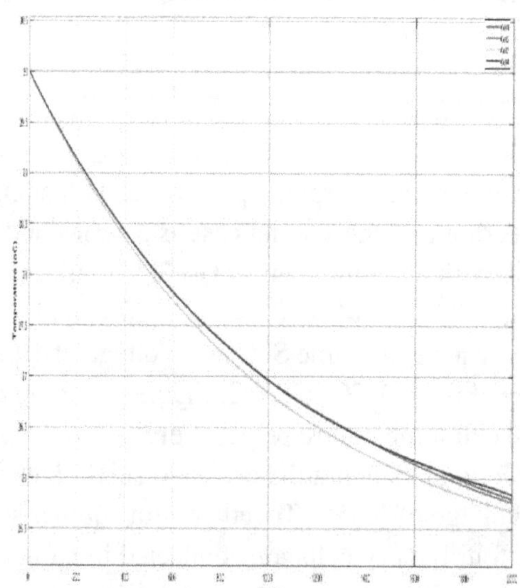

Figure 6. SoC response of cells

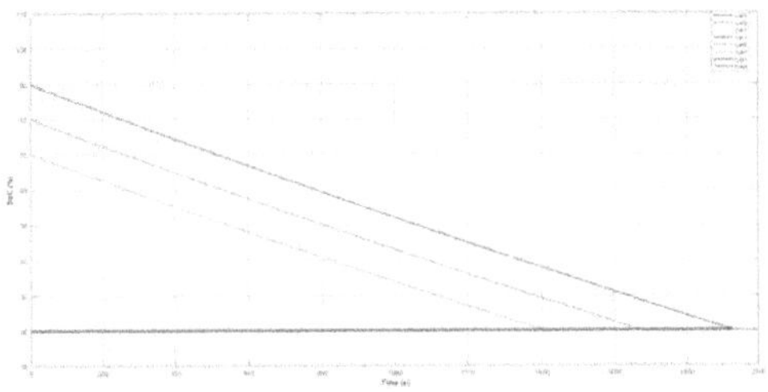

Figure 7. Voltage response of Cells

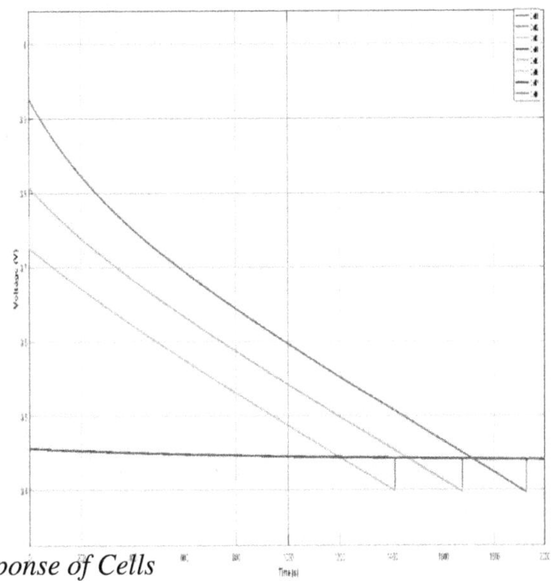

Figure 8. Temperature response of Cells

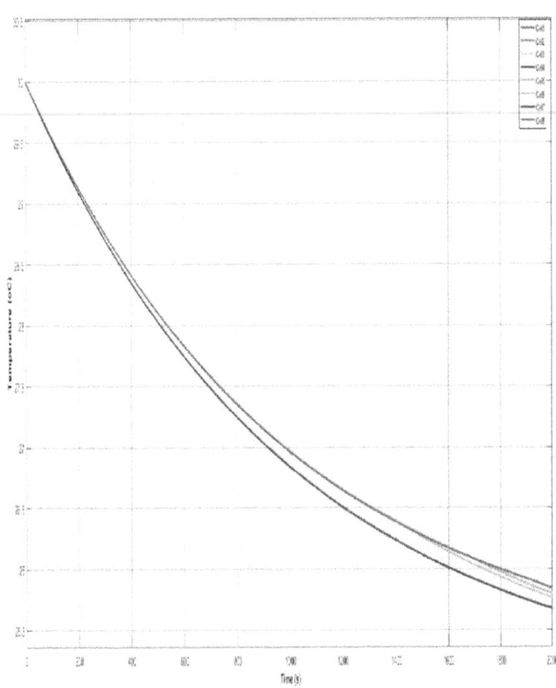

SoC is 70% for cell3 and cell7 the SoC is 20% for cell4 and cell8 the SoC is 90%. When the cells are balanced, their SoC and voltage drop to 19.96% and 3.441V, respectively, and their temperature is lowered to 25°C from 30°C. Figure 7, Figure 8, and Figure 9 show the SoC response, Voltage response, and cell temperature respectively.

CONCLUSION

In this study, a passive cell balancing method for battery cells coupled in series and series parallel was proposed. This passive cell balancing technique was performed on lithium cells which are widely used in industries, and vehicles. The passive cell balancing method produces great results and is perfect for low-power applications due to its low heat loss through a resistor. Based on the findings, it was determined that using the passive cell balancing technique, the four battery cells with varying states of charge and voltage were linked in series and obtained the same state of charge (19.97% state of charge) and voltage of 3.441V for four cells. Eight cells, four of which are connected in series and four of which are connected in parallel with the other series-connected cells, are balanced passively using the same method. All eight cells reached 3.441V of voltage and a 19.96% state of charge following the application of the passive cell balancing procedure. At 1930 seconds, all cells are balanced by employing an IGBT switching operation to lose energy through a resistor.

REFERENCES

Aizpuru, U., Iraola, J., Canales, M., Echeverria, M., & Gil, I. (2013). Passive balancing design for Li-ion battery packs based on single cell experimental tests for a CCCV charging mode. *2013 International Conference on Clean Electrical Power (ICCEP),* 93-98.

Arachchige, B., Perinpanayagam, S., & Jaras, R. (2017). Enhanced prognostic model for lithium ion batteries based on particle ðlter state transition model modiðcation. *Applied Sciences (Basel, Switzerland), 7*(11), 1172. doi:10.3390/app7111172

Barsukov & Instruments. (n.d.). *Battery Cell Balancing: What to Balance and How*. Academic Press.

Cao, J., Schofield, N., & Emadi, A. (2008). Battery balancing methods: A comprehensive review. In *IEEE Vehicle Power and Propulsion Conference, VPPC 08* (pp. 1-6). IEEE.

Daowd, M., Omar, N., Van Den Bossche, P., & Van Mierlo, J. (2011). Passive and active battery balancing comparison based on MATLAB simulation. In *Vehicle Power and Propulsion Conference (VPPC)* (pp. 1-7). IEEE. 10.1109/VPPC.2011.6043010

Diao, W., Xue, N., Bhattacharjee, V., Jiang, J., Karabasoglu, O., & Pecht, M. (2018, January). Active battery cell equalization based on residual available energy maximization. *Applied Energy, 210*, 690–698. doi:10.1016/j.apenergy.2017.07.137

Duan, Caisheng, Zongzheng, Jianfei, Wang, Snyder, & Jiang. (2018). A Solar Power-Assisted Battery Balancing System for Electric Vehicles. *IEEE Transactions on Transportation Electrification, 4*(2).

ESCTUFF4U. (n.d.). ESCTUFF4U for electronics engineering. *ESCTUFF.* https://www.ecstuff4u.com/2018/01/advantages-and-disadvantages-of-igbt.html

Guran, I. C., Perişoară, L. A., Florescu, A., & Săcăleanu, D. I. (n.d.). 4-Cell Passive Battery Management System for Automotive Applications. *IEEE 27th International Symposium for Design and Technology in Electronic Packaging (SIITME).* 10.1109/SIITME53254.2021.9663604

Hemavathi, S. (2020). Overview of cell balancing methods for Li-ion battery technology. *Energy Storage,* e203.

Huang, S.-C., Tseng, K.-H., Liang, J.-W., Chang, C.-L., & Pecht, M. G. (2017). An online SoC and SOH estimation model for lithium-ion batteries. *Energies, 10*(4), 512. doi:10.3390/en10040512

Lawson, B. (n.d.). A software conðgurable battery. In *Proc. EVS26 Int. Batter. Hybrid Fuel Cell Electr. Veh. Symp.* (*vol. 1*, pp. 252–263). IEEE.

Lin, J.-C. M. (2017). Development of a new battery management system with an independent balance module for electrical motorcycles. *Energies, 10*(9), 1289. doi:10.3390/en10091289

Loniza, E., Situmorang, J.A., Kusuma, D.D.A., Cahyadi, A.I., & Wahyunggoro, O. (2016). Passive balancing of battery lithium polymer using shunt resistor circuit method. *Adv Sci Technol Soc., 1755,* 90011-1-90011-6.

Mestrallet, F., Kerachev, L., Crebier, J.-C., & Collet, A. (2014, June). Multiphase Interleaved Converter for Lithium Battery Active Balancing. *IEEE Transactions on Power Electronics, 29*(6), 2874–2881. doi:10.1109/TPEL.2013.2276152

Moore, S., & Schneider, P. (2001). A Review of Cell Equalization Methods for Lithium Ion and Lithium Polymer Battery Systems. In *Proceedings of the SAE 2001 World Congress.* SAE Mobilus. 10.4271/2001-01-0959

Negrohol, A., Kaleg, S., & Nugrhol, A. (2017). Passive Balancing Battery Management System using MOSFET Internal Resistance as Balancing Resistor. *International Conference on Sustainable Energy Engineering and Application (ICSEEA).*

Omariba, Z. B., Zhang, L., & Sun, D. (2019). *Review of Battery Cell Balancing Methodologies for Optimizing Battery Pack Performance in Electric Vehicles.* IEEE. doi:10.1109/ACCESS.2019.2940090

Pinto, C., Barreras, J. V., Schaltz, E., & Araujo, R. (2016). Evaluation of Advanced Control for Li-ion Battery Balancing Systems Using Convex Optimization. *IEEE Transactions on Sustainable Energy, 7*(4).

Qi, G., Li, X., & Yang, D. (2014). A control strategy for dynamic balancing of lithium iron phosphate battery based on the performance of cell voltage. In *Proc. IEEE Conf. Expo Transp. Electriðc. Asia–Paciðc (ITEC Asia–Paciðc)* (pp. 1–5). IEEE. 10.1109/ITEC-AP.2014.6941212

Quan, O., Chen, J., Zheng, J., & Fang, H. (2018). Optimal Cell-to-Cell Balancing Topology Design for Serially Connected Lithium-Ion Battery Packs. *IEEE Transactions on Sustainable Energy, 9*(1).

Shukla & Patel. (2022). Battery Management System by Passive Cell Balancing for Electric vehicle. *2022 2nd International Conference on Power Electronics & IoT Applications in Renewable Energy and Its Control (PARC).*

Supanat, A., Manop, M., & Nisai, F. (2019). On a Design of Adjustable Passive Balancing Circuit Using PWM Technique for Li-Ion Battery. *Research, Invention, and Innovation Congress (RI2C).*

Vezzini, A. (2014). *Lithiium-ion battery management. In Lithium-Ion Batteries Advance and Applications*. Elsevier.

Vulligaddala, VVernekar, SAdusumalli, SEle, VBrandl, MM.B., S. (2020, May). A 7-Cell, Stackable, Li-Ion Monitoring and Active/Passive Balancing IC With In-Built Cell Balancing Switches for Electric and Hybrid Vehicles. *IEEE Transactions on Industrial Informatics*, *16*(5).

Wei, L., Jie, Lv., Song, W., & Feng, Z. (2017). Study on passive balancing characteristics of serially connected lithium-ion battery string. In *IEEE 13th International Conference on Electronic Measurement & Instruments*. IEEE. 10.1109/ICEMI.2017.8265862

Yusof, M. S., Toha, S. F., Kamisan, N. A., Hashim, N. N. W. N., & Abdullah, M. A. (2016). Battery cell balancing optimisation for battery management system. In *International Conference on Mechanical, Automotive and Aerospace Engineering*. Semantic Scholar.

Zheng, L., Zhu, J., Wang, G., Lu, D. D.-C., McLean, P., & He, T. (2017). Model predictive control based balancing strategy for series-connected lithium- ion battery packs. In *Proc. 19th Eur. Conf. Power Electron. Appl. (EPE ECCE Eur)* (pp. 1–8). IEEE. 10.23919/EPE17ECCEEurope.2017.8099189

Zheng, Y., Ouyang, M., Lu, L., Li, J., Han, X., & Xu, L. (2014). On-line equalization for lithium-ion battery packs based on charging cell voltages: Part 1. Equalization based on remaining charging capacity estimation. *Journal of Power Sources*, *247*, 676–686. doi:10.1016/j.jpowsour.2013.09.030

Chapter 4
Design and Evaluation of Battery Management Systems for Electric Vehicle Applications With Harris Hawk Optimized Algorithm (HHOA) Based on Fuzzy Logic Controller

Shanmukha Naga Raju Vonteddu
https://orcid.org/0000-0002-7874-5956
Jawaharlal Nehru Technological University, Kakinada, India

Ravindra Kollu
University College of Engineering, Kakinada, India

Prasanthi Kumari Nunna
University of Petroleum and Energy Studies, India

ABSTRACT

Utilizing electric vehicles (EVs) that don't pollute the atmosphere is one way to solve environmental pollution. This chapter as per the author proposes Harris Hawk optimized algorithm (HHOA) based on fuzzy logic controller with battery management system in electric vehicles. The suggested system is made to alter and regulate the current values of battery characteristics like battery voltage, current, power usage, and state of charge (SOC). To estimate the model temperature dependency, at various temperatures, the heat parameters can be found. Then, these temperature-dependent parameters are extracted for model decoupling. With offline data, the provided model is verified to attain good reliability over a wide range of temperatures and charge states. Extended Kalman filter (EKF) is used to monitor and control BMS operation.

DOI: 10.4018/978-1-6684-6631-5.ch004

INTRODUCTION

The transportation of the future is electric vehicles. It is essential to create more efficient EVs because of the expanding market for EVs (Electric Vehicles) and the dropping price of fossil fuels. In an EV, batteries serve as the main storage device. A crucial component of any EV is a battery management system. It is made up of several electrical and electronic circuits that have been configured to monitor and get the most out of a battery system, including different converter and inverter circuits. The sensors in the BMS keep track of the cell conditions, which are then utilized to compute the SOC and carry out numerous tasks. Since batteries are the most crucial component of all, accurate modelling of battery storage is crucial (Ananthraj & Ghosh, 2021). As lithium-ion battery technology has advanced, using lithium-ion batteries to power electric vehicles has become a viable option (EVs). Lithium-ion batteries have the following advantages over Nickel-metal hydride and lead-acid batteries: Low self-discharge, good coulombic efficiency, high specific energy, large capacity, and extended cycle life. Lithium-ion batteries are now regarded by automakers and research organizations as the most dependable form of motor vehicle energy (Zhang et al., 2020). It uses batteries for electricity. Battery power for high performance. The "drive performance" and "high dependability" trade-off for EVs should be pleasant. The battery used in EVs is distinct from other situations where it is typically charged and discharged with a big current and operates in a hazardous environment (Chen et al., 2012).

The BMS in EVs consists of a large number of circuits, parts, power electronics, sensors, actuators, diodes, capacitors, inductors, transformers, switches, converters, and safety equipment, all of which are managed by a large number of algorithms, models, and control signals. The creation of suitable algorithms for BMS has been the subject of extensive study. Model-based methods and intelligent methods are the two approaches used in BMS the most frequently. The complex, dynamic, and nonlinear features of lithium-ion batteries can be addressed by the intelligent algorithms (Martinez et al., 2017). The resistors, capacitors, and other model parameters are chosen by directly fitting the external current voltage responses. The lack of consideration for internal electrochemical factors, however, may reduce the accuracy of state prediction when a battery ages or its load changes. The pseudo-two-dimensional (P2D) electrochemistry-based models, on the other hand, can get around these issues.

Each of these features corresponds to an internal electrochemical process physically (physical and chemical phenomena)that occur while the battery is in use (Vaideeswaran et al., 2019). EV lithium batteries were the main focus of the BMS's development. Fuzzy logic was used in the development of this BMS to represent user knowledge and enhance the EV's performance and autonomy (Qureshi et al., 2021). The efficiency of EVs affects the precise value of State of Charge and State - of - health tracked by the BMS used in a particular EV. The State of Chrgea and State - of - health of a battery can be tracked using a variety of both traditional and modern methods. The design and modelling of the Universal Battery Management System, which monitors battery attributes such load voltage, load current, energy usage, and SOC, uses real-time physical restrictions and predefined parameters to verify the proposed BMS (Dost & Sourkounis, 2014).

Overview Of Battery Management System In Electric Vehicles

General Battery Management System

A battery management system (BMS)'s duty is to provide that the energy stored in the battery that operates the current source is used as efficiently as possible and that the danger of battery damage is kept to a minimum. This is accomplished by keeping an eye on and managing how the battery is charged and discharged. The monitor and control features of the BMS include intelligence. According to the description, the monitor's functions include measuring things like charger status, battery voltage and load activity. Based on these measured variables, the control functions determine how the battery is charged and discharged. The design of these monitoring and control features should guarantee optimum battery utilization and minimize the possibility of battery damage.

Figure 1. A general Battery Management System (BMS)

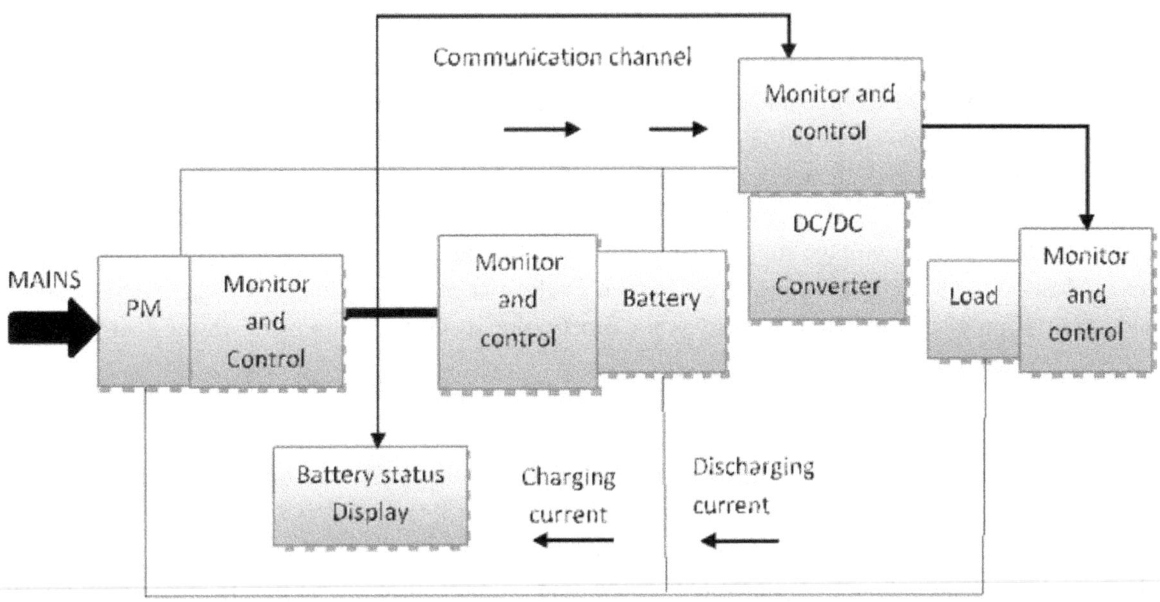

The many components can share the intelligence required for the BMS. The division of intelligence is a crucial component of a BMS' design. Cost is the primary determining factor in this case. Intelligence can be implemented via specialized battery management ICs (Hossain Lipu et al., 2021). There is a wealth of relevant background knowledge and examples available. A communication channel is used to transmit measured variable and parameter values as well as control instructions between the various components of the BMS. This channel can take the form of a bus that is managed by a specific protocol or only one wire that controls a Pulse Width Modulation (PWM) switch. Figure 1 shows the general BMS's structure. The addition of "Monitor and Control" blocks to each component of the system represents the division of intelligence. A Battery Status Display is also controlled by the BMS shown in Figure.1.A Light-Emitting Diode (LED) that indicates a "battery low" situation is an example. A liquid crystal display (LCD) can also be used to show the battery status. This includes the State-of-Charge (SOC) and battery condition.

A) The Power Module

By transforming the PM's major responsibility is to re-create electrical energy from mains electricity that may be utilized to recharge the battery. The mains could be replaced with alternative energy sources like a solar cell or vehicle battery. In certain situations, such as when the battery is low, the PM can also be utilized to directly portable power source. The PM may be built into the portable apparatus itself, as in shavers for example, or it may be a separate device, like a travel charger (Wu et al., 2011). The energy conversion process' efficiency must be good enough, particularly in the latter scenario, because lower efficiency results in higher internal temperatures for the portable device and, consequently, for the battery. The capacity of the battery will reduce over time when exposed to high temperatures.

B) A DC/DC converter

When voltage across the load does not match the voltage required, a DC/DC converter in a portable instrument is primarily used to connect a charge to the different system components. Either too low and too high battery voltages are possible. Of course, DC/DC up-conversion is required in the first scenario. When the battery voltage is greater than the maximum permissible voltage source for the load, the second scenario asks for DC/DC down conversion. When a DC/DC converter is used, the load design can be optimized for the lowest voltage supply instead of the entire output voltages of the battery degradation curve. Most often, this will result in a larger burden. Performance (Mihet-Popa & Groza, 2015). Second, using more supply voltage than necessary when functioning a system component wastes energy.

C) The Load

The primary function of the load is to convert the electrical energy from the battery that will allow it to continue out its functions. There are numerous varieties of portable goods, and hence numerous varieties of loads. Various supply voltages are required for the majority of portable products. As technology advances, the voltages change regularly. Due to the decreasing feature size of ICs, the maximum power supply at which an IC may run without failing is decreasing. On the other hand, some analog operations will continue to demand a higher supply voltage than a digital system requires. components, for examples because the analogue function requires a particular signal-to-noise ratio (Quantmeyer & Liu-Henke, 2013).

Block Diagram of Battery Management System

A. Measurement Block

The primary purpose is to record the relevant data, includes ambient temperature, cell voltages, currents, and temperature. The processing of measured values then involves converting them to digital signals. Although using monitor at the unit level is expensive, it has the benefit of providing cell balance at its minimum point.

B. Block of the Battery Algorithm

Its primary purpose is to estimate the Charge of Soc and SOH using data from the measurement block (State of Health). A battery's actual capacity is calculated as a percentage of its entire maximum load and is known as the state of charge (SOC).The remaining distance an EV can travel can be calculated using SOC, which serves as a fuel gauge. Temperature and charge-discharge cycles also affect SOC (Zhang et al., 2016). Therefore, the method employed should likewise take these parameters into account. The risk of overcharging and undercharging is also reduced with the aid of SOC estimation. Charge dumping brought on by regenerative braking may result in overcharging of cells (Lee, 2010). To avoid causing harm to the battery in certain circumstances, the BMS should monitor and regulate this.

C. Block for Capability Estimation

This block establishes the upper limit charged and discharged current at any given time depending on the State of Charge and State of Health values. After that, the result is delivered to the ECU (Electronic Control Unit) (Xiong, 2020). The Battery Management System may therefore controlled how the battery is charged and discharged and do other battery-related testing.

D. Cell Equalization Block

There will be variations in the characteristics of different cells since no cell is flawless. Therefore, the level battery charge may varies from cell to cell even when charging. The major responsibility of this modules is to keep the State of Charge and cell voltages in check in order to guard against system damage. This approach states that if one cell has an undervoltage, the another using an algorithm, cells are employed to charge it, decreasing energy waste caused by dissipative cell equilibrium (Xiong et al., 2011).

E. Block for Thermal Management

To prevent any harm to the cells, it keeps a tight eye on the temperatures of the cells. A air conditioning system and a water heater receive the output, and they work in tandem to maintain the temperature within acceptable operating ranges. Additionally, it sends a warning to the Electronic Control Unit (ECU) if the temperature rises abnormally because this could be harmful. The user in addition to permanently ruining the batteries (Gao et al., 2021).

THE LITHIUM-ION BATTERY

Lithium-ion Batteries Diagram of the P2D Model

Rechargeable lithium-ion batteries are especially suitable applications for Electric Vehicles because of their increased energy density and longer life cycle. The following is the formula to use to Identify the battery's volume:

Figure 2. Block diagram of Battery Management System

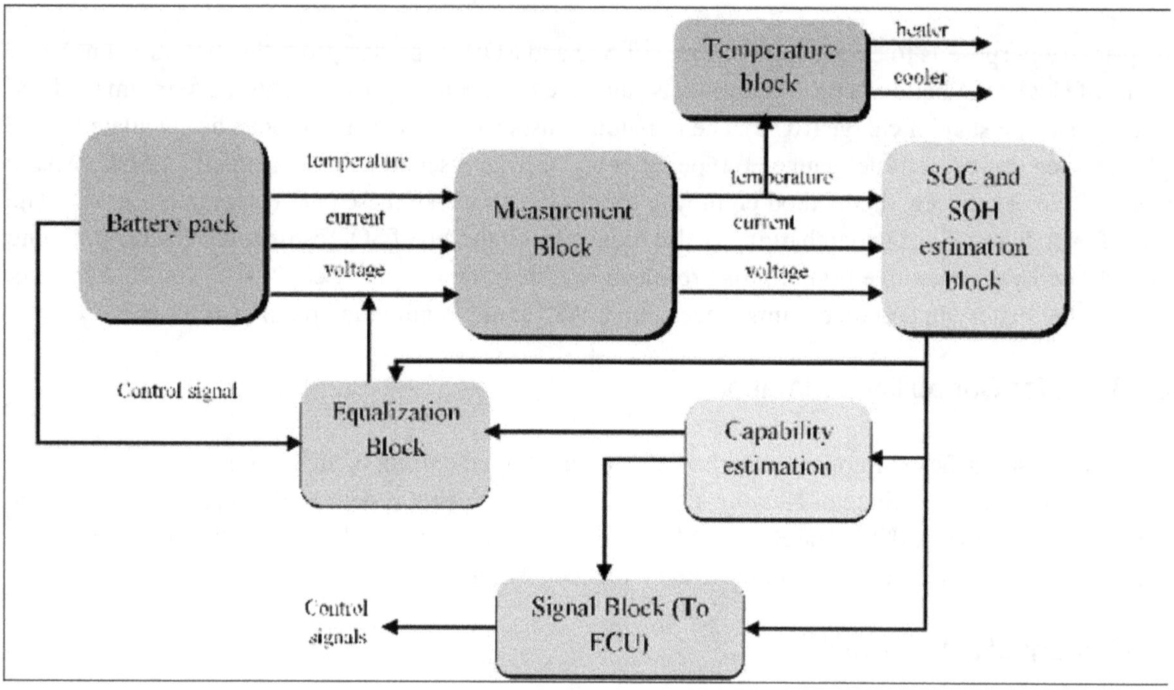

$$\text{Volume of battery} = \frac{(Device\,longevity)(Average\ power)}{Energy\,density}$$

The negative and postive electrodes of a lithium-ion battery are separated by a separator.which is a rechargeable battery. LiPF6, LiBF4, LiCIO4, and other electrolytes are utilised in lithium-ion batteries.

Figure 3 displays a Diagrammatic Structure of the P2D Model for a LiCoO2 battery. two phases and three areas (solid phase and electrolyte phase)make up the battery's internal structure (positive, negative, and separator). Spherical particles are used to represent the solid phase, and the diffusion of Li-ions moving in a particle's r-radius is used to explain the solid-phase diffusion process. The Li-ion diffusion and migration along the current battery thickness, or in the x-direction, is what is meant by the electrolyte-phase diffusion process. The discharge procedure essential requirements. Lithium ions diffuse from the interior to the exterior of the negatively charged dynamic particles. Additionally, electrochemical processes taking place on the aera of the negative active particles cause Li-ions to be discharged into the electrolyte (Sreedhar & Karunanithi, 2021).

Lithium ion then diffuse in that direction and pass through the separator to the positive electrode. Lithium ion diffuse into the interior of the positive dynamic particles as a result of electrochemical reactions taking place on their surface. As this is happening, electrons are travelling from the negative collector to the positive collector, creating the external circuit's electric current (Huang et al., 2020).

A positive electrode's electrochemical reaction

$$LiCo\ O_2 \xleftrightarrow[discharge]{charge} Li_{(1-x)}\ \mathrm{CO}\ O_2 + xLi^{+} + xe^{-}$$

Figure 3. Diagrammatic Structure of the P2D Model

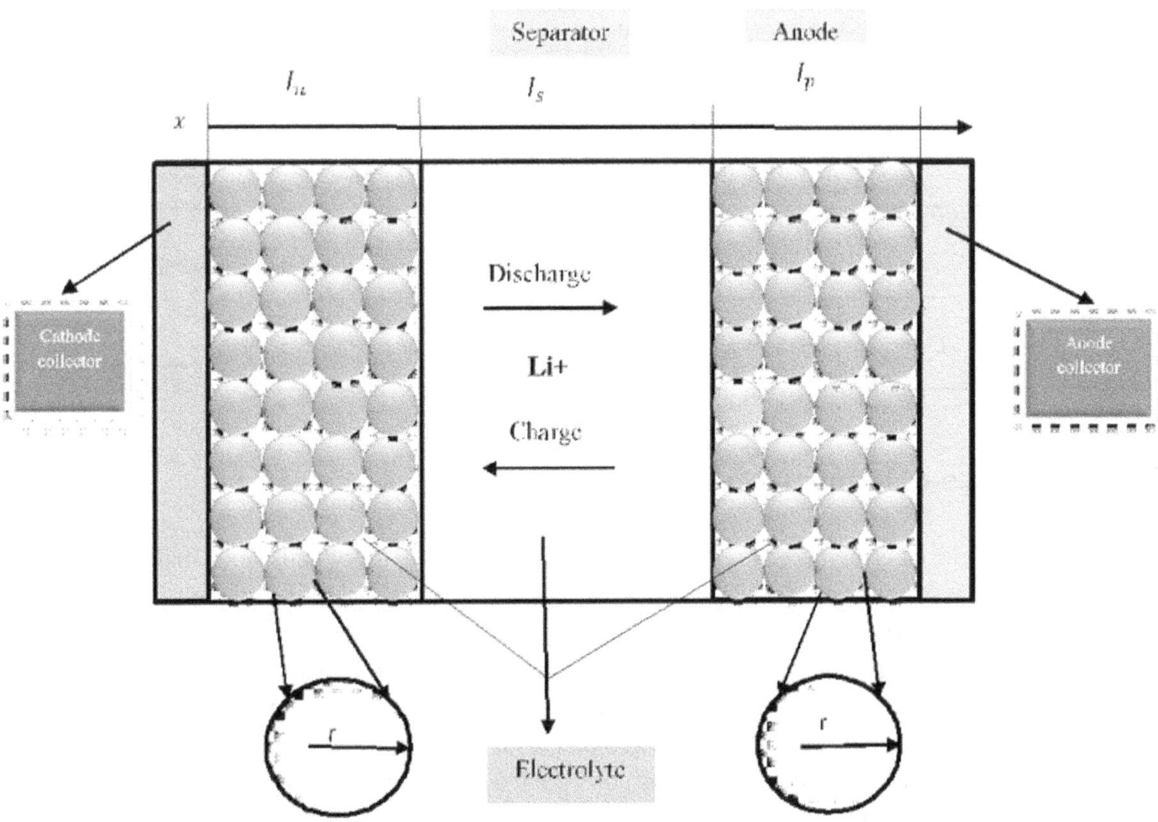

A negative electrode's electrochemical reaction

$$6C + xLi^{+} + xe^{-} \xleftrightarrow[discharge]{charge} Li_{x}C_{6}$$

The P2D model's logical implication allows for the establishment of a number of algebraic equations and partial differential equations that characterized the properties of the battery from an electrochemical standpoint.

(1)The active material, negative terminal, and separating all have charge equilibrium equations.

(2) Li-ions in the separation battery diffusion formulae for the electrode surface, negative electrode,
(3) The positivity, negative, and separator parts of the battery Ohm's law equation.
(4) The equation for the diffusion of solid-phase Lithium ion in the negative and positive electrode.
(5) A Butler-Volmer kinetic equation describes the surface of the functional particles in each electrode.

Table 1. Essential Lithium-ions Battery Components and Reactions.

Components	Functions	Materials
Negative electrode	Lithium ions move into the electrode as the battery drains, and they leave while the battery energized.	Powdered lithium metal oxide
Positive electrode	Lithium ions leave the positive electrode while the battery de-energized, and they return when the battery energized.	Powdered graphitic carbon
Electrolyte	The electrolyte used for the movement of lithium ions but not electrons among the negative and positive electrode	Organic solvents and lithium salts
Separator	Short circuits among both the anode and cathode are avoided by the separator, which restricts the amount of lithium ions that can travel through its pores.	microscopic pores in membranes

Characteristics of Lithium-ion Batteries

The cathode, anode, electrolyte, and separator are the lithium-ion battery cells' four main component, just like other kinds of batteries. Table 1 shows these crucial elements together with their roles, common materials, and the electrochemical reactions in lithium-ion batteries' operating principles.

Lithium ions and diffusion may typically be accommodated on the positive electrode, which has a major problem on the battery's overall performance, combination compounds, lithium manganese oxide, lithium iron phosphate, and lithium cobalt oxide (usually nickel-cobalt-manganese, nickel-cobalt-aluminum)made of several metal oxides are currently the most often utilized positive electrode materials.

A mixture composed of a binder, an organic solvent, and carbon-based substance is frequently used as the negative electrode's active ingredient. The paste is placed as a thin layer on a copper base. Additionally, the negative electrode of several contemporary fast-charging battery types is made of lithium nanoparticles (Li4Ti5O12).

The ions are delivered by the electrolyte, which is typically a mixture of the organic solvents and a lithium salt solution with high permittivity. The electrolytes should be able to respond to the potent redox reaction that takes place during energized and de-energized while remaining highly stable to the active ingredient.

Only lithium ions are used to travel via the separator, which is between the negative and positives electrodes to divide the electrons. Typically, it is constructed from a microporous polyethylene or polypropylene film

CONTROL-ORIENTED ELECTROCHEMICAL MODEL'S DERIVATION

In this section describes an electrochemical technique with the laser dynamics that is control-oriented. The basis for the expression on the simplified isothermal electrochemical model's transfer-function type, which provides accurate estimations of the terminal voltage and internal electrochemical states of the cell. Here, we go through the electrochemical model in more detail and extend its use to a wider range of temperatures.

As seen in Fig. 3, the fundamental structure is known as a P2D Structure. The particles electrode thickness and radius (r) are used to indicate the two independent directions (x). The conservation of species and the charge in the electrolytes and electrodes are regulated by four PDEs. At the solid-electrolyte interface, the overpotential is predicted by the Bulter-Volmer equation.

Under boundary circumstances, lithium ion battery conservation at the electrode is:

$$\frac{\partial c_s}{\partial t} = \frac{D_s \partial}{r^2 \partial r}\left(r^2 \frac{\partial c_s}{\partial r}\right) \tag{1}$$

$$\partial c_s \,|_r = 0, D_s \frac{\partial C_s}{\partial r} I_r = R_s = -\frac{j}{F} \quad (2$$

where j is the charge transfer current density and C_S is the solid coefðcient. Laplace transformation can be used to determine the equation of motion of D_s with reference to j.

$$\frac{C_s(S)}{J(s)}\,|_r = R_s = \frac{R_s}{FD_S}\left(\frac{\tanh\left(\frac{R_s\sqrt{s}}{D_s}\right)}{\tanh\left(\frac{R_S\sqrt{S}}{D_S}\right) - \frac{R_S\sqrt{S}}{D_S}}\right) \tag{3}$$

The equation (3) is simplified at both the negative and positive electrodes using the second Pade approximation.

$$\frac{C_s(s)}{J(S)} = -\frac{\frac{3}{R_S, neg}}{\frac{a_s F}{35} R^2 s, \frac{neg}{D_S, neg} S^2 + a_s FS} - \frac{\frac{2R_s, neg}{7D_S, neg}}{\frac{a_s F}{35} R^2 s, \frac{neg}{D_s, neg} s + a_s F}(X = 0) \tag{4}$$

$$\frac{C_s(s)}{J(S)} = -\frac{\frac{3}{R_S, pos}}{\frac{a_s F}{35} R^2 s, \frac{pos}{D_S, pos} S^2 + a_s FS} + \frac{\frac{2R_s, pos}{7D_S, pos}}{\frac{a_s F}{35} R^2 s, \frac{pos}{D_s, pos} s + a_s F}(X = L) \tag{5}$$

Lithium ion conservation in the electrolyte phase results in

$$\frac{\partial(\varepsilon_e c_e)}{\partial t} = \frac{\partial}{\partial x}\left(D_e^{eff}\frac{\partial}{\partial x}c_e\right) + 1 - \frac{t^0 +}{F}j \tag{6}$$

$$\frac{\partial c_e}{\partial x}\Big|_x = 0 = \frac{\partial c_e}{\partial x}\Big|_x = L = 0 \tag{7}$$

where $D_e^{eff} = D_e \varepsilon_e^{1.5}$ is the electrolytes' effective diffusion coefficient. Similar calculations can be used to determine the increases in the amount between the ion concentration and current density:

$$\frac{C_e(s)}{J(S)} = \frac{3\left(K' + 2\delta +\right)^2}{\delta_+ \varepsilon ep\left(3k'^2 + 10k'\delta_+ + 10\delta^2 +\right)s + \left(12k' + 24\delta_+\right)D_{e,pos}^{eff}}(x = L) \tag{8}$$

$$\frac{C_e(s)}{J(S)} = \frac{3\left(K' + 2\delta +\right)^2}{\delta\_\varepsilon ep\left(3k'^2 - 10k'\delta\_ + 10\delta^2 -\right)s - \left(12k' - 24\delta\_\right)D_{e,neg}^{eff}}(x = 0) \tag{9}$$

where $k'=2\delta s_{ep+}\delta-{}_{\_}\ \delta+;{}_{k}= -2\delta sep+_{\delta-}\ \delta+$, Equations (10) and (11) and (12) and (13) respectively describe how charge is conserved in the electrodes and electrolyte.

$$\sigma^{eff}\frac{\partial\varphi s^2}{\partial x^2} - a_s j = 0 \tag{10}$$

$$-\sigma_-^{eff}\frac{\partial\varphi_s}{\partial x}\Big|_x = 0 = \sigma_+^{\frac{eff\,\partial_s}{\partial xx}|_x = L = \frac{I}{A}.\frac{\partial\varphi_S}{\partial XX}|_X = \frac{\delta_{=\partial\varphi S}}{\partial XX}|_X = \delta_{+\delta_{sep}} = 0} \tag{11}$$

$$k_-^{eff}\,\partial\varphi^2{}_{\frac{e}{\partial x^2} + \frac{k_d^{eff}}{c_e.o}.\partial c^2{}_{\frac{e}{\partial x^2} + a_s j = 0}} \tag{12}$$

$$\frac{\partial\varphi}{\partial x}\Big|_x = 0 = \frac{\partial\varphi_e}{\partial xx}\Big|_x = L = 0.\frac{k^{eff}\partial\varphi_e}{\partial xx} + \frac{k_d^{eff}\partial c_e}{\partial xx}\Big|_{x-} = \frac{k^{eff}\partial\varphi_e}{\partial xx} + \frac{k_d^{eff}\partial c_e}{\partial xx}\Big|_{x-} \tag{13}$$

where φs is the potential of the electrode, $\sigma ef^{f=\sigma}\varepsilon s$ is the effective electrodes conductivity. φe is $_t$he potential of electrolytes $k^{eff} = k\varepsilon_e^{1.5}$ is the effective ionic conductivity $k_d^{eff} = \frac{2RTK^{eff}}{F}\left(t_+^0 - 1\right)$ is the ionic conductivity of effective diffusion. The electrolytes potential difference among the anode and cathode electrodes can be expressed mathematically as follows:

$$\varphi_e^{\sim}(s) = -\frac{I(S)L_n}{2AK_-^{eff}} - \frac{I(S)L_m}{2AK_{sep}^{eff}} - \frac{I(s)L_P}{2AK_+^{eff}} + \frac{2RT\left(1-t_+^0\right)}{C_e, o^F}\left(C_e(L,s) - C_e(o,s)\right) \quad (14)$$

The following sentence describes how the intercalation current and overpotential are related:

$$\eta(s) = \frac{RT}{Fi_o\left(\alpha_a + \alpha_c\right)}J(s) \quad (15)$$

where $\alpha a_+ \alpha c$ $_a$re the cathode and anode transfer coefficient respectively.

$$i_0 FK\left(C_s.max - C_S.surf\right)^{\alpha a} C_s.suf^{\alpha a} C_e^{\alpha^c}$$

is the exchangecurrent density. The terminal voltage of cell is expressed as:

$$V_{cell}(s) = \left(U_p\left(C_s.\frac{pos(s)}{p.max}\right) - U_n\left(C_s.\frac{neg(s)}{n.max}\right)\right) - \left(\eta_+(s) - \eta_-(s)\right)$$

$$-\varphi_e^{\sim}(s) - R_f I(S) \quad (16)$$

where R_f is the cell internal resistance. U_p and U_n represent the positive and negative electrodes' respective open circuit voltages. What is transient voltage? $U_d = V_{cell} - V_{OCV}$, and the voltage-current 2nd order transfer function, excluding the OCV, is given as

$$\frac{U_s(s)}{I(s)} = R_f + \frac{RT}{AF\left(\alpha_a + \alpha_c\right)}\left(\frac{1}{a_s.p_o^i.p^\delta +}\right) + \left(\frac{1}{a_s.n_o^i.n^\delta +}\right) + \left(\frac{L_n}{2Ak_-^{eff}} + \frac{L_m}{2AK_{sep+}^{eff}}\right.$$

$$+\frac{L_p}{2AK_+^{eff}}\frac{2RT\left(1-t_+^o\right)^2}{AC_e.0F^2}\left[\frac{3(k^{\cdot} + 2\delta_+}{\delta_+\varepsilon ep(3k^{\cdot 2} + 10\delta^2_{+)s+\left(12K^{\cdot}+24\delta_+\right)D_{e.pos}^{eff}}} + \frac{3\left(k - 2\delta_-\right)^2}{\delta_-\varepsilon en\left(3k^2 - 10k\delta_-\right)s - \left(12 - 24\delta_-\right)D_{eneg}^{eff}}\right] \quad (17)$$

$$U_d.\frac{ECM(S)}{I(S)} = R_o + \frac{R_1}{1+R_1C_1s} + \frac{R_2}{1+R_2C_2s} \tag{18}$$

The thermal effect on electrochemical temperature-dependent parameters is modelled using the Arrhenius law:

$$\varnothing(T) = \varnothing_{ref} \exp(\frac{E_a}{R}\left(\frac{1}{T_{ref}} - \frac{1}{T}\right)) \tag{19}$$

Based on Equation, the transfer-function type electrochemical-thermal model is developed.

PERFORMANCE ASSESSMENT COMPARED TO OTHER BATTERY PACKS

Table 1 compares the traits of Li-ion batteries with various power backup techniques based on operational conditions. It has a better power density (110–160 Wh/Kg), according to the comparison. Lithium-ion batteries are appropriate for EV applications since efficient vehicles need the highest power and energy density.

Li-ion batteries also have a wide operating temperature range, which is an added benefit for EV applications. For EVs, the weight of the batteries is a big consideration. Li-ion batteries are consequently lighter and smaller than traditional batteries.

Table 2. Performance of li-ion batteries with various types of batteries

Classification of Battery	Lead Acid	Ni MH	Zn Br	Ni-Cd	Li Ion
Energy Density (Wh/Kg)	30-40	70-100	40-80	35-50	110-165
Power Density	180	250-1000	-	150	1800
Temperature (c)	-20-65	-20-65	-20-65	-40-65	-20-65
Running Voltage (v)	2.1	1.26	1.68	1.26	3.7
Self-Discharge	Less	More	Less	Moderate	Less
Energy Efficiency	75	80	85	85	85

Table 2 compares the characteristics of Lithium-ion batteries with those of other batteries. As was described in earlier articles, Li-ion batteries have linear discharge capacities when compared to other batteries.

BMS BATTERY STATE ESTIMATION

The energy storage device estimate states take into account the state of charge (SOC), state of health (SOH), and state of operation (SOF). Each state's specifics are given.

Estimated State of Charge

The fuel device of a typical fuel vehicle is analogous to the battery PackState of SOC of a new electric vehicle. State of Charge, one of the important energy management factors, is essential to optimize power management Vehicles, increasing capacity of battery and energy utilization, over discharging of batteries, preventing overcharging and s assuring the safety and lengthy service life of batteries, and improving battery capacity and utilization. Under the assumption that the battery's static capacity is present, this section will go into further detail about the SOC estimate method.

In the event that the characteristics of the battery structure are thoroughly examined, the design SOC estimate techniques have a lower offline operation period and produce usable findings. However, they have a time-consuming and expensive parameterization technique in addition to a Due of the computationally challenging computation, online execution is slow. The performance of model-based methods degrades under uncertainties in battery model parameters, as well as noise and ageing impacts, so a special technique for SOC estimation and model identification with noise effect based on Frisch mechanism bias compensating recursive least squares was also proposed (FBCRLS) (Lokhande et al., 2020).

The experiment findings showed how well the suggested method reduced bias noise during model identification, increasing the accuracy of SOC estimation. This led to the construction of a BMS's SOC and capacity estimate framework uses dual estimation and several time scales. Recursive vector-type least squares was used to evaluate the parameters of the online battery model (VRLS). the fusion of the Kalman filter (KF)-based SOC estimation with the RLS-based capacity estimator was used to perform the dual estimation.

The battery's state of charge refers to how much energy is kept there. The battery's open circuit voltage is used to measure it. As a potential voltage that is immeasurable, this voltage must be deduced from the terminal voltage. Values between 0 and 1 roughly describe the status of charge. It is possible to say 0% to 100% at the same time. Where 1 or 100%, respectively, denotes the permitted open circuit voltage's upper limit and 0 or 0%, its lower limit.

As most battery have a less voltage decrease up to 10% State of Charge, which is immediately followed by a rise in voltage, there isn't always a straight correlation between the charges and the voltage.

$$\delta Q \sim \delta uoc \tag{1}$$

In light of this, it is necessary to compute the soc in proportion to the charges (Q).

$$S_{soc} \sim Q \tag{2}$$

Figure 4. Classification of SOC estimate techniques

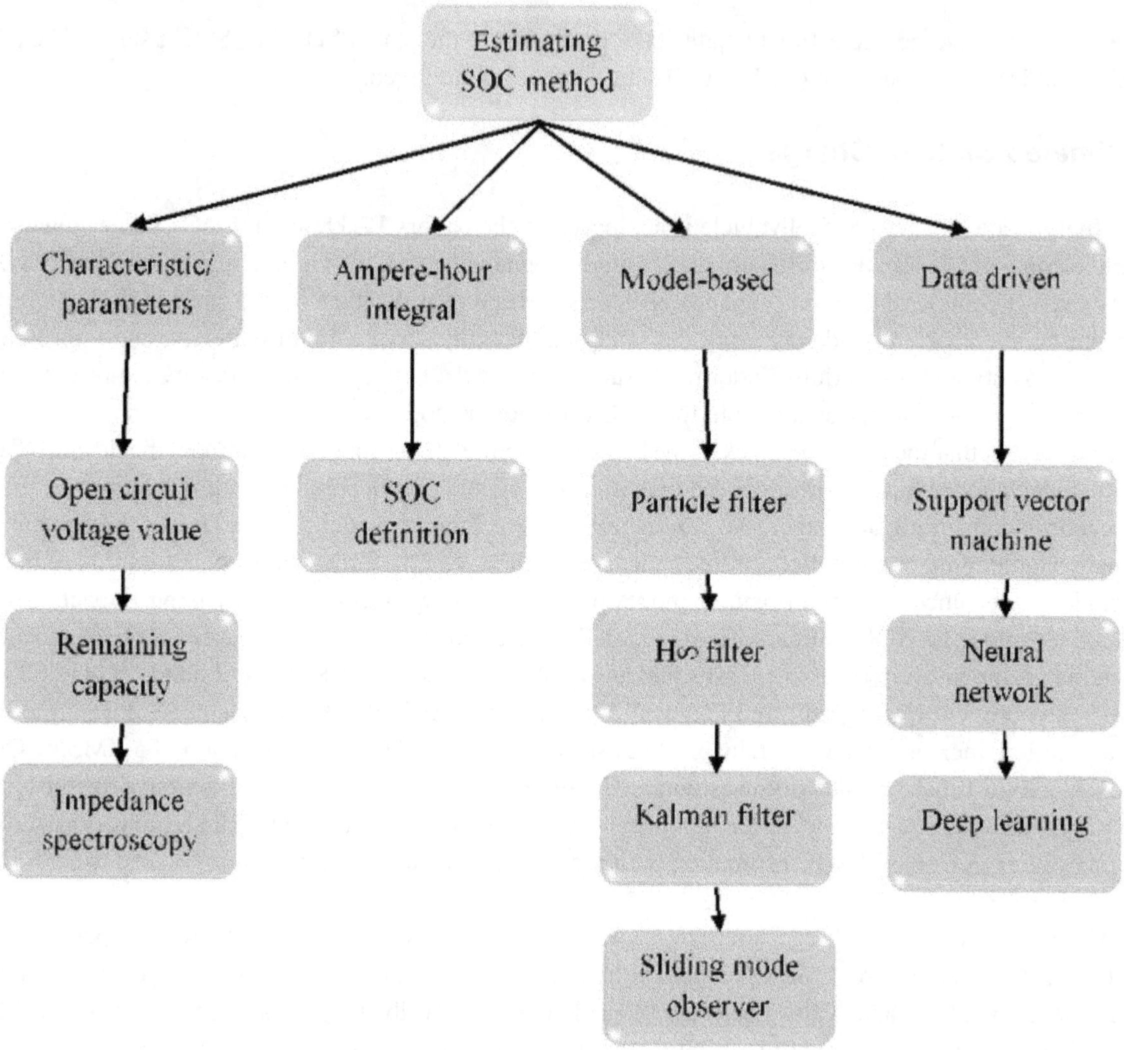

Classification of SOC Estimation Methods

Vehicle operating circumstances are harsh and unexpected, and the electrochemical reaction mechanism and battery stage are complex and challenging to determine. Since it is a concealed state, it is quite challenging to collect correct SOC. Four categories of battery SOC estimating techniques may be identified. The estimation technique using defining parameters.

- Establishing a real-time calculation of the parameter values and calibration of the battery's SOC.
- Creating an offline relationship among the characteristic parameters and SOC.

The discharge experiment procedure, which is thought to be the simplest method for figuring out the battery State of Charge (SOC) can be used to figure out the available capacity. However, it is challenging

Figure 5. Capacity - State of Health – correlation

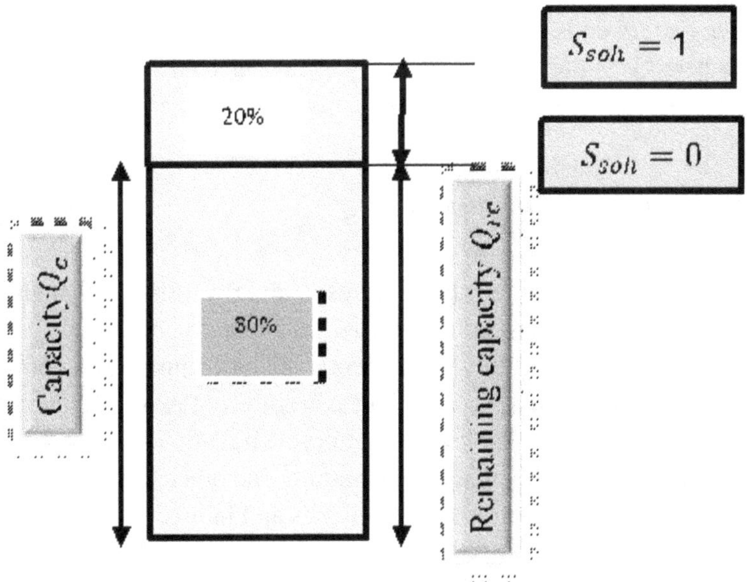

to estimate the capacity via steady current discharge for an extended period of time while really driving, which makes this approach only relevant in certain circumstances.

To some particular settings, like laboratories. The electrochemical workstation must achieve the impedance under various SOCs, determine the mapping between the State of Charge and the parameter sand calibrate State of Charge by consulting tables in order to use the electrochemical impedance spectroscopy method (Muralidharan et al., 2020).

State of Health Estimation

Battery deterioration results in a reduction in the power storage and quick energized de-energized capabilities of batteries. The quantitative index used to assess how much the batteries have aged is called State of Health. The precise SOH value is necessary for an accurate SOC estimation. Since the known State of Health can only serve as a reference, it is not practical to estimate the SOC using this data.

The battery's level of health needs to be determined based on its age. The capacity Qrc remaining is represented by the condition of health. The maximum and minimum open voltages are typically not directly influenced, but instead, the voltage will increase or decrease for a specific amount of energized or de-energized. The output is also represented by the range of values 0% to 100%.

A drop in capacity of more than 20% typically results in unfavourable operation because the frequency of recharging rises to undesirable levels, the process stability is compromised, or the process cannot be completed. This results in the definition that follows;

Let Qc represent a new battery's capacity. So that $S_{soh} = 1$ displays this ability. In addition, a residual capacity of $Q_{rc} = 0.8 \cdot$ Qc leads to $S_{soh} = 0$.

Between the two constraints, the run is proportional.

$S_{soh} \sim Q_{rc}$

The capacity of the battery does not directly decrease with age. For the first 150 to 250 cycles, this typically increases, where Qs is the nominal capacity. Between 10% and 20% of the nominal voltage may be included in this rise. If not cropped, this typically has no impact on the Ss oh and results in an inflated number (1.1 or 1.2).

Classification of SOH Estimation Methods

Resistance and capacity of batteries value, which, correspondingly, reflect power and energy capacities, are the two SOH metrics that are most frequently used. In general, a new battery is thought to have a 100% SOH. As demonstrated in Fig. 3.2.1, experimental analytic approaches and model-based methods can be used to classify SOH estimation techniques. Methods of experimental analysis can directly extract some characteristics indicating the level of battery degradation by examining the experimental data that has been collected on current, voltage, temperature, and other variables. Experimental analysis techniques can be split into direct measurement techniques and indirect analysis techniques depending on the many health indicators used. To achieve the SOH calibration, The parameters related to health must be calculated using model-based methods. Consequently, it could be divided into adaptive data using various estimating strategies.

Design and Optimization Harris Hawk Optimized Algorithm -HHOA Based on Fuzzy Logic Controller

There are many difficulty levels involved in EV battery development and optimization. In terms system engineering, lithium-ion battery-related problems are mostly at the market, levels of the battery system and battery response units. The Harris Hawk Optimized Algorithm (HHOA) of EV battery design primarily manifests itself at the level of the battery system .To obtain the best overall performance of battery goods, the best battery technology and structure must be adopted in conformity with the unique usage requirements of EVs. An eco-design for the EV Battery Management System (BMS) optimization HHOA

An Electric Vehicles battery pack's design and development is a challenging engineering problem. The design team must efficiently arrange the battery cells on the vehicle's chassis, taking into mind the EV battery system's mass, stability, cost, heat dissipation, and other characteristics in order to achieve a suitable total stored energy. The total design of the EV battery pack necessitates a multidisciplinary understanding of materials, electrochemistry, electrical and electronic equipment, mechanical engineering and thermal engineering. If the environmental component is included, its complexity will rise. The EV battery pack with HHOA eco-design. EV battery structure HHOA eco-design enhancement (Fig 7).

By reducing the possible optimization design risk, it is advisable to support design in carrying out their eco-design responsibilities and guarantee the legitimacy of the eco design a decision-making process. A conceptual model of the product, an optimization design model, and performance analysis modules are all included in the framework. According to the product conceptual model, the basic packaging for cells, modules, and packs, as well as the techniques for thermal management and the control system for charging and discharging batteries, make up the majority of an Electric Vehicles battery pack's overall design .After the conceptual design scheme has been chosen in accordance with the fundamental In order to study the battery performances, a variety of analysis modules, such as product structure model,

Figure 6.SOH estimate methods are classification

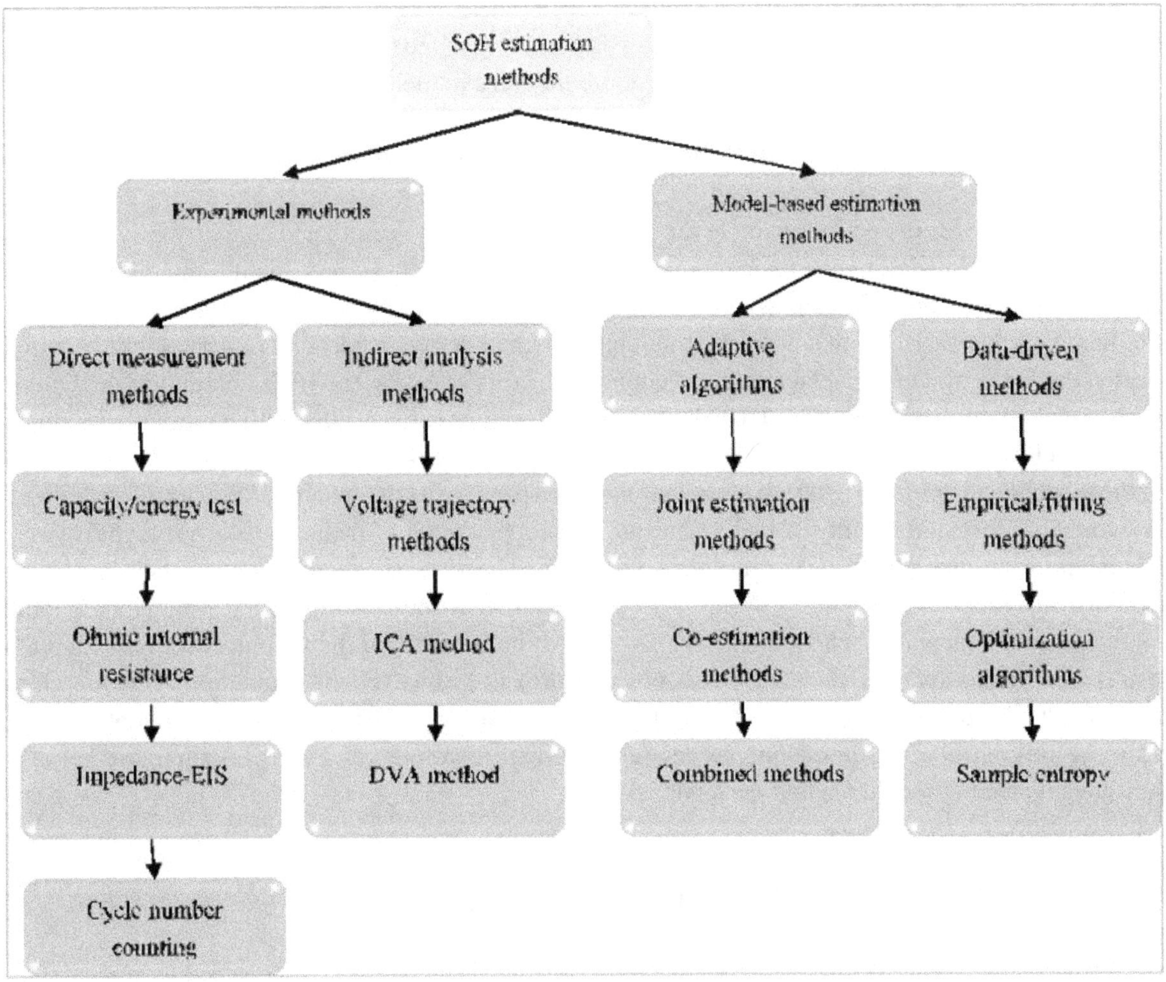

physical research model, electrolyte model of rechargeable batteries, and LCS model, can be built. For all the structure discussed in the most crucial performance analysis module is the product structural model, and it can serve as a vital input for other analysis modules. By entering pertinent product information entered into these modules, battery capacity response data, such as model, size, mechanical performance and environmental performance, can be generated. The module for implementing the examination of the environmental impact of Electric Vehicles battery system the cornerstone of eco-design is the LCS model, in particular. The three most important elements. The design constraints, objective functions, and design variables that make up an optimization design model are chosen by performance analysis modules. The ultimate solution technique can be reached by determining the accurate values for the design variables for Electric vehicles battery system.

While taking into account, To increase the optimized design efficiency of Electric Vehicles batteries, the optimization framework uses an open architecture to account for the complexity and variety of Electric Vehicles battery design and development.. Always, how difficult an optimization task is will depend on the quantity of design variables. The HHOA optimized design's design space is evaluated, and the optimization problem is solved more challengingly the more design factors that are included in the optimization. According to the actual needs of development, the scale for the design analysis and the objectives for the performance analysis can be dynamically changed. If a more thorough investigation is needed, the designers can additionally include more modules in their optimization design.

Fuzzy Logic Controller for the EV Complete Model in-HHOA

In order to reduce energy consumption in the ESS, fuzzy control was used, and change the system's default attitude after modelling the three probable EVs. For this, the fuzzy sets, inputs, and outputs of the fuzzy system are defined. The inputs to the fuzzy system are the energy demand and SOC which enable it to take multiple variables into account simultaneously and deliver feedback. To achieve the required results, the fuzzy system must take in the account the factors affecting the EV's energy use in HHOA. Harris hawk Optimized Algorithm-HHOA Based on Fuzzy Logic Controller with battery management system in electric vehicles. The output of the fuzzy system matches the optimum energy demand based on the input data. In order to determine whether there is a need, a fuzzy set named "%Ahorro" is defined. Due to the uniform uncertainty of fuzzy sets and the fact that none of their inputs are straightforward, it is advised to employ soft, not semi-linear, functions. The system's regulations are then laid down after that. All inputs and outputs were initially combined to take into account every eventuality that the controller could encounter. The inference rules are then built based on the graphical analysis, and the results are filtered to create the algorithm's best performance. the cycle of simulation. This loop lists the elements that are relevant and crucial for the decision-making process. The parameters that are most active in the control decision-making process are the fuzzy controller and the ESS simulator because they are situated between them and the feedback loop.

BATTERY MANAGEMENT SYSTEM APPLICATION

The effectiveness of the battery and BMS has a significant impact on the performance of the EV. to fulfil the demands of the EVs' dynamic performance, safety, economy, and environmental friendliness.

i) Battery Electric Vehicle

BEVs are new energy vehicles that only store their energy in batteries. In order to drive the motor and move the vehicle forward, the battery transfers the battery transfers electric power to the motor. The battery's capacity has a direct impact on how far a BEV can travel. The vehicle's driving range will improve with the battery capacity, which will also result in an increase in the battery's volume and mass. As a result, the decision of the appropriate battery for a BEV relies on several design objectives, driving situations, and road conditions.

ii) Hybrid Electric Vehicle

A hybrid energy vehicle, or HEV, combines an internal combustion engine with an electric drive system to provide propulsion. The battery serves as the primary energy source for the HEV's electric propulsion system in general. Instead of having a big capacity, a HEV battery must be able to supply the car with

Figure 7. Harris hawk Optimized Algorithm -HHOA in BMS

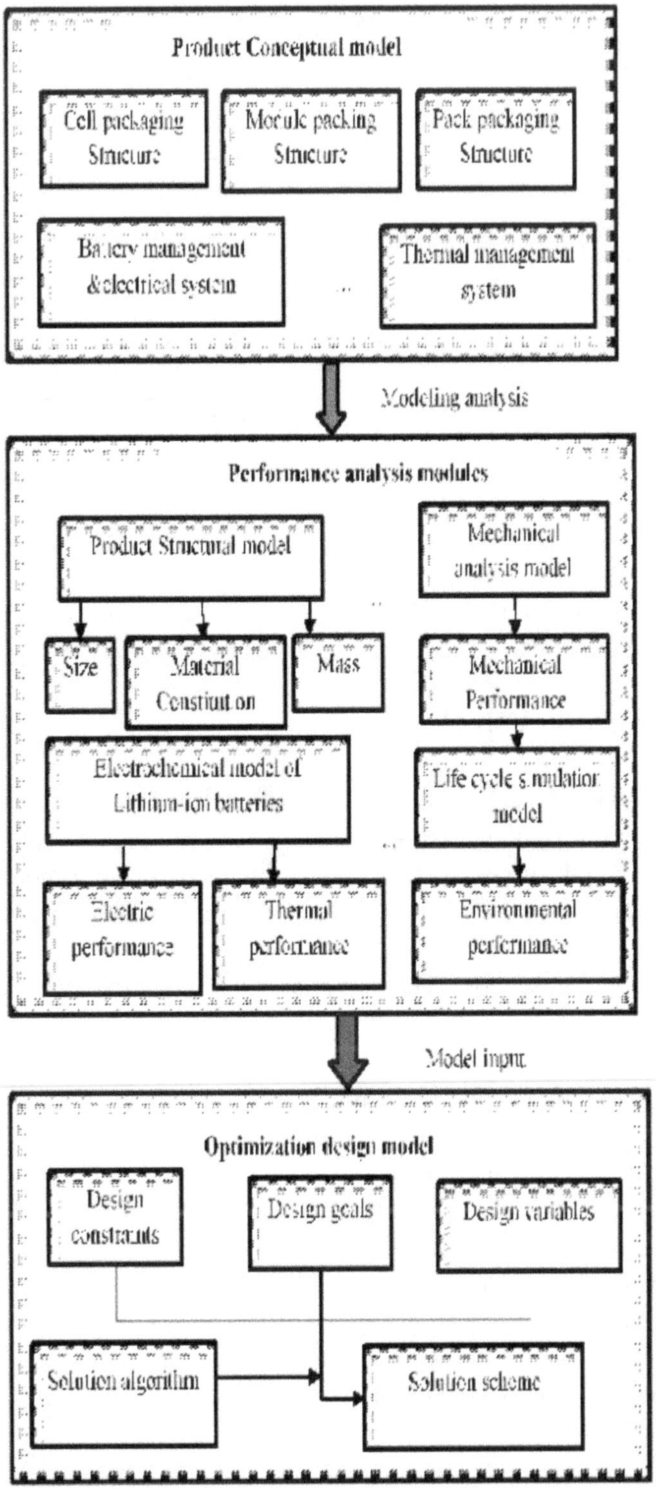

Table 3. Optimization results of design schemes

Name	Scheme 1	Scheme 2	Scheme 3
Number of modules in the pack (parallel, series)	1, 26	1, 25	1, 25
Number of cells in a module (parallel or series	2, 5	2, 6	3, 4
Positive porous electrode of Thickness [um]	89	80	74
Positive electrode coating Width in [mm]	135	145	130
Positive electrode coating Length [mm]	290	297	295
A Cells number of positive current collectors	28	27	33
Cell thickness [mm]	14	11	14
Pack length, width, height [mm]	1446, 692, 172	1395, 706, 180	1486, 702,168
Stored energy[kwh]	61.0	66.1	61.4
Pack mass[kg]	321	348	337
Electric Vehicles driving range[km]	327	352	328
Total environmental impact value	5523	5978	5998

the high power it needs right away. HEVs, as opposed to BEVs, have three distinct structural configurations: series, parallel, and series-parallel, each of which has its own set of battery design specifications.

iii) Plug-in Hybrid Electric Vehicle

The PHEV is a form of HEV that may be charged by the electrical grid. It contains operating modes for BEV and HEV, which are just BEV and HEV combined. In addition to being able to travel dozens of kilometres in BEV mode and the necessary distances on highways in HEV mode, these cars frequently have a high output capacity in a low SOC range. Because of this, PHEV battery requirements for energy density and power density are comparable to those of BEV batteries and HEV batteries, respectively. In reality, because PHEV batteries use more energy, they are more expensive than BEV and HEV vehicles.

SIMULATION RESULTS AND DISCUSSION

The proposed BMS's operation with different battery types, such as Li-ion batteries with the same capacity, is simulated. In order to contrast the performance of the suggested Battery Management System, the simulation work is initially done with the identical RL load and SOC value. Using MATLAB/Simulink, several instantaneous battery parameter values, consisting of voltage, current, energy cost, and SOC, are analyzed, and their performance characteristics are displayed as graphs. This comparison shows that lithium ion batteries perform the best across the board for most battery parameters. These results show that Overall, the lithium ion battery shows the best performance characteristics.

Figure 8. Flowchart for the complete EV model

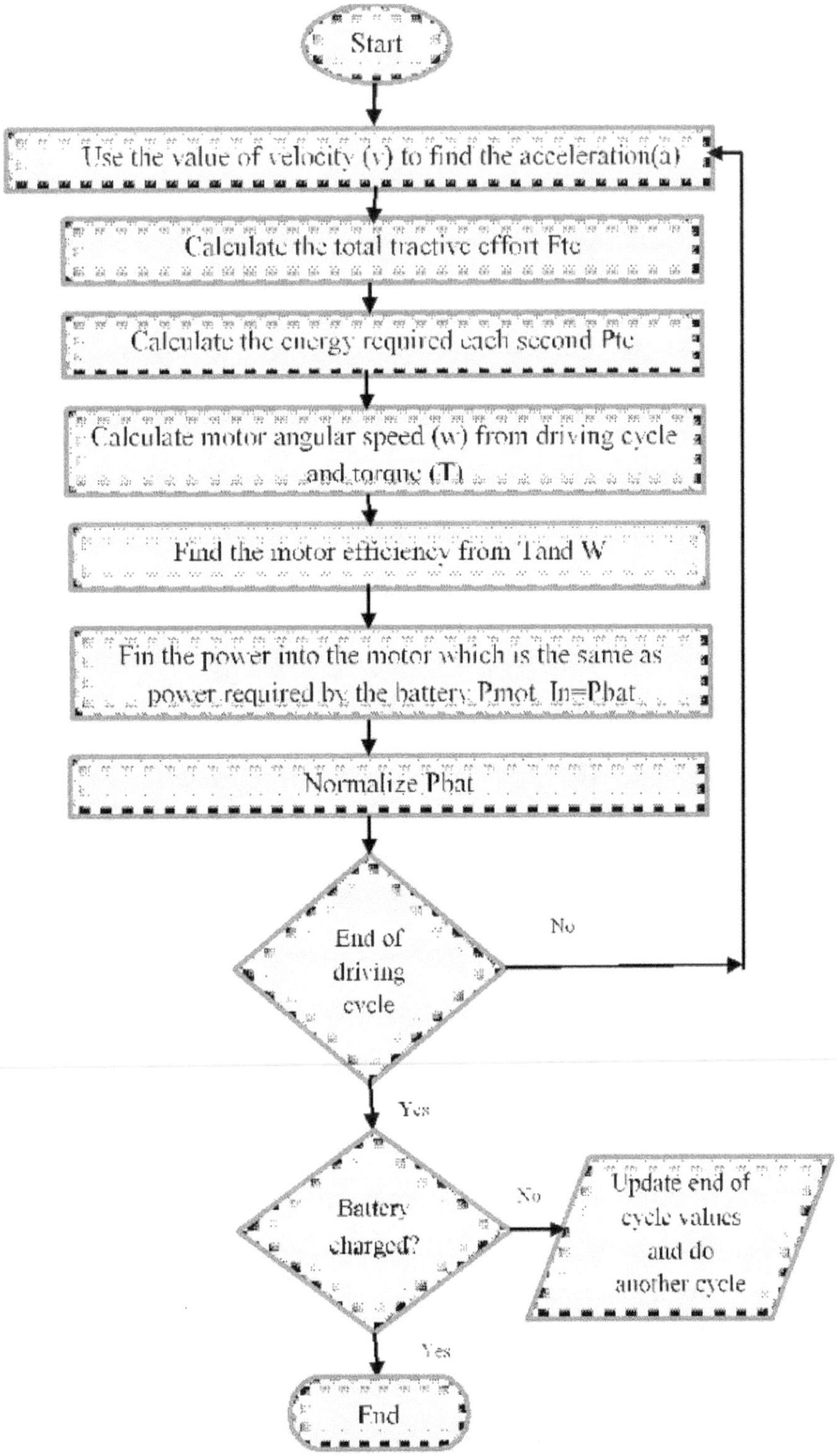

Figure 9. BMS measured Estimated Voltage Curve

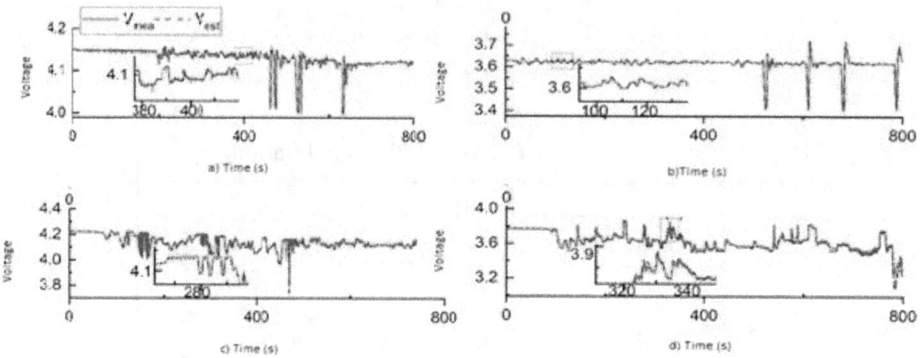

Figure 10. Fuzzy states defined for the variable "%Ahorro","Energy demand", variable "SOC"

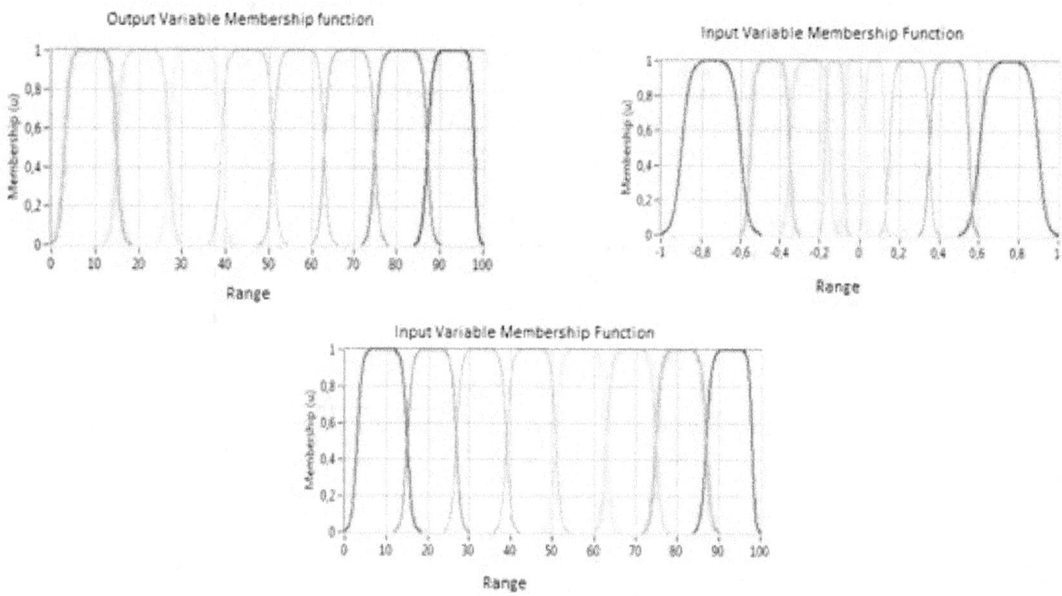

Figure 11. Characteristics of Li-Ion battery

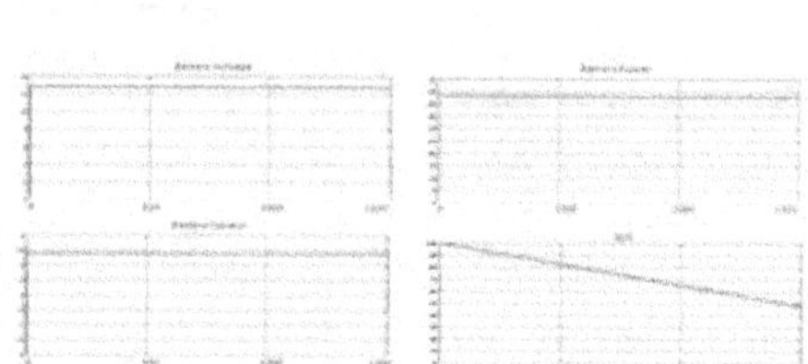

CONCLUSION

This paper developed a simulink battery model of a lithium-ion battery. This model was developed utilizing a variety of mathematical and physical components in simulink and simscape. The model does the data collection and SOC successfully (State of Charge),and SOH functions (State of Health). A control-oriented transfer-function type model was made using the original P2D model after it had been simplified using the Pade approximation. This model was then utilized to include a complicated model into an onboard BMS. The generated model's structure and complexity were comparable to those of a second order ECM, which substantially aided the process of analytically solving the problem. System for managing batteries. Using MATLAB/Simulink, several instantaneous battery parameter values, such as voltage, current, power consumption, and SOC, are examined, and their performance characteristics are displayed as graphs. This BMS was developed using fuzzy logic to model the methodologies for improving the autonomy and performance of the EV.The output of the FLC allows saving stored energy according to the demand at each time step of the simulation. This comparison shows that lithium ion batterie sperform the best across the board for most battery parameters.

REFERENCES

Ananthraj & Ghosh. (2021). Battery Management System in Electric Vehicle. *2021 4th Biennial International Conference on Nascent Technologies in Engineering (ICNTE)*. doi:10.1109/icnte51185.2021.9487762

Chen, X., Shen, W., Vo, T. T., Cao, Z., & Kapoor, A. (2012). An overview of lithium-ion batteries for electric vehicles. *2012 10th International Power & Energy Conference (IPEC)*, 230–235. doi:10.1109/ASSCC.2012.6523269

Dost & Sourkounis. (2014). *Battery management system realisation for electric vehicles*. doi:10.1109/MED.2014.6961456

Gao, Y., Zhu, C., Zhang, X., & Guo, B. (2021). Implementation and evaluation of a practical electrochemical- thermal model of lithium-ion batteries for EV battery management system. *Energy*, *221*, 119688. Advance online publication. doi:10.1016/j.energy.2020.119688

Hossain Lipu, M. S., Hannan, M. A., & Tahia, F. (2021). Intelligent algorithms and control strategies for battery management system in electric vehicles: Progress, challenges and future outlook. *Journal of Cleaner Production*. Advance online publication. doi:10.1016/j.jclepro.2021.126044

Huang, X., Acharya, A. B., Meng, J., Sui, X., Stroe, D-I., & Teodorescu, R. (2020). Wireless Smart Battery Management System for Electric Vehicles. *2020 IEEE Energy Conversion Congress and Exposition (ECCE)*, 5620–5625. doi:10.1109/ECCE44975.2020.9236279

Lee, J. (2010). Battery management technology for an electric vehicle. *Proceedings of the 16th ACM/IEEE international symposium on Low power electronics and design - ISLPED '10*, 361. doi:10.1145/1840845.1840923

Lokhande, J. S., Daigavhane, P. M., & Sarkar, M. (2020). A Critical Approach Towards a Smarter Battery Management System for Electric Vehicle. *2020 4th International Conference on Trends in Electronics and Informatics (ICOEI)*, (48184), 232–235. doi:10.1109/ICOEI48184.2020.9142869

Martinez, D. A., Poveda, J. D., & Montenegro, D. (2017). Li-Ion battery management system based in fuzzy logic for improving electric vehicle autonomy. *2017 IEEE Workshop on Power Electronics and Power Quality Applications (PEPQA)*, 1–6. doi:10.1109/PEPQA.2017.7981677

Mihet-Popa, L., & Groza, V. (2015). Battery management system test platform developed for electric vehicle applications. *2015 IEEE 9th International Symposium on Intelligent Signal Processing (WISP) Proceedings*, 1–6. doi:10.1109/wisp.2015.7139163

Muralidharan, A., Sreelekshmi, R. S., & Nair, M. G. (2020). Cell Modelling for Battery Management System in Electric Vehicles. *2020 Third International Conference on Smart Systems and Inventive Technology (ICSSIT)*, 558–564. doi:10.1109/ICSSIT48917.2020.9214253

Quantmeyer, F., & Liu-Henke, X. (2013). Modeling and identification of lithium-ion batteries for electric vehicles. *2013 International Symposium on Electrodynamic and Mechatronic Systems (SELM)*, 19–20. doi:10.1109/SELM.2013.6562959

Qureshi, K. N., Alhudhaif, A., & Jeon, G. (2021). Electric-vehicle energy management and charging scheduling system in sustainable cities and society. *Sustainable Cities and Society*, *71*, 102990. Advance online publication. doi:10.1016/j.scs.2021.102990

Sreedhar, R., & Karunanithi, K. (2021). Design, simulation analysis of universal battery management system for EV applications. *Materials Today: Proceedings*. Advance online publication. doi:10.1016/j.matpr.2020.12.136

Vaideeswaran, V., Bhuvanesh, S., & Devasena, M. (2019). Battery Management Systems for Electric Vehicles using Lithium Ion Batteries. *2019 Innovations in Power and Advanced Computing Technologies (i-PACT)*, 1–9. doi:10.1109/i-PACT44901.2019.8959965

Wu, Y., Liao, X., Chen, W., & Chen, D. (2011). A Battery Management System for electric vehicle based on Zigbee and CAN. *2011 4th International Congress on Image and Signal Processing*, 2517–2521. doi:10.1109/cisp.2011.6100781

Xiong, He, Guo, & Ding. (2011). *Modeling for Lithium-Ion Battery used in Electric Vehicles*. doi:10.1016/j.proeng.2011.08.540

Xiong, R. (2020). *Battery Management Algorithm for Electric Vehicles*. doi:10.1007/978-981-15-0248-4

Zhang, C., Liu, Y., Qian, Y., & Bao, H. (2020). An optimization framework of electric vehicle (EV) batteries for product eco-design. *Procedia CIRP, 90*, 366-371. doi:10.1016/j.procir.2020.01.081

Zhang, X., Wu, J., & Kang, G. (2016). SOC estimation of Lithium battery by UKF algorithm based on dynamic parameter model. *2016 13th International Conference on Ubiquitous Robots and Ambient Intelligence (URAI)*, 45–950. doi:10.1109/URAI.2016.7734116

Section 2

Electric Vehicles: Modelling and Control

Chapter 5
Electric Vehicles and Their Types

Vigneshwaran A.
Hindusthan College of Engineering and Technology, India

Naveen Kumar K. A.
Hindusthan College of Engineering and Technology, India

ABSTRACT

Electric vehicles are the only way to reduce greenhouse toxic gases. Electric vehicles, including battery vehicles, hybrid vehicles, plug-in hybrid vehicles, fuel cell electrical vehicles, are the most commonly used transports recently. There are some major obstacles for EV to overcome before totally replacing IC engine vehicles including EV configurations, battery energy sources, electrical machines, charging techniques, optimization techniques, impacts, trends, and possible directions of future developments. Currently, the battery technologies are moving from the lead-acid batteries to the lithium-ion. Moreover, different standards are available for charging process of the EV, including the power control and battery energy management proposals.

1. INTRODUCTION

Electric vehicles (EVs) are being developed on land (road and rail vehicles), on water surfaces (ship and submarine underwater vessels), and on-air surfaces (aircraft and spacecraft). The electric vehicle is the future of transportation, also known as an electric vehicle. First introduced in the 19th century, electric cars took almost 100 years for the internal combustion engine to become the dominant mode of motor vehicle propulsion. For propulsion, the electric vehicle consists of one or more electric motors or propellants. Electric cars are stored in batteries. The energy sources are a charging system, a solar panel, and a generator that converts fuel into electricity.

Electric cars are starting to be driven as a prescription due to technological developments and emerging technologies in the 21st century, focusing mainly on renewable sources.

DOI: 10.4018/978-1-6684-6631-5.ch005

1.1. Key Components of Basic Electric Vehicles

Battery (Electric Auxiliary Battery): In an electric vehicle, the auxiliary battery supplies electricity to the components of the vehicle.

Charging port: The vehicle can be connected to an external power source via the charging port, which is charged by the battery charge.

DC/DC Converter: This device converts the DC voltage of the high-voltage battery into the low voltage DC required to power the vehicle's components and charge the auxiliary battery.

Electric Traction Motor: The car uses battery power to drive the wheels. Some vehicles use engine generators that are responsible for starting and restarting the vehicle.

Onboard Charger: It takes AC power from the ignition port and converts it to DC current to supply the portable battery. It connects to the charger to charge the battery while monitoring the battery characteristics such as voltage, current, temperature and flow rate.

Power Electronics Controller: This device controls the flow of electrical energy supplied from the traction battery to control the speed and torque generated by the electric traction motor.

Thermal System (Cooling): This system maintains the proper temperature of the engine, electric motors, electronics and other components.

Traction battery: It stores the electricity consumed by the electric traction motors.

Transmission (Electric): The transmission transmits mechanical power from the electric traction motor to drive the wheels.

1.2. Types of Electric Vehicles

Based on working technologies and storage types, electric vehicles are divided into four types

- Battery Electric Vehicle (BEV)
- Hybrid Electric Vehicle (HEV)
- Plug-in Hybrid Electric Vehicle (PHEV)
- Fuel Cell Electric Vehicle (FCEV)
- Extended Range Electric Vehicle (ER-EV)

1.3. Advantages and Disadvantages of Electric Cars

1.3.1. Advantages of Electric Cars

As a consumer, EVs are a great way to save gas. However, there are many reasons to invest in electric vehicles with advanced technologies.

I. No gas required

Electric vehicles are fully charged with the electricity they provide, so you don't have to buy gas again. Driving a gas car when gas prices are at an all-time high can leave a hole in your pocket.

It costs an American about 15 cents per mile to drive a gas vehicle, while many electric cars cost 5 cents per mile. Electricity is cheaper than petrol.

If most people install a few solar panels to charge their car from the garage, the cost can be significantly reduced, saving the cost of powering the whole house. With an EV, you can avoid spending $2,000 to $4,000 a year on gas bills.

II. More acceptable

Charging an electric car is easy. No need to go to the gas station to fill up your car before you go! Electric cars can also be charged through a simple household socket.

III. Storage space

The fuel in this car is quite cheap and many new cars offer big discounts from the green government. Electric cars are a great way to save money in your personal life.

IV. No emissions

The biggest advantage of electric vehicles is eco-friendly certification. Electric vehicles are 100% eco-friendly because they run on electric motors.

It is a clean source of energy that does not release any toxic gases or fumes into the environment. Gasoline hybrids are better than hybrids because they produce less exhaust gas. You contribute to a healthier and greener climate.

V. Reputation

Electric cars are becoming more and more popular. According to Wikipedia, it is about three times more efficient than an internal combustion engine car. As it becomes more and more popular, all new models are released, giving us more options in the future.

VI. Drive safely

Electric cars go through a trial-and-error process just like other gasoline-powered vehicles. Electric vehicles are safer, have more traction, and are very stable on the road in the event of an accident.

In the event of an accident, the airbag pulls and releases the battery power. It will not cause serious injury to you or other occupants of the vehicle. The absence of flammable fuels and gases minimizes the possibility of an explosion.

VII. Cost-effectiveness

You've heard of the blockbuster electric car. However, as technology has advanced, both costs and maintenance have come down.

Mass production of batteries and tax breaks further reduce costs and make them more profitable. Contact your tax professional for more information on state or federal tax credits.

VIII. Low maintenance

Because electric vehicles operate as electrically powered motors, many maintenance tasks for motors, internal combustion engines, or gas engines generally do not require lubrication.

Other expensive engineering jobs are a thing of the past. This reduces the maintenance of these vehicles. You don't have to go to the gas station as often with it as with a regular gas car.

IX. Reduce noise pollution

Electric cars are very quiet and reduce noise. Electric motors provide high acceleration and smooth movement over long distances. Many EV owners report annual savings in the tens of thousands of dollars.

X. Battery Life and Cost

Batteries are an integral part of an electric car. Most EV batteries are lithium-ion batteries, which are increasing in price every year.

The total capacity of the lithium battery should be suitable for 300-500 cycles. A good battery can last up to 10 years. As technology advances, battery costs are expected to decrease.

XI. Simple operation

In the automotive world, electric cars are the way to drive. Commercial EVs only come with long gearing and they don't allow you to buy problems like big money.

This eliminates the need for a table to prevent. That is, the gas pedal, brake pedal and steering wheel

1.3.2. Disadvantages of Electric Vehicles

While the positive evidence is pretty clear, there are some downsides that everyone should consider before making the next big investment in an EV. Here's why:

I. Recharge point

Electric filling stations are still in the development phase. Many places you visit every day do not have electric stations for your car. That is if you go on a long trip or visit family in the country or the suburbs, it won't cost you anything. Find a charging station. You can get stuck there.

But before charging stations become more common, have a map of the places you live and frequent. This way you can charge your new EV when needed.

II. Fixed initial investment

Electric cars are relatively new, so you might be surprised by the expected retail price of an electric car. Most affordable brands cost around $30,000 to $40,000.

If you ask for an extra option, you can pay up to $80,000. $50,000 to $50,000 more.

III. Electricity is not driving

If these options are not carefully considered, electric vehicles can cause problems with your electric bill. If you haven't done enough research when considering an electric car, you may be making an unwise investment.

Electric usually requires a high charging function. This will negatively affect your monthly electricity bill.

IV. Short run and speed

Electric cars have limited range and speed. Most of these vehicles give about 50 to 100 thousand miles. It is not currently available for long-term travel but will be improved in the future.

V. Long charging time

Gasoline vehicles take minutes to refuel, while electric vehicles take 4-6 hours, sometimes a day, to fully charge.

Therefore, dedicated power plants are needed over a longer period of time. As a result, the investment of time and necessary planning is very difficult for some.

Some tools can speed up the charging time. But again, it would be another investment. So think about yourself.

VI. Silence is lost

The silence can be a little off beat because people like to bang in the background. However, electric cars are quiet and can sometimes cause accidents.

VII. Ordinary 2 seats

Most electric vehicles available today are small and only have two seats. Not all family members are suitable, and it can be a bit uncomfortable for a third person to ride with the other two passengers.

VIII. Replace the battery

Almost all electric vehicle batteries need to be replaced every 3 to 10 years, depending on the battery type and purpose. Inertia is not good for cities.

Cities that already suffer from severe power shortages are not suitable for EVs, as they need electricity to charge. Excessive electricity consumption limits their daily energy needs.

IX. Fewer options

The electric vehicle market is expanding today and shows no signs of slowing down. However, it is clear that there is less opportunity to choose and customize the aesthetics of the EV.

In addition, traditional tea can often be customized. That may change over time, but it's a loss for many.

X. Minimal pollution

Electric cars are also not 100% zero emission, but they do indirectly produce a small amount of pollution. The power and energy to make the attack does not always come from renewable energy sources.

2. BATTERY ELECTRIC VEHICLE (BEV)

A battery electric vehicle works with the help of a battery and a motor to drive the vehicle. Battery electric vehicles are the most used electric vehicles recently. The concept is basically simple and easy to implement. It consists of a reachable battery for its storage and reuse, a controller (control module), and an electric motor.

Battery: The battery is mainly used to accumulate electricity further used for the electric motor for propulsion. The battery is recharged from electricity via plugs and now explores wirelessly. Recharging the battery at charging stations and at home.

Controller: The controller simply controls the engine power, vehicle speed, forward and reverse. This is commonly known as a "two-quadrant controller" - forward and reverse. It is generally preferable to use regenerative braking, a suitable form of frictionless braking, to recover energy. Additionally, when the controller allows it to regenerate in the forward and backward directions known as a "four-quadrant controller"

2.1. BEV Architecture and Main Components

Figure 1. Architecture and main components of a battery electric vehicle (BEV).

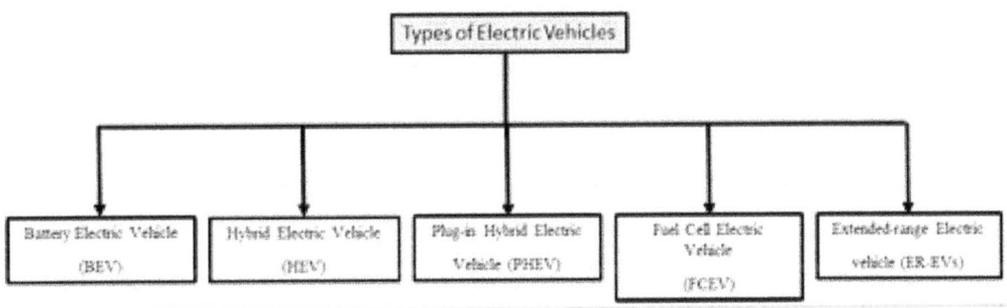

2.2. BEV Components

- Electric motor
- Inverter
- Battery
- Control module
- Drive the train

2.3. Principles of BEV Work

- The power source is converted from a DC battery to an AC battery.

- The accelerator pedal sends a signal to the controller, which changes the speed of the car by changing the frequency from outside to inside the car.
- The engine shifts and shifts gears.
- When the brakes are applied or the electric vehicle slows down, the engine turns on the generator and powers the battery.

Figure 2. Battery Electric Vehicle (BEV) operation.

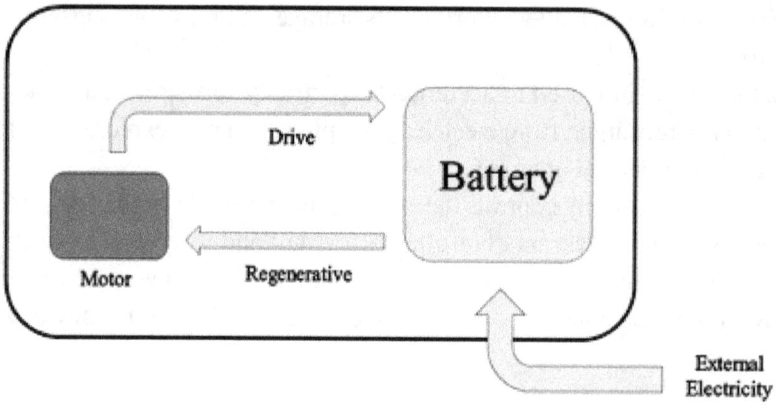

3. HYBRID ELECTRIC VEHICLE (HEV)

Hybrid electric vehicles have more than one power source. Power can be batteries, bicycles and so on. The power source may be an engine, fuel, cell, motor, capacitor, depending on configuration, voltage of the motor, or two or more components used. Hybrid vehicles combine electric motors with combustion engines (ICEs) to save energy and reduce pollution, using any type of vehicle. Ingredients are usually divided into several types of products and similar products. In a hybrid car, the engine is driven, and the generator is driven by an electric motor. Similar hybrids, engines and electrical systems are paired to drive the car. Hybrids can use less fuel in city driving. Ice continues to work very well at first. Similar hybrids can use less fuel in highway driving and perform better in ICES at higher speeds. Hybrids are further divided into mid-range models, hybrids, and hybrid engines. It's an electronic device. For a plug-in hybrid car, it can be more or equal to a battery. As it is used, the length of the vehicle and the rear load increase, in the electrical system.

3.1. With the Help of an Electric Motor

The additional power provided by the electric motor in an HEV can dwarf the internal combustion engine. When the vehicle is stationary, charging batteries can cause additional load and reduce engine braking power. These features combine to reduce fuel consumption without compromising performance.

3.2. Regenerative Braking

HEVs cannot be driven externally to charge the battery. Instead, he rebuilt used car cylinders and rebuilt the internal combustion engine. The car uses the electric motor as a generator and stores it in the battery to replenish energy lost during braking.

3.3. Fuel Saving System Design

HEVs can be hybrids or mild hybrids, and full hybrids can be produced in series or in parallel.

A mild hybrid (also called a mild hybrid) uses a battery and an electric motor to power the car, starting the engine when the car is stopped (at a traffic light or on the road). Riches. Light mixers cannot run on electricity alone. These cars are generally more expensive than full hybrids, but offer lower fuel efficiency than full hybrids.

A full hybrid has a larger battery and more efficient electric motors, allowing the vehicle to travel shorter distances and lower speeds. This vehicle is more expensive than a mild hybrid, but has better fuel economy.

There are many ways to confuse electric motors and motors. There are series hybrid vehicles and parallel hybrid vehicles.

3.4. A Series of Hybrid Vehicles

It is a hybrid electric vehicle with a battery source. The engine is usually connected to a generator to supply electricity to the battery. The system can be set up so that it acts as a generator to supply the engine and the engine. In this case, the size of the battery can be reduced, but the size of the generator and motor must be increased. The electrical output of a conventional hybrid car system is:

1) Switch the output switch to DC to charge the battery.
2) DC/AC converter motor vehicle for power. The vehicle has a DC-DC converter to charge the 12V battery. In addition, electric air conditioners require inverters and related power systems.

3.5. Parallel Hybrid Vehicles

Hybrids are cheaper and can generate motors, batteries and electricity from the existing grid. However, hybrid vehicles require the same complexity. Similar hybrids are available in different designs, depending on the motor/generator segment and the engine. Parallel hybrid vehicles can use engines and electric motors separately or in combination. The Toyota Prius and Honda Insight are examples of similar hybrid systems on the market. Figure: 4 shows a schematic diagram of the parallel hybrid drive system.

Figure 3. Propulsion system of a serial hybrid vehicle.

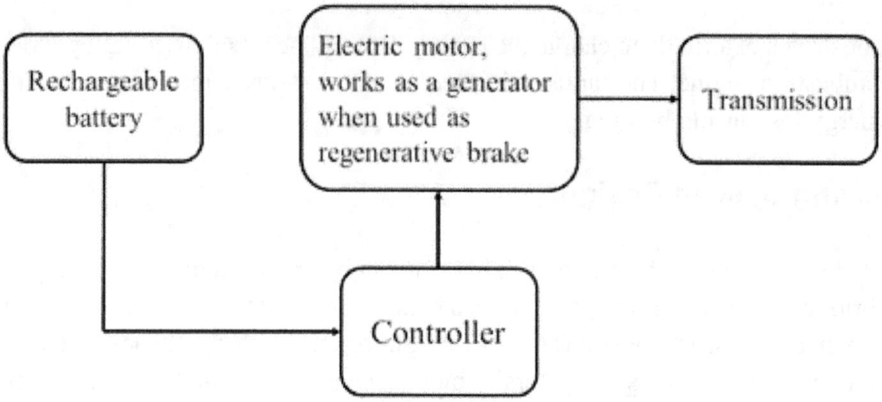

Figure 4. parallel hybrid vehicle propulsion system.

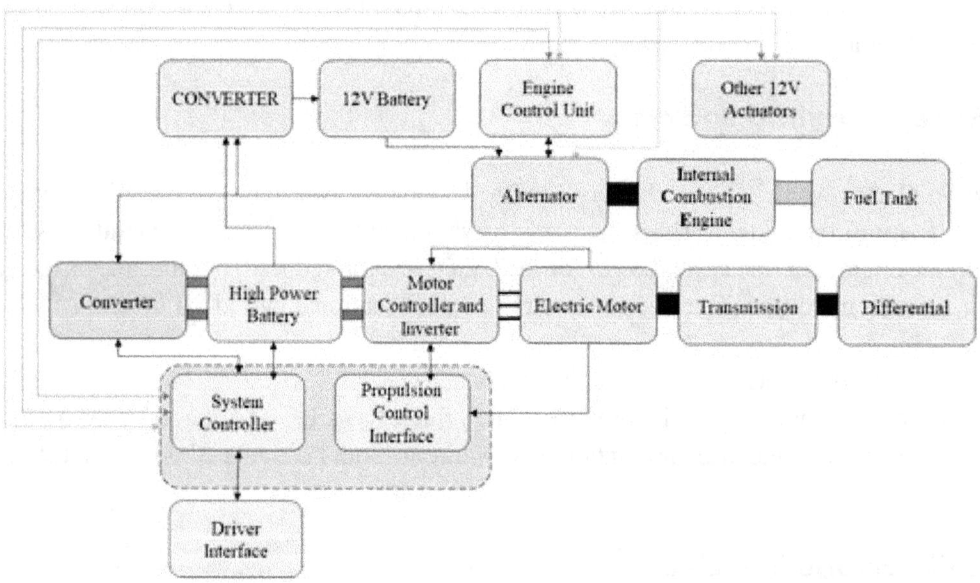

3.6. HEV Architecture and Main Components

Figure 5. Architecture and main components of HEV.

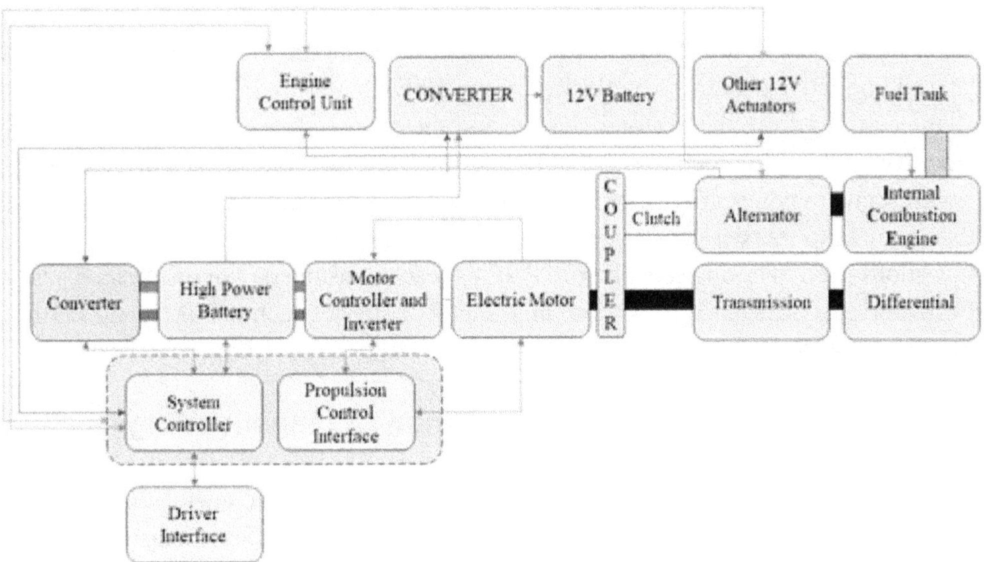

3.7. HEV Components

- Engine
- Electric motor
- Battery with controller and inverter
- Fuel tank
- Control module

3.8. Principles of HEV Work

A hybrid generator engine is one or more electric motors that run on energy stored in a battery. Hybrid electric vehicles cannot be plugged in to charge the battery. Instead, the battery is powered by a regenerative brake system and an internal combustion engine. Too much power delivered by an electric car leads to engine inefficiency. The engine battery can be used to power the accessory load and reduce the slip and tear of the engine lock. Combined, these features improve fuel economy without compromising performance.

4. PLUG-IN HYBRID ELECTRIC VEHICLE (PHEV)

PHEVs have been recognized by industry and academia as well as government agencies around the world as an advanced hybrid vehicle technology. PHEVs have battery capacity that can be charged externally, allowing them to go farther than conventional electric HEVs, with better mileage per gallon (MPG). The battery can be charged by using the nearest exit or parking lot. PHEVs have improved electricity consumption because the batteries are charged at night.

A typical PHEV architecture configuration is shown in Figure 6. Many companies are trying to convert existing HEVs to PHEVs as a viable technology to increase HEV efficiency. Automakers are also considering and preparing to bring PHEVs to the commercial market. These upgrades increase energy efficiency by adding active batteries or replacing existing HEV batteries. In either case, the storage space stores all the electricity from the external charger and the regenerative brake and transfers electricity to the engine. Network wiring requires an AC-DC power converter (PFC) and a digital programmable controller with accurate power curves for high power loads. A two-way DC-DC converter with a duty curve is also required to transfer power between the battery and the drive motor.

Many questions need to be answered before PHEVs reach consumers. For example, if you use high-quality PFC batteries at the same time, choose high-quality batteries such as NiMH and Li, the safety of use in automotive batteries, thermal management and cell balancing are important issues.

4.1. Driven by an Electric Motor and an Internal Combustion Engine

A PHEV has an internal engine and an electric motor that harnesses the energy stored in the engine. PHEVs generally have more batteries than hybrid electric vehicles. It offers a reasonable range (about 15-60 miles or more for current models) on electricity alone.

When driving in the city, most of the PHEV's power comes from the energy storage device. For example, PHEV light vehicle drivers can drive at full power, plug in,Leave their car overnight and prepare all the other electronics for the next day. The combustion engine in the car drives the car, and when the battery is depleted, the vehicle accelerates, or the heating and air conditioning conditions are under heavy load. Other large PHEVs do the opposite. Internal combustion is used to move the engine to and from the manufacturing site, which controls the power of vehicle components and the air conditions in the cabin during manufacturing.

4.2. Refuelling and Driving Options

The PHEV battery can be charged by external power, internal heating or adaptive braking. When braking, the electric car acts as electricity, using energy to charge the battery and restore lost energy.

PHEV fuel consumption depends on battery life. Like when a car stops while charging, fuel consumption is roughly comparable to a hybrid electric vehicle of the same size. If the vehicle is driving below the net electricity limit and is plugged in for charging between trips, only electricity is available. So, the best way to maximize the benefits of electricity is to keep your vehicle charged at all times.

4.3. Construction of a Fuel-Efficient System

In addition to motor power and batteries, there are various methods such as motors and engines. The two main structures are parallel and series. Some PHEVs can operate in a parallel or series configuration with a transmission that can switch between them depending on the driver's profile.

Parallel plug-in hybrid operation in which the engine and electric vehicle are connected to the wheels in a hybrid configuration. Electric motors and motors can drive the wheels directly.

A typical plug-in hybrid uses only one electric motor to drive the wheels. It is made from the combustion of electric engines in electric cars. These vehicles are often referred to as remote electric vehicles. An electric car almost always drives the wheels, but if the battery dies on the road, the car can be swapped for something like a hybrid.

4.4. PHEV Architecture and Main Components

Figure 6. PHEV architecture and main components.

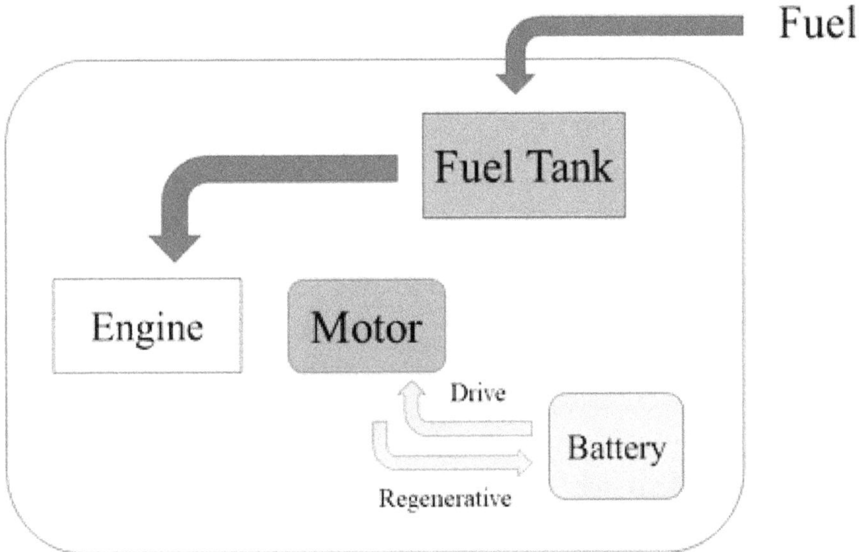

4.5. PHEV Components

- Electric motor
- Engine
- Inverter
- Battery
- Fuel tank
- Control module
- Battery charger (if the model is on board)

4.6. Principles of PHEV Work

PHEVs are usually all-electric and run on electricity until the battery runs out. Some models go into hybrid mode when highway speeds are reached (usually 60 or 70 mph). When the battery is discharged, the engine starts, and the vehicle operates as a plug-in hybrid.

In addition to connecting to an external power source, the PHEV's battery can be powered by the internal combustion engine or adaptive braking. When braking, the electric car acts as a generator that uses energy to charge the battery. The car is powered by an electric motor. As a result, smaller engines can use more power, resulting in better fuel economy.

Figure 7. Principles of PHEV work.

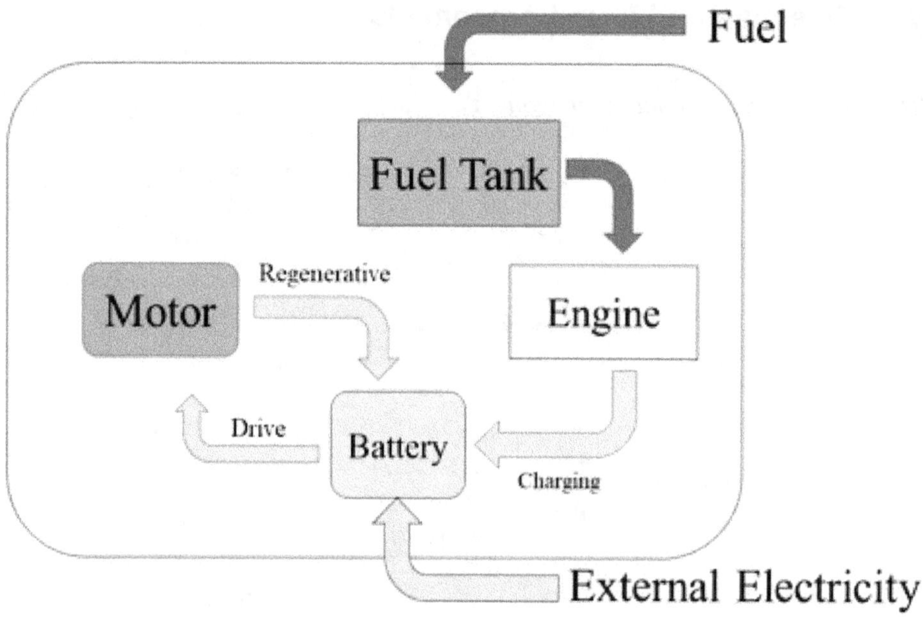

5. FUEL CELL ELECTRIC VEHICLE (FCEV)

A fuel cell vehicle (FCV) runs on hydrogen. Internal combustion vehicles are more efficient than conventional vehicles, they do not emit smoke, they only produce steam and hot air. FCEV and hydrogen infrastructure must be developed first. U.S. The Department of Energy is leading research to provide affordable, environmentally friendly, and safe hydrogen vehicles. Under the World Act of 1992, hydrogen is considered an alternative fuel for cars.

FCEVs use a propulsion system similar to electric cars, where energy stored in the form of hydrogen from a fuel cell is converted into electricity. Unlike conventional combustion vehicles, these vehicles do not emit harmful exhaust gases. Other benefits include increasing energy efficiency and growth through economic diversification.

FCEVs always run clean and run on fuel in the vehicle's tank. Like a conventional combustion engine car, it can run for four minutes and has a range of more than 300 miles. FCEVs have other advanced technology to improve efficiency, such as a braking system that recovers energy lost during braking and recharges the battery.

5.1. FCEV Architecture and Main Components

Figure 8. FCEV architecture and main components.

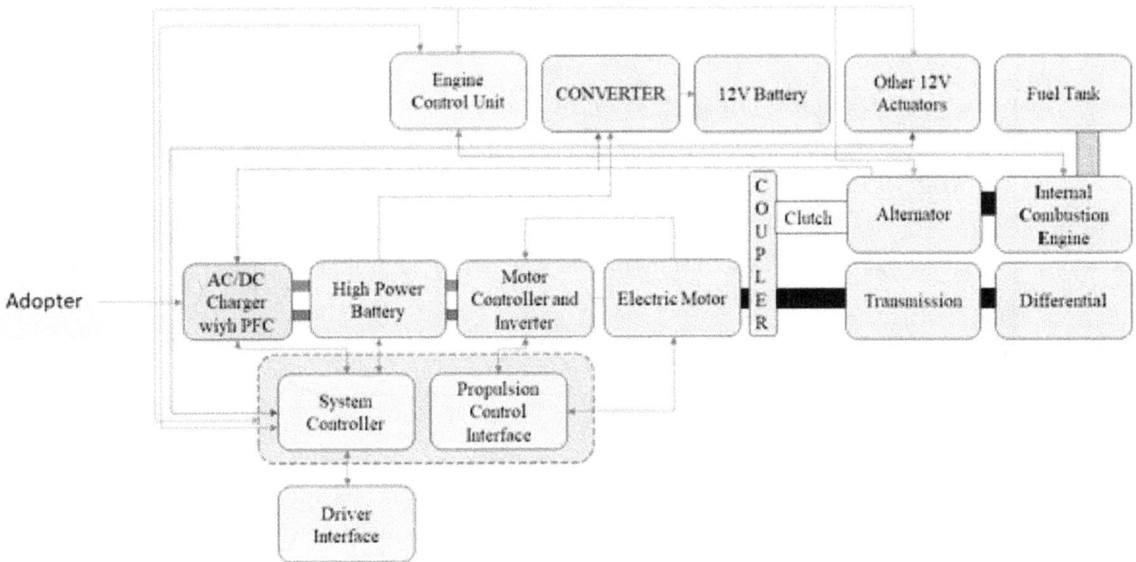

5.2. FCEV Components

- Electric motor
- Fuel cell reservoir
- Hydrogen tank
- Battery with converter and regulator

5.3. Principles of FCEV Work

The polymer electrolyte membrane (PEM) fuel cell is the most common fuel cell for vehicles. In a PEM fuel cell, an electrolytic membrane is placed between the anode (cathode) and the cathode (cathode). Hydrogen is produced at the anode and oxygen (from the atmosphere) at the cathode. By splitting hydrogen molecules into protons and electrons by the electric field of nuclear reactors. Protons cross the membrane at the cathode.

The electrons are captured by an external circuit (electrical motor) and combine with protons at the anode side, where protons, electrons, and oxygen form water molecules.

6. EXTENDED RANGE ELECTRIC VEHICLE (ER-EV)

An electric vehicle (EV) is a type of vehicle designed to replace an internal combustion engine (ICE). Electric cars have electric motors that run on electricity stored in batteries.

Figure 9. Principles of FCEV work.

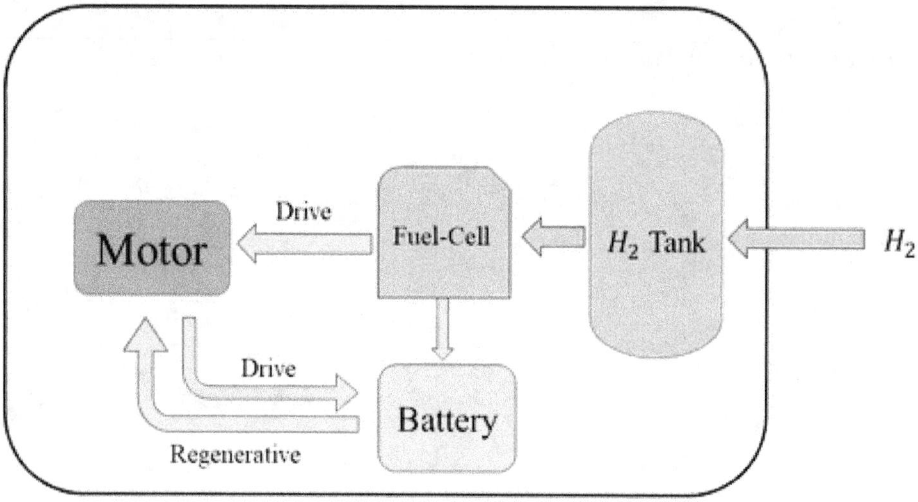

EVs now use new technologies. These companies are taking various approaches to develop electric vehicles. Advanced electric vehicles (e-REV) typically use an auxiliary power unit (APU) to charge the battery. E-REVs are also known as plug-in hybrid vehicles.

An APU is a small internal combustion engine used to power a vehicle's electric motors. This compact machine has a 5–10-liter fuel tank. The capacity of the APU is equal to or greater than the capacity of the battery. If the APU voltage is less than the battery voltage, the battery operation will fail. A typical APU is a small power turbine. It provides low charging, but that's different for an EV battery.

If the battery voltage is too low (5-10%), the battery is charged to start the APU. This happens spontaneously without the intervention of the driver. Additional APU charging eliminates longer charging times and the need for fast charging stations. This allows Eve to move forward and allows her to move forward.

Electric cars can reduce or reduce fossil fuel consumption, eliminate oil dependence, and eliminate greenhouse gas emissions. Compared to conventional vehicles, it can convert energy into speed with lower fuel consumption and improve overall energy efficiency. Driving an electric car instead of a car with an internal combustion engine has several important advantages.

Lower energy costs per mile for electric vehicles: Charge less at night than during peak hours.

Reduction of fossil fuel and carbon emissions: Electric cars emit little or no emissions.

The overall efficiency of converting energy into motion: Electric cars use less waste heat and less energy to travel the same distance.

Renewable energy charging: Vehicles can be charged using solar or other renewable energy sources.

6.1. Electric Cars and Driving Restrictions

ICE vehicles have a typical mileage of about 300 miles. So, I expect consumers to like the cars. The average range of an electric vehicle today is about 250 miles. It is necessary for manufacturers to have ample power so that they can continuously improve their technology.

One of the biggest obstacles for fleet managers in the use of electric vehicles is the issue of distance. This happens when the electric car batteries are dead when there are few stations nearby. Passive anxiety is the fear of being caught in a car without violence or crime.

An electric vehicle will reduce electric externalities. For example, low temperatures are available to reduce wear and tear in an electric car. Without autonomy for long-distance driving, the driver of an electric car may feel limited.

However, recent advances in electric vehicle design and technology have allowed the range to be extended. This helps address short-term options for first-generation EVs. No pain in the basketball player, but also to provide the ability to charge batteries while the vehicle is in motion.

6.2. Architecture and Main Components of ER-EV

Figure 10. Architecture and main components of ER-EV.

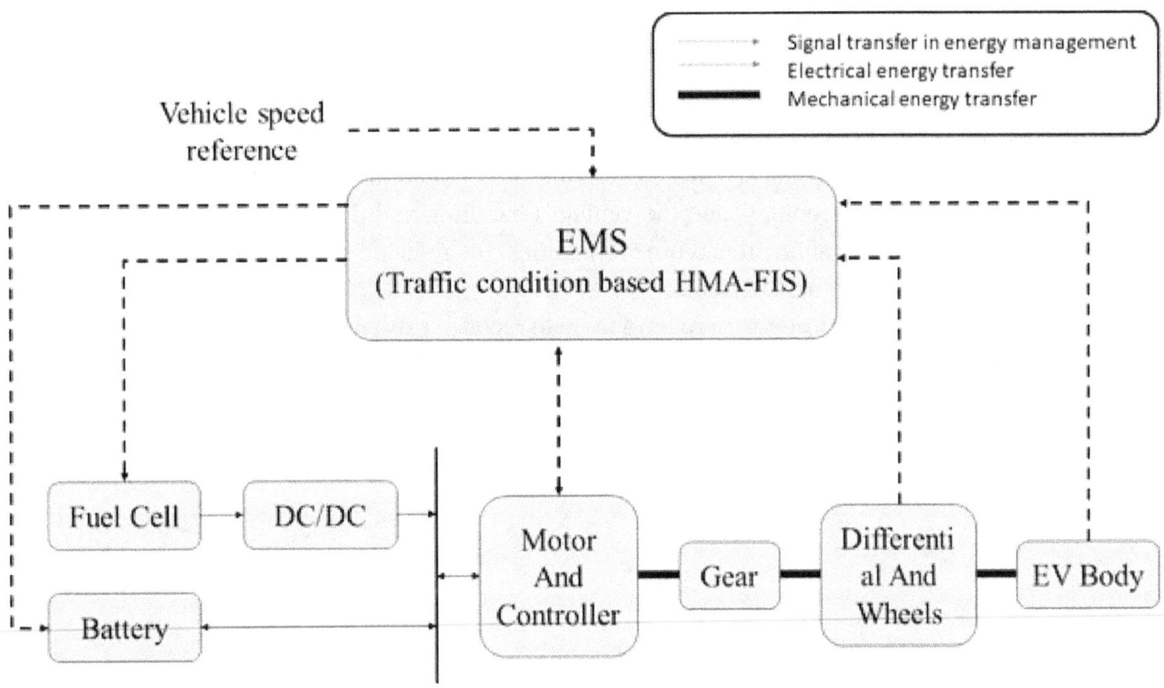

6.3. FCEV Components

- Traction engine
- Engine
- ISG
- Battery pack
- Battery with converter and regulator

6.4. Principles of FCEV Work

An expanded range of electric vehicles bridges the gap between the plug-in and pure electric variants. EREVs run on electricity only, but in many cases use an internal generator and are continuously powered by conventional fuel. Simply put, this generator/auxiliary power unit extends the range of the EV by keeping the motor/battery running when needed. Long-range electric cars cannot run on sustainable fuels. However, when the device is running and powering electrical components, it continuously emits carbon dioxide. Since the vehicle runs on electricity, all the initial benefits of owning an electric vehicle apply to this type of vehicle as well. For example, two of these benefits are acceleration and better performance.

7. CONCLUSION

Many technologies will be implemented in new generation vehicles. There are many technical challenges to overcome, especially in the area of car cameras. Weight, size, and cost dimensions must be overcome to achieve the desired power and performance. Productivity, reliability, safety, and reliability are other concerns, while price is of utmost importance.

The barrier to the adoption of high-performance EVs is economic, not technical. Hybrid and plug-in hybrid vehicles are more expensive. This calculation is an equation of the vehicle's added cost, the manufacturer's average fuel economy, and the vehicle's handling and dynamic performance, the total internal energy used by entertainment and other amenities, the reduced emissions, and finally the savings in food. Original device. Includes fabric

Power electronics and rotary engines are advancing to reduce costs and improve performance. Power transmission and engine rotation problems are common in BEVs, HEVs, PHEVs, FCEVs, and ER-EVs. There are also reductions in the power consumption of electronic and electric motor drive systems for hybrids and plug-in hybrid vehicles based on internal combustion engines.

Chapter 6
Modelling and Simulation of E-Bikes Driven by BLDC Motors

Maladhi D.
Vellore Institute of Technology, Chennai, India

Selvasundar K,
Vellore Institute of Technology, Chennai, India

Maheedhar Marabathina
Vellore Institute of Technology, Chennai, India

Kripalakshmi T.
Vellore Institute of Technology, Chennai, India

Deepa T.
Vellore Institute of Technology, Chennai, India

Angalaeswari S.
https://orcid.org/0000-0001-9875-9768
Vellore Institute of Technology, Chennai, India

Subbulekshmi D.
Vellore Institute of Technology, Chennai, India

ABSTRACT

Recent advancements in e-vehicle technology are becoming an emerging area of research in electrical and chemical engineering. Many studies are going on in the field of battery technology and charging stations, which will take the country to the next level of its infrastructure. Locomotives operated through these technologies offer efficiency and a pollution free environment. An e-bike is an electric bicycle incorporated with an electric motor. Though renovations in the battery technology are a prime factor, usage of appropriate electric motor is yet another important aspect for wider and smoother speed operations. This work discusses the brushless direct current (BLDC) motor, which is used for propulsion. This work presents an e-bike model developed in Simulink/Simscape. The performance of the vehicle at different conditions is observed and studied through simulation. The variations in parameters like speed of vehicle, current drawn by motor, modulated current, and voltage waveforms at different conditions are discussed in this work.

INTRODUCTION

E-bikes, commonly referred to as electric bicycles, are a workable answer to the world's energy dilemma since they can replace vehicles with internal combustion engines (combustion engines) to meet the

DOI: 10.4018/978-1-6684-6631-5.ch006

requirement for emission-free medium transportation. E-bikes are a substitute form of transportation for people and goods, especially over short and medium distances. E-bikes may move at speeds that are about equivalent to those of urban traffic while also using energy for movement that is quite similar to that needed for moving people and cargo. The e-bike powered by a brushless direct current (BLDC) motor is the subject of this study.

The modelling of the BLDC motor and its supporting elements, including the buck converter, inverter, and hall sensors, serves as the first step in the process. Next, a model of the longitudinal dynamics of an e-bike will be developed. Using Matlab's Simulink / Simscape Tool, the entire system and its subcomponents will be modelled and simulated (software). Within the Simulink platform, Simscape facilitates the creation of physical system models.

A survey of the literature is conducted to learn the longitudinal dynamics and BLDC motor modelling equations for e-bikes. Additionally, several literatures are researched to determine the parameter values that were utilised to simulate and model the specific component.

MODELLING OF BRUSHLESS DC (BLDC) MOTOR AND INVERTER

BLDC Motor Construction

Motors that use brushless DC (BLDC) technology are quickly gaining prominence. Consumer electronics, automotive, aerospace, consumers, the medical field, industrial automation, and instrumentation are just a few of the industries that use BLDC motors (Babu et al., 2012). BLDC motors do not need brushes for rectification, as the name suggests. Instead, they are corrected electronically. A faster range, high dynamics, high efficiency, extended service life, and noise-free operation are a few of them. Another is speed enhancements for torque characteristics. Additionally, the high torque to motor size ratio makes it beneficial in situations where weight and space considerations are crucial.

Synchronous motors include BLDC motors. This indicates that the magnetic fields produced by the stator and the rotor have the same rotational frequency. Induction motors typically have "slip," which BLDC motors do not. There are single-phase, two-phase, and three-phase BLDC motor options. The stator's design calls for an equal number of turns. Three-phase motors are the most well-liked and frequently used of these.

Stator

A BLDC motor's stator consists of stacks of steel laminations with windings inserted into slots which are axially cut along the innermost periphery. It is often shaped like an induction motor, but the windings are placed differently.

Three stator windings are often connected in a star pattern in BLDC motors. Each of these windings is made up of a number of coils joined together to form a winding. A winding is created by inserting one or more coils into the slots and connecting them together. To provide an even number of poles, these windings are placed evenly across the stator's periphery. Sinusoidal and trapezoidal motors are the two types of stator windings. This differentiation is made in order to produce different sorts of back electromotive force. It relies on the coupling of stator coils (EMF).

Each winding of a BLDC motor produces a voltage known as back Electromotive Force which, in accordance with Lenz's Law, opposes the applied voltage provided to the stator windings. This reverse EMF's polarity runs counter to the voltage's activated direction. Three elements in particular influence back EMF:

1) Angular velocity of the rotor
2) Magnetic field generated by rotor magnets
3) The number of turns in the stator windings

In addition to the back EMF, the corresponding types of motor also exhibit sinusoidal and trapezoidal changes in the phase current.

Rotor

The rotor is built of a permanent magnet and has alternate North (N) and South (S) poles in pairs ranging from two to eight. The appropriate magnetic material is selected to manufacture the rotor based on the desired magnetic field density in the rotor. Permanent magnets are often created using ferrite magnets. Magnets made of rare earth alloy are growing in favour as technology develops. Although ferrite magnets are cheap, they have a low flux density for a specified volume, which is a huge drawback.

The alloy material, however, allows the rotor to compress further for the equal torque since it has a high magnetic density per volume. Additionally, compared to ferrite magnets, these alloy magnets provide more torque for the same size motor. Examples of rare earth alloy magnets include those made of boron, neodymium, neodymium, samarium cobalt and ferrite.

Hall Sensors

A BLDC motor's commutation is electronically controlled in contrast to a brushed DC motor. The stator windings should be activated sequentially in order to turn the BLDC motor. Understanding which winding will be powered after the energising sequence depends on knowing the position of the rotor. Hall effect sensors built within the stator are used to find the position of the rotor. The BLDC motor drive circuit is shown in Figure 2.1.

Principle of Operation

In each commutation sequence, one of the windings receives positive, the second winding receives negative power and the third winding receives no power at all. Torque is generated when the magnetic field from the stator coils interacts with the magnetic field from the permanent magnets. Torque is expected to be greatest when the two fields are perpendicular to one another and to decrease when they are moving in unison. It is necessary for the rotor's motion to cause the windings' magnetic field to change and catch up with the stator field so that the motor can continue to spin. Six-Step Commutation refers to the process of supplying power to the windings.

Figure 1. BLDC Motor Drive

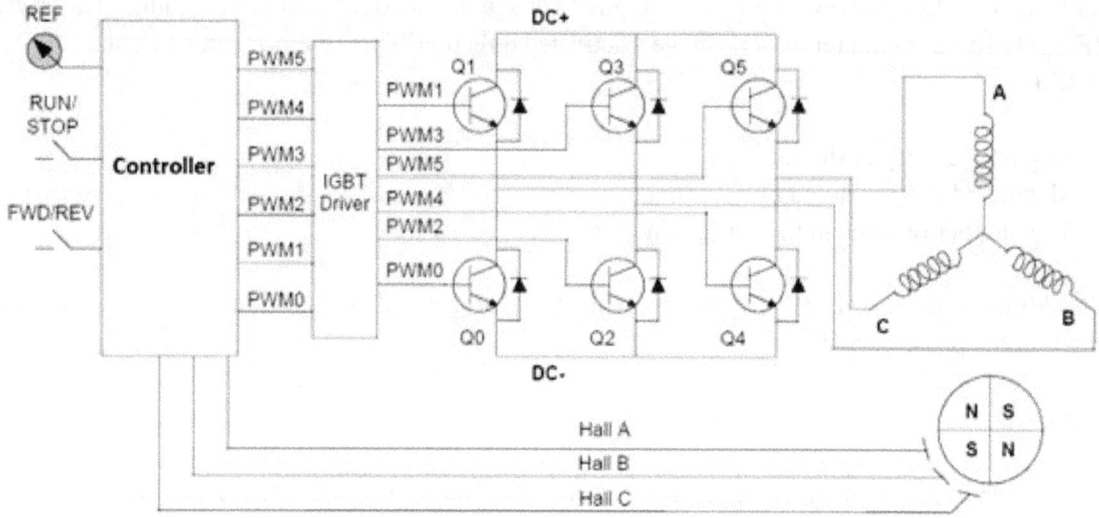

Commutation Sequence

One of the Hall sensors undergoes a state change every 60 electrical degrees of rotation. As a result, an electrical cycle involves six steps. In synchronous, the phase current switching needs to be updated every 60 electrical degrees. The rotor's mechanical rotation might not entirely fit into one electrical cycle, though. The rotor pole pairs define how many electrical cycles must be repeated to complete a mechanical revolution. One electrical cycle is completed for each pair of rotor poles. Thus, the number of rotor pole pairs equals the number of electrical cycles or revolutions.

Measuring the motor's real speed allows for closed-loop speed control. Calculated is the difference between the set speed and the actual speed. The speed error can be amplified and the PWM duty cycle can be dynamically changed using a Proportional + Integral + Derivative (PID) controller.

MATHEMATICAL MODELLING OF BLDC MOTOR

A trapezoidal distribution of the flux is found in a PMBLDC motor (Rao et al., 2012). Given the trapezoidal flux distribution, it makes sense to create a mathematical model of the system. An R-L back emf stator circuit coupled to a three-phase star is assumed to be the motor. Since it is assumed that there are three identical phases and that the rotor has zero inductance with angle, the mutual and self-inductances of the three phases are also equal. The following voltage equations for phases a, b and c shed light on the balancing constraints imposed by the stator currents ia, ib and ic and stator induced emf's e_a, e_b, e_c.

$$V_a = i_a R_s + (L - M)\frac{d}{dt} i_a + e_a \rightarrow \tag{1}$$

Figure 2. Hall sensor output and Back EMF waveforms

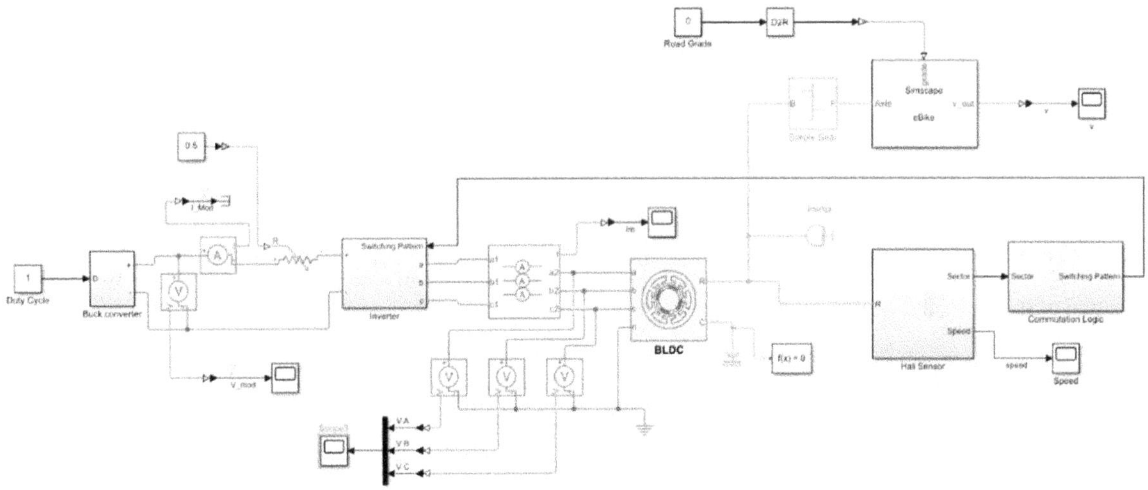

$$V_b = i_b R_s + (L - M)\frac{d}{dt} i_b + e_b \rightarrow \quad (2)$$

$$V_c = i_c R_s + (L - M)\frac{d}{dt} i_c + e_c \rightarrow \quad (3)$$

Where, R_s – Stator resistance and L, M – Self-inductance and Mutual inductance of phase a-b-c.
Simplifying equations (1), (2), & (3)

Figure 3. Output torque and phase current waveforms

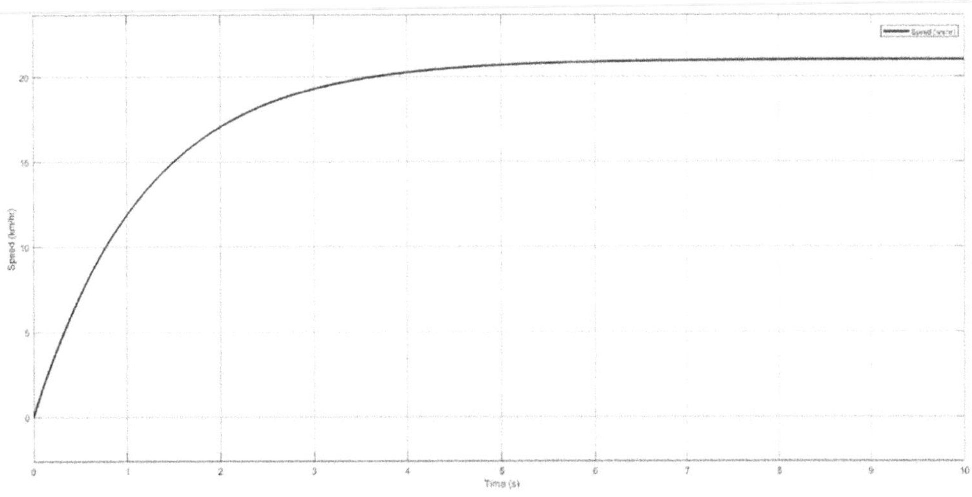

$$\begin{bmatrix} V_a \\ V_b \\ V_c \end{bmatrix} = R_s \begin{bmatrix} 1 & 0 & 0 \\ 0 & 1 & 0 \\ 0 & 0 & 1 \end{bmatrix} \begin{bmatrix} i_a \\ i_b \\ i_c \end{bmatrix} + (L-M) \begin{bmatrix} 1 & 0 & 0 \\ 0 & 1 & 0 \\ 0 & 0 & 1 \end{bmatrix} \frac{d}{dt} \begin{bmatrix} i_a \\ i_b \\ i_c \end{bmatrix} + \begin{bmatrix} e_a \\ e_b \\ e_c \end{bmatrix}$$

The trapezoidal induced emf in a PMBLDC motor have a peak value of Ep, which is expressed as,

$E_p = \lambda \mathrm{p}_{\omega}\mathrm{r}$

The total electromagnetic torque is given by

$T_e = (e_a i_a + e_b i_b + e_c i_c) / \omega \mathrm{r}$

The instantaneous induced emf of phase a can be written as

$e_a = f_a(\theta \mathrm{r}_) \lambda \mathrm{p}\omega_r$

Where, fa($_\theta$r) $_-$ Rotor position which is dependable for inducing emf in phase a and λp - $_p$eak mutual flux linkage of stator winding.

The total electromagnetic torque in terms of rotor position is given as

Te= λp[f$_a$(θ_r)ia + $_f$b($\theta r_)$ib $_+ f_c(\theta \mathrm{r})_c$]

R$_o$tor speed

$$\omega_r = \frac{1}{J}\int (T_e - T_L - B\omega_r)\,dt$$

Where, J – moment of inertia and TL _ load torque.

$$\frac{d\theta_r}{dt} = \frac{P}{2}\omega_r$$

Where, P – poles and Θ_r – rotor position.

SIMSCAPE MODELLING OF BLDC MOTOR AND INVERTER

Simscape Modelling of BLDC Motor

Figure 4. Simscape block of BLDC motor

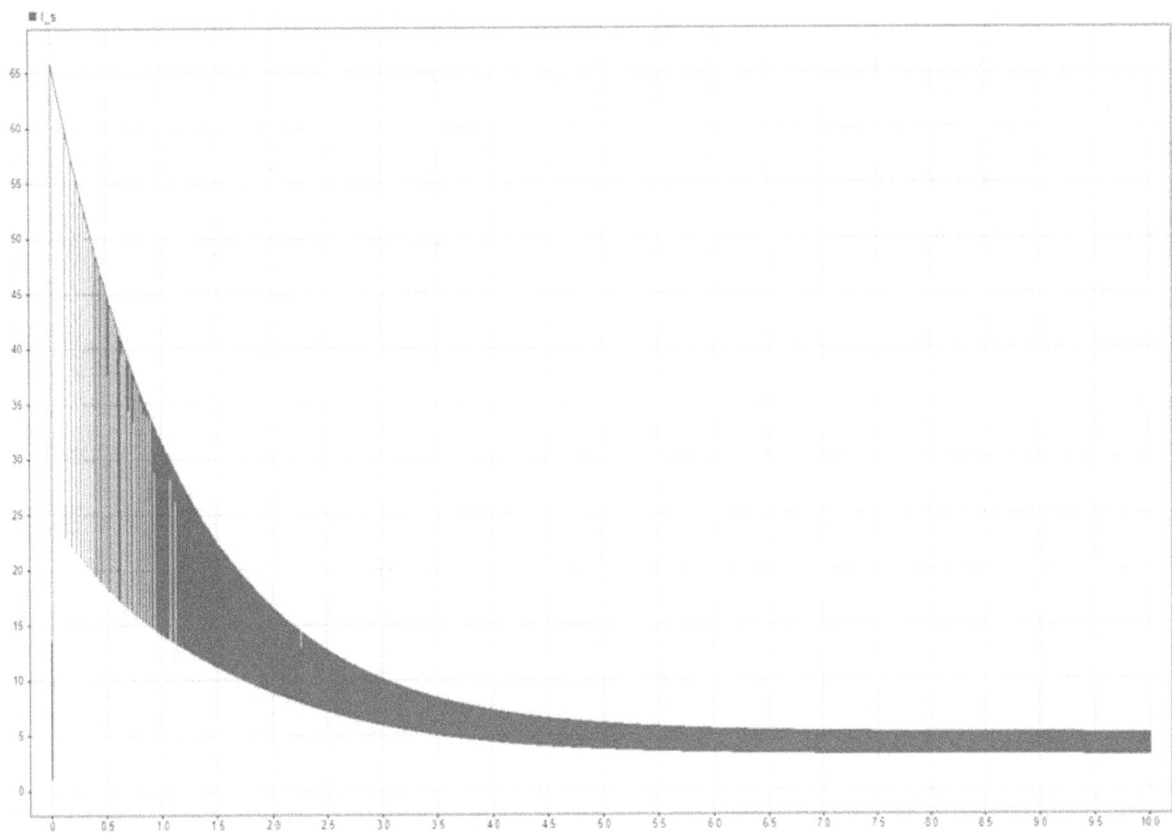

Simscape Modelling of Hall Sensor

A rotary motion sensor is used to measure the rotor angle and angular velocity of a BLDC motor. Figure 2.5 shows a subsystem representing the operation of a Hall sensor. This subsystem outputs sector information that is used to turn on/off the corresponding IGBT device in the inverter.

Simscape Modelling of Inverter

Three phase converter consists of IGBT power semiconductor switches. Converter is driven by commutation sequence signals.

Table 1. Parameters used in BLDC model

Settings	Parameter	Value
Rotor	Windind type	Wye-wound
	Back Emf profile	Perfect trapezoid – specify maximum flux linkage
	Maximum permanent magnet flux linkage	0.18 Wb
	Rotor angle over which back emf is constant	120/4 deg
	Number of pole pairs	4
	Rotor angle definition	Angle between the a-phase magnetic axis and the q-axis
Stator	Stator parameterization	Specify Ls, Lm, and Ms
	Stator self-inductance per phase, Ls	1e-4 H
	Stator inductance fluctuation, Lm	0 H
	Stator mutual inductance, Ms	1e-5 H
	Stator resistance per phase, Rs	0.1 Ω
	Zero sequence	Include

Figure 5. Hall sensor modelled using Simulink and Simscape components

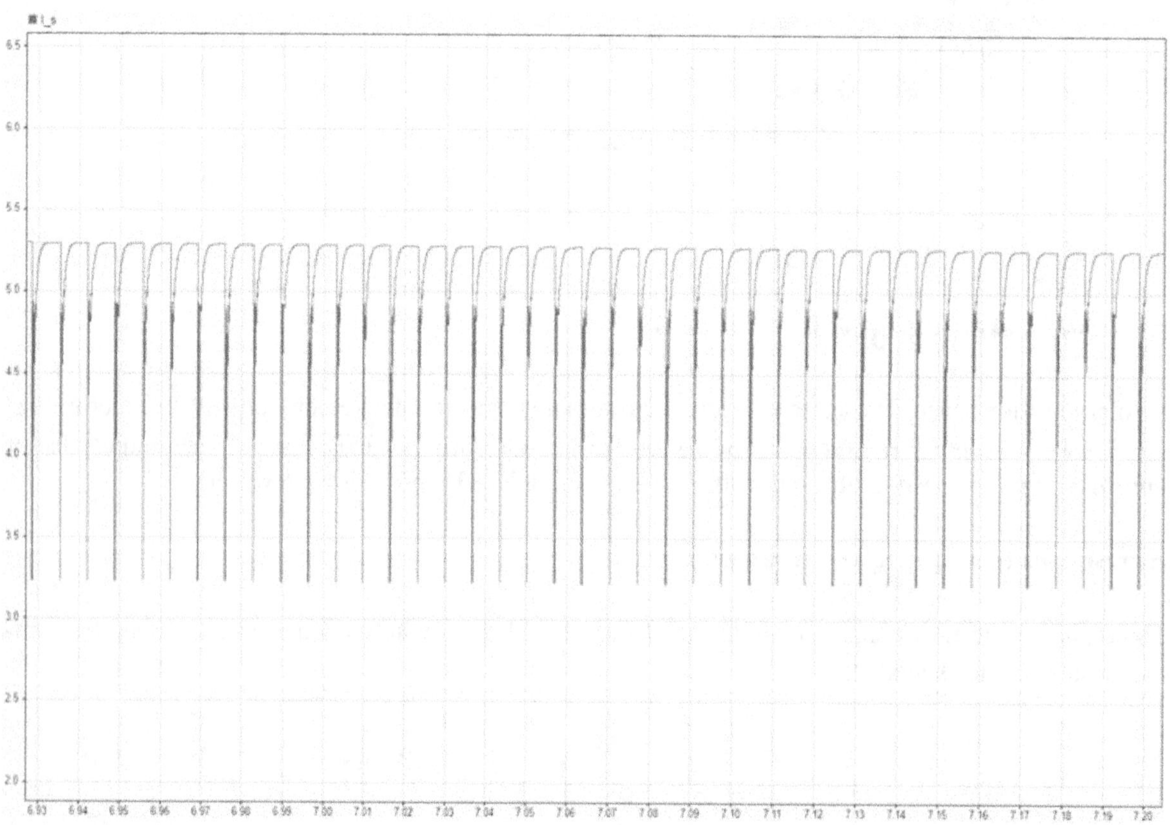

Figure 6. Inverter modelled using Three phase converter

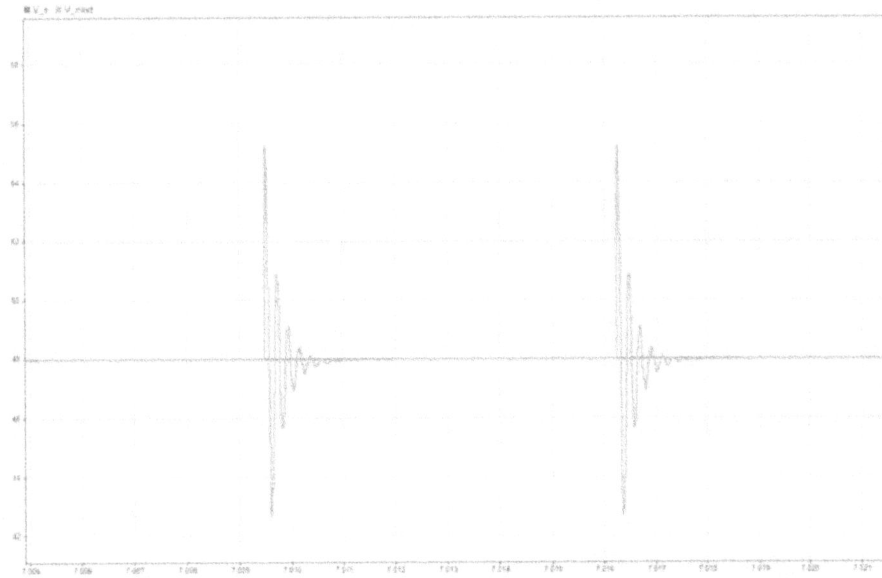

Table 2. Parameters used in Converter (three phase) block

Settings	Parameter	Value
Switching Devices	Switching device	IGBT
	Forward voltage, Vf	0.8 V
	On-state resistance	0.001 Ω
	Off-state resistance	1e-5 1/Ω
	Threshold voltage, Vth	0.5 V
Integral Diodes	Integral protection diode	Diode with no dynamics
	Forward voltage	0.8 V
	On resistance	0.001 Ω
	Off conductance	1e-5 1/Ω

Buck Converter

PWM Generator is used to drive the Buck converter. Switching frequency of 1000 Hz is used.

Figure 7. Buck converter modelled using Simulink / Simscape components

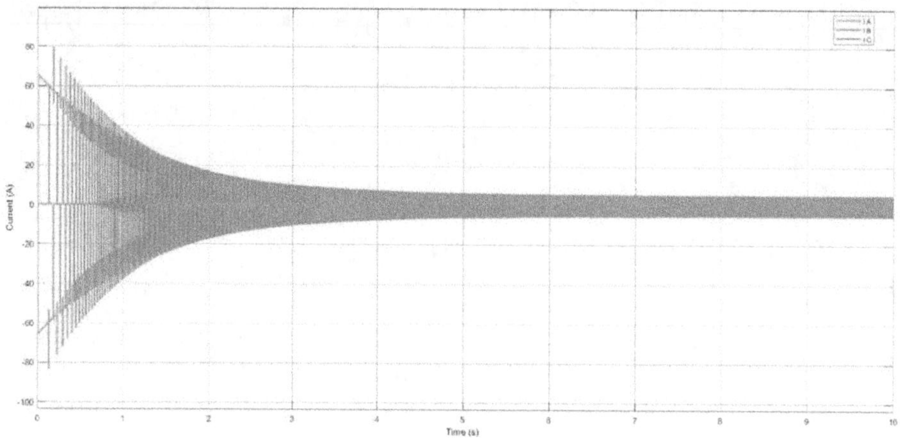

MODELLING OF E-BIKE DYNAMICS

Introduction to e-Bike

E-motorcycle is as an electric powered bicycle additionally known as a energy-motorcycle or booster motorcycle (Hung & Lim, 2019; Mclean et al., 2021). It is a bicycle with an included electric powered motor that may be used for propulsion. The rider has the selection of the use of the motor help as and

Table 3. Parameters used in Buck Converter block

Settings	Parameter	Value
Switching Device	Switching Device	MOSFET
	Drain-source on resistance	0.001 Ω
	Off-state conductance	1e-5 1/Ω
	Threshold voltage	0.5 V
Diode	Model dynamics	Diode with no dynamics
	Forward voltage	0.8 V
	On resistance	0.001 Ω
	Off conductance	1e-5 1/Ω
LC filter	Inductance	220 μH
	Inductor series resistance	0 Ω
	Capacitance	10 μF
	Capacitor effective series resistance	0.25 Ω
Snubbers	Snubber	RC snubber
	Snubber resistance	0.1 Ω
	Snubber capacitance	1e-7 F

whilst required, to journey uphill or on a strenuous lengthy road. This is done via a controller. The motor compensates the specified energy and this manner the journey will become smooth. In this work, e-motorcycle that is completely powered with the aid of using BLDC motor is taken into consideration for modelling and simulation.

Figure 8. Forces acting on e-Bike (Abagnale et al., 2015)

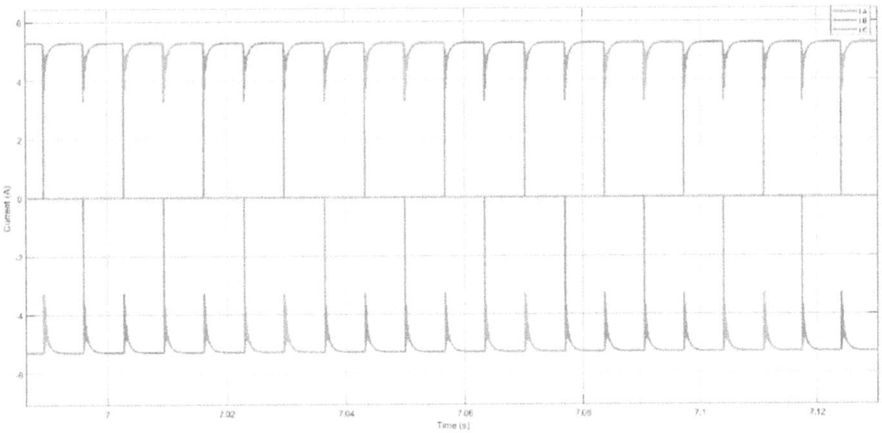

Mathematical Modelling of e-Bike

The longitudinal dynamics of e-Bike can be expressed as (Abagnale et al., 2015),

$$M\frac{dv}{dt} = F_t - \left(F_w + F_{rr} + F_a\right)$$

in which,

Aerodynamic drag force $F_w = 0.5A_sC_w\rho a_{irv}2$
Rolling resistance force $F_{rr} = C_{rr}Mgcos(\alpha)$
Gradient resistance force $F_a = Mgsin(\alpha)$
Driving force F_t

Where,
M – total mass of e-Bike including rider
v – longitudinal velocity of e-Bike
A_s – reference area of e-Bike
C_w – aerodynamic drag coefficient
C_{rr} – rolling resistance coefficient
ρa_{ir} – density of air
g – acceleration due to gravity

α – slope of the road

Figure 9. Simscape component of eBike

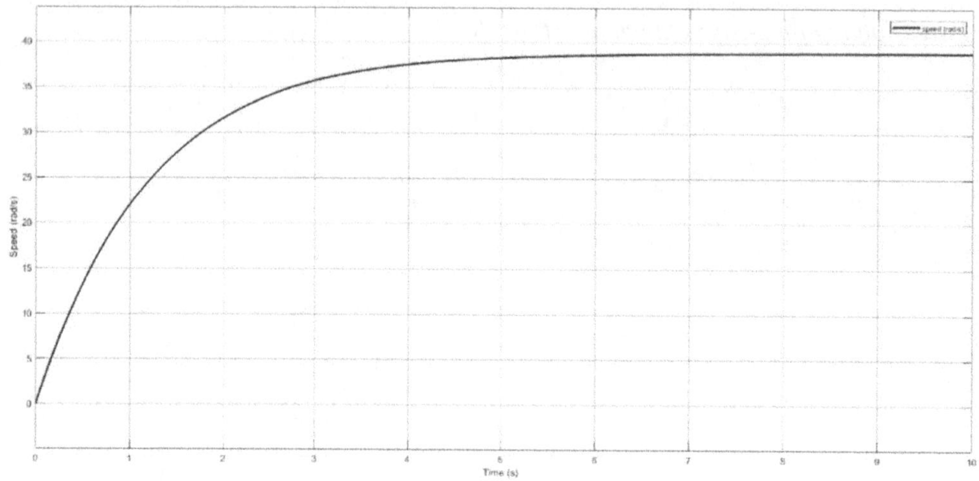

Table 4. Parameters used in eBike simscape component

Parameter	Value
Vehicle mass	90 kg (Rider = 75 kg + ebike = 10 kg)
Tire radius	0.3 m
C-rr	0.0025
A_s	0.509 m^2
C_w	0.76
P_air	1.2 kg/m^3

RESULTS AND DISCUSSION

Modelling of e-Bike Driven by BLDC Motor (Open loop)

Open loop model of e-Bike driven by BLDC motor is modelled (Sathiyan, 2018) and simulated using Simulink/Simscape. Here, constant duty cycle is used to power the buck converter. BLDC motor is electronically commutated by three phase inverter circuit. BLDC motor drives the e-Bike tires through reduction gear.

The modelled e-Bike can reach the maximum speed of 21 km/hr at 0° road grade (slope). e-Bike draws maximum current of 66A during its starting and reaches 5.3 A during running (steady-state). BLDC motor is powered by 48V voltage source. The source voltage is modulated by buck converter using Pulse Width Modulation (PWM) technique.

Figure 10. Open-loop model of e-Bike

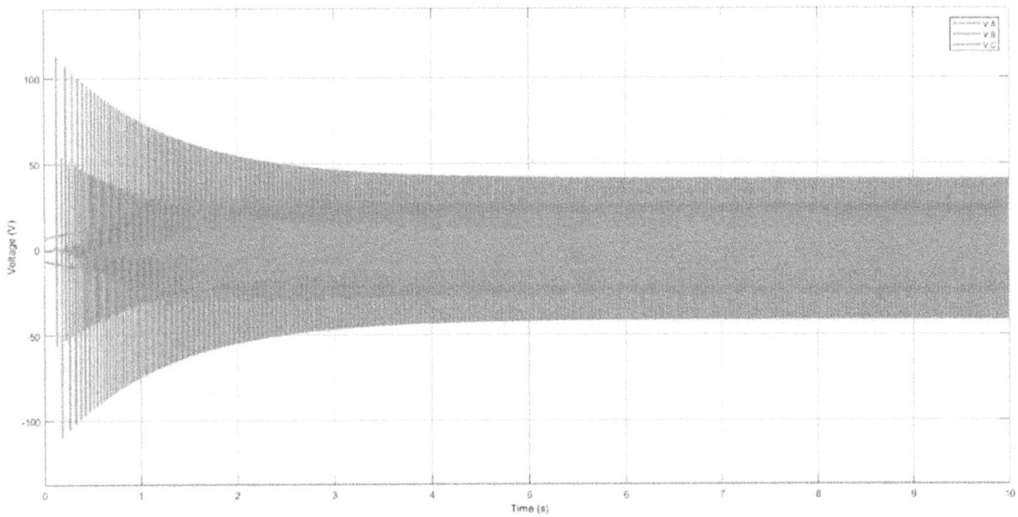

Figure 4.7 & 4.10 illustrates the trapezoidal waveform of EMF and current.

The maximum speed of the vehicle is slightly reduced to 19.6 km/hr when the road grade (slope) is at 2° but the current drawn by the motor is 9.4 A at steady state.

Closed Loop Control of e-Bike Driven by BLDC Motor

Speed control of e-Bike is done using PID controller. Speed of the vehicle is used as the feedback signal. In this Discrete PI controller with parameters Proportional (P) =1.2 and Integral (I) = 0.1 is employed. Output saturation is set with limits 0 and 1. PID controller is connected to buck converter circuit.

Figure 11. Speed of e-Bike

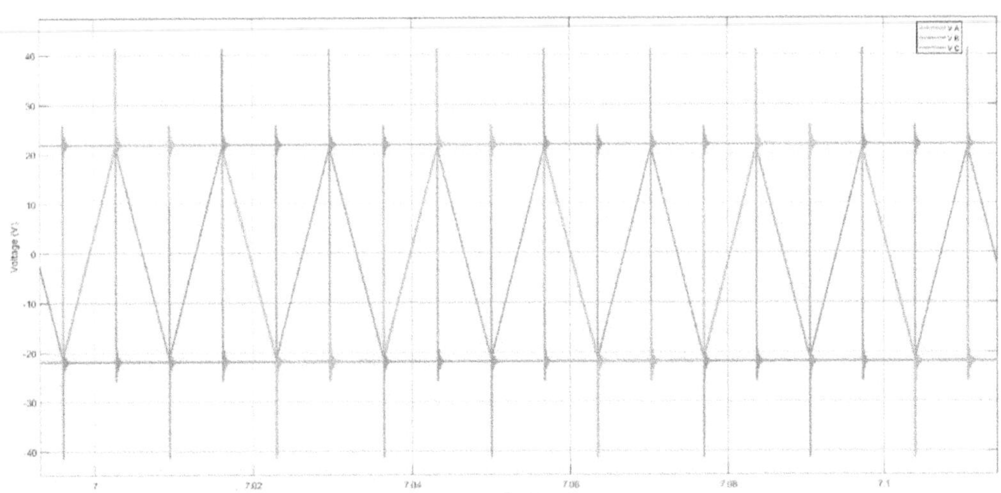

Figure 12. Source Current profile

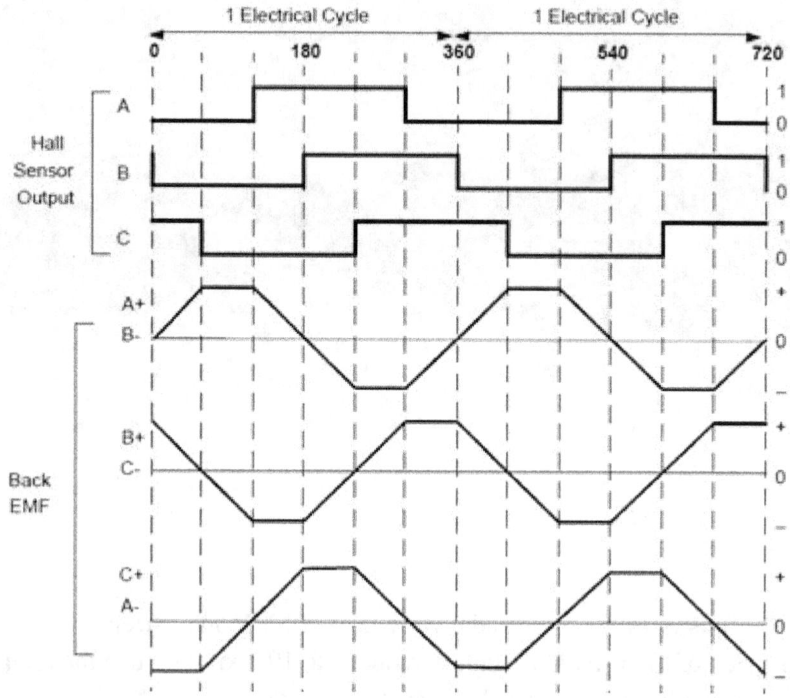

During starting, duty cycle is at 1. Once the motor reaches the reference speed, duty cycle is reduced to 0.7.

Figure 13. Current draw from source at steady state running

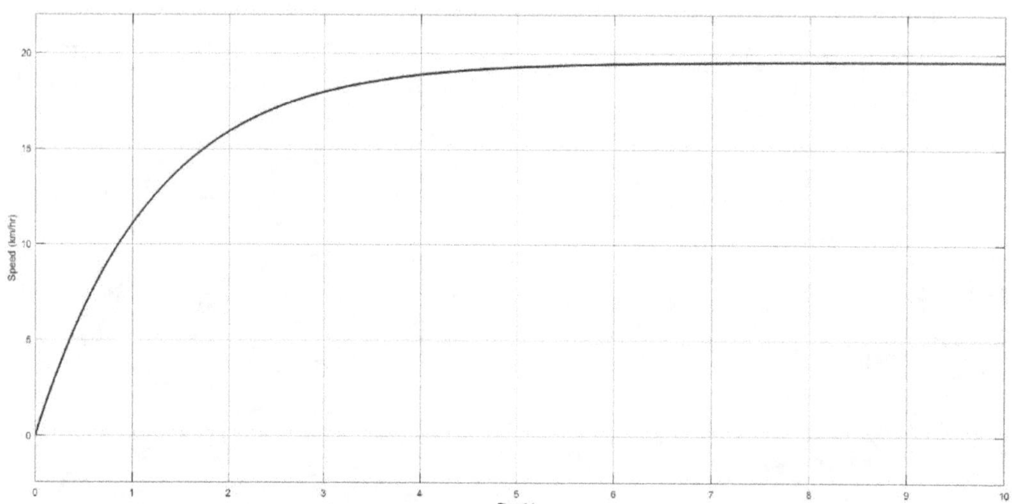

Figure 14. Source vs. modulated voltage

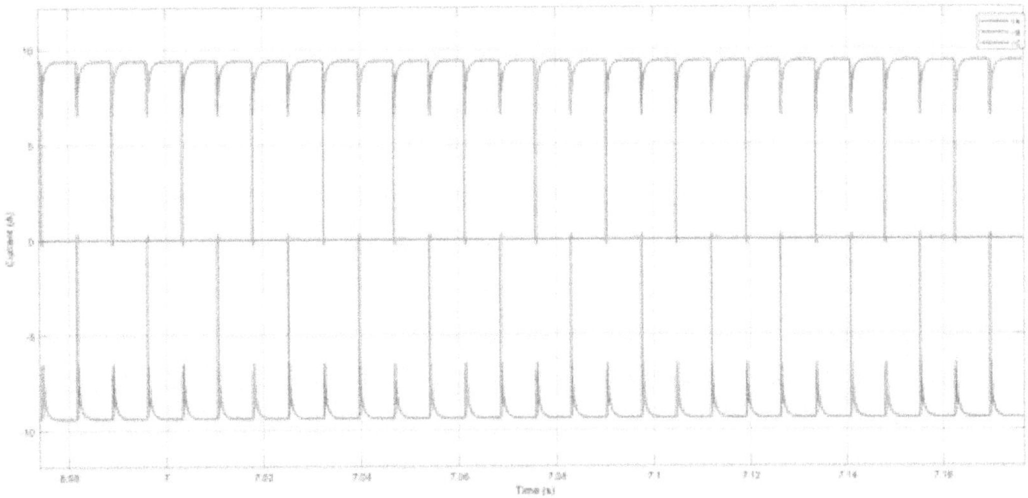

Performance of e-Bike at Different Speed and Road Grade

Performance of eBike is studied with different speed reference and road grade. Speed reference is given as step signal and constant speed of 10 and 15 km/hr is given as input. Signal builder block is used to create variable road grade between 0° and 5° degree slope.

Figure 4.21 and 4.22 illustrates how the current drawn by the system varies at different conditions (Thejasree & Maniyeri, 2019). When the slope of the road is increased from 0° to 5°, the current drawn by the motor is increased (increase of 2.5 A) steadily till it reaches the steady state

Figure 15. Current profile – drawn by motor

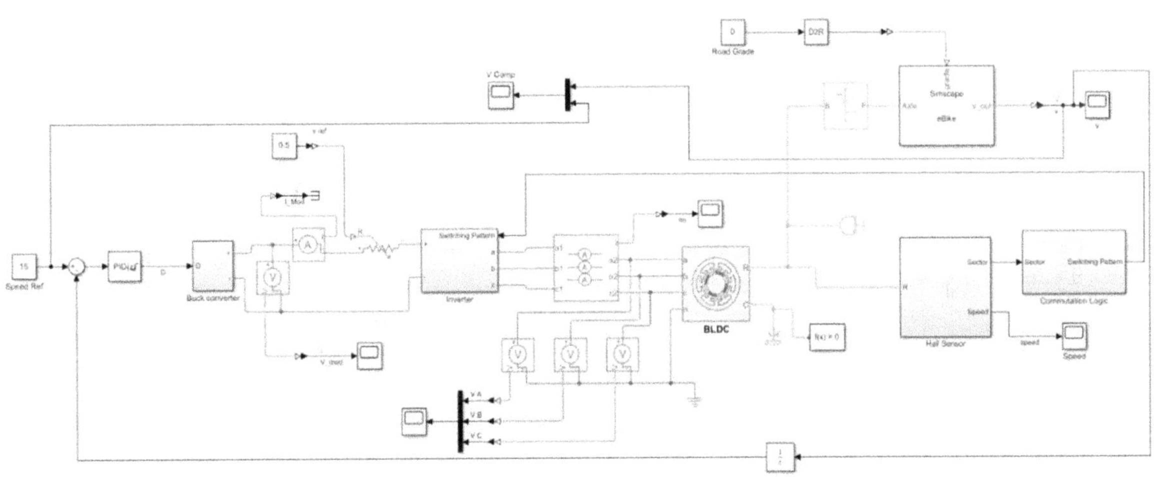

Figure 16. Current drawn by motor at steady state

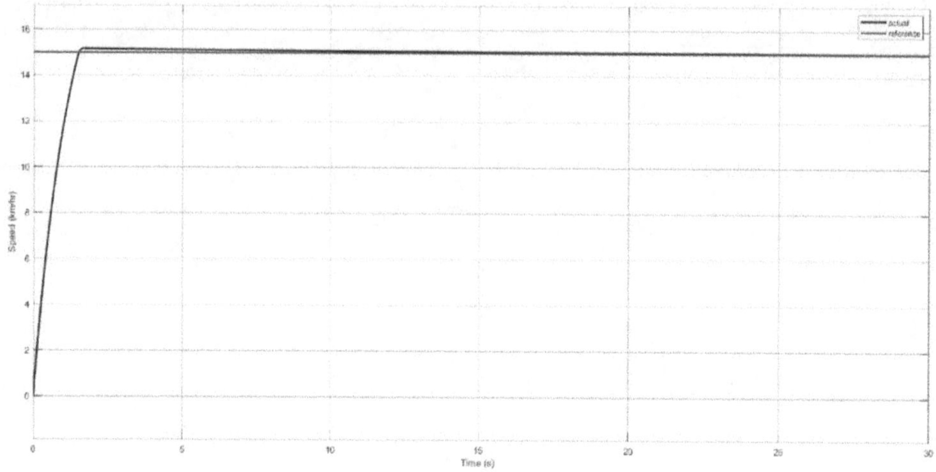

CONCLUSION

A detailed modelling of BLDC motor and its subcomponents and dynamics of e-Bike is done. Both the open loop and closed loop response of the system is discussed in detail. Closed loop control is implemented using PID control algorithm. Performance parameters like speed of vehicle, current drawn by motor at different reference speed and diverse road grade (slope) is studied in detail. The current drawn by the BLDC motor increases when speed and road grade increases. The maximum speed attained by modelled e-Bike is 21 km/hr, which is well below the permitted speed of 25 km/hr (based on Indian speed limit for non-licensed vehicles).

Figure 17. Speed of motor

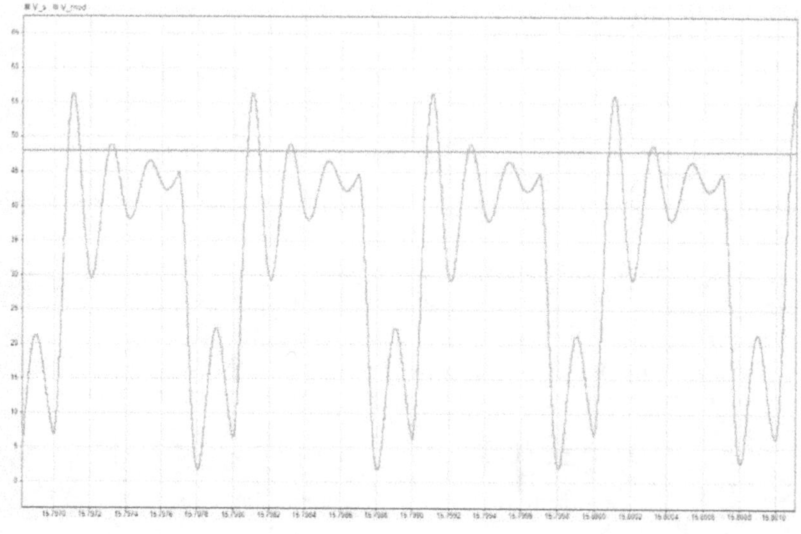

Figure 18. Voltage profile of BLDC motor

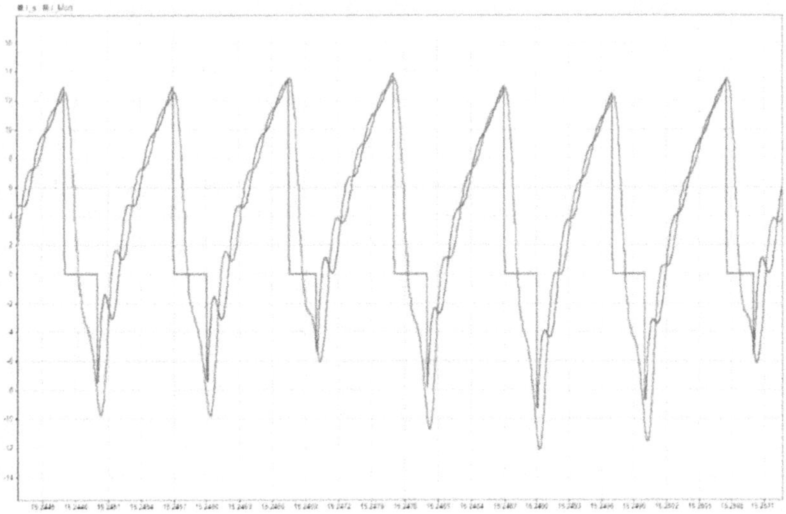

Figure 19. Voltage of BLDC motor at steady state

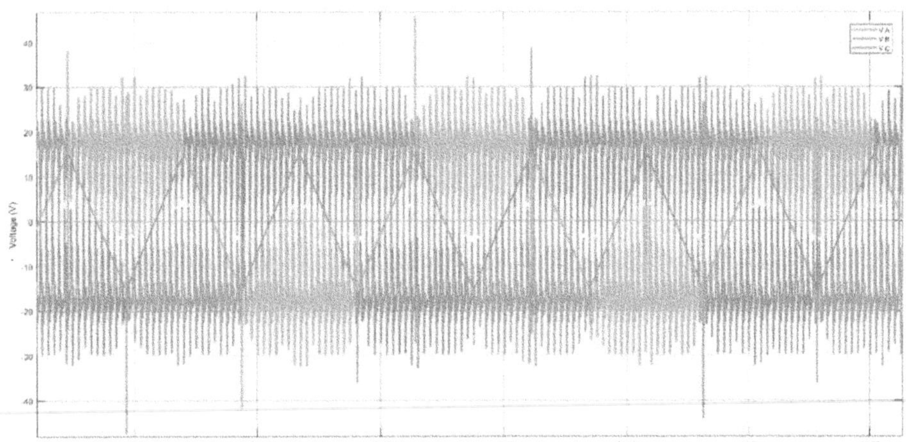

Figure 20. Speed of e-Bike at 2° road grade

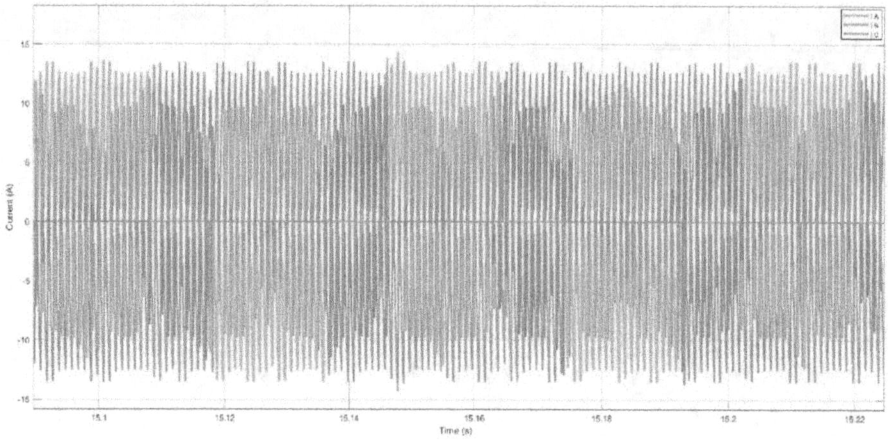

Figure 21. Current drawn by motor at 2° road grade (steady state)

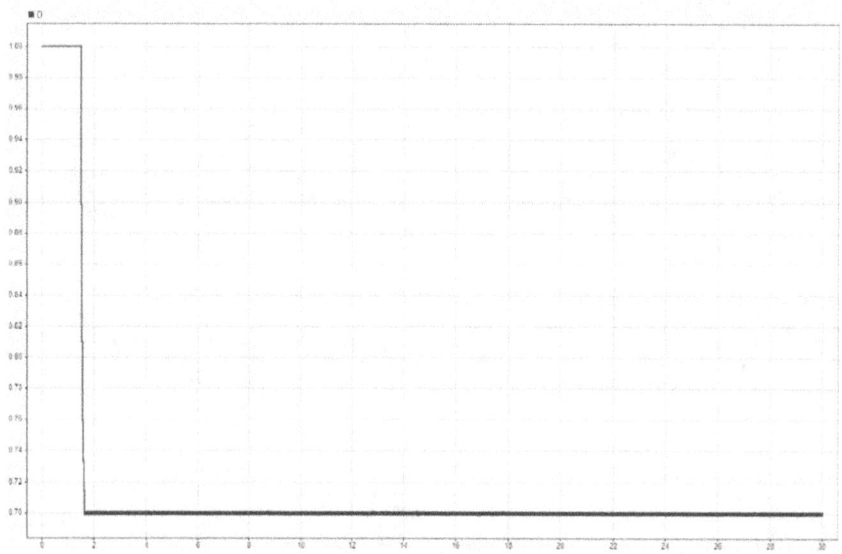

Figure 22. Closed loop model of e-Bike

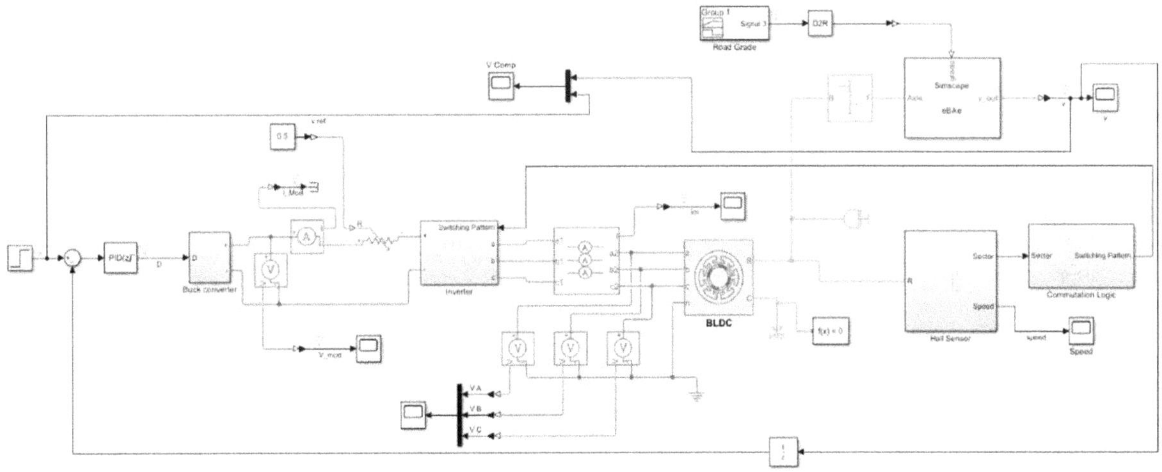

Figure 23. Speed of vehicle (Reference Speed = 15 km/hr)

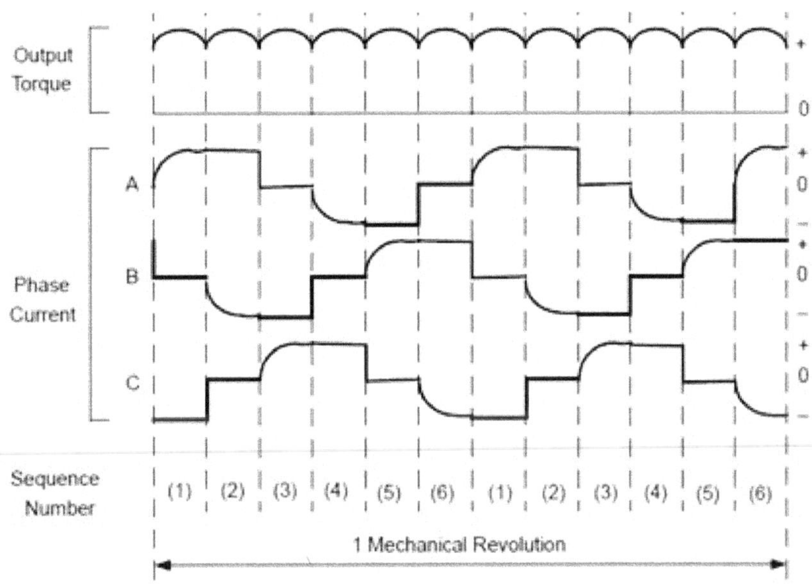

Figure 24. Source vs. modulated voltage (steady state)

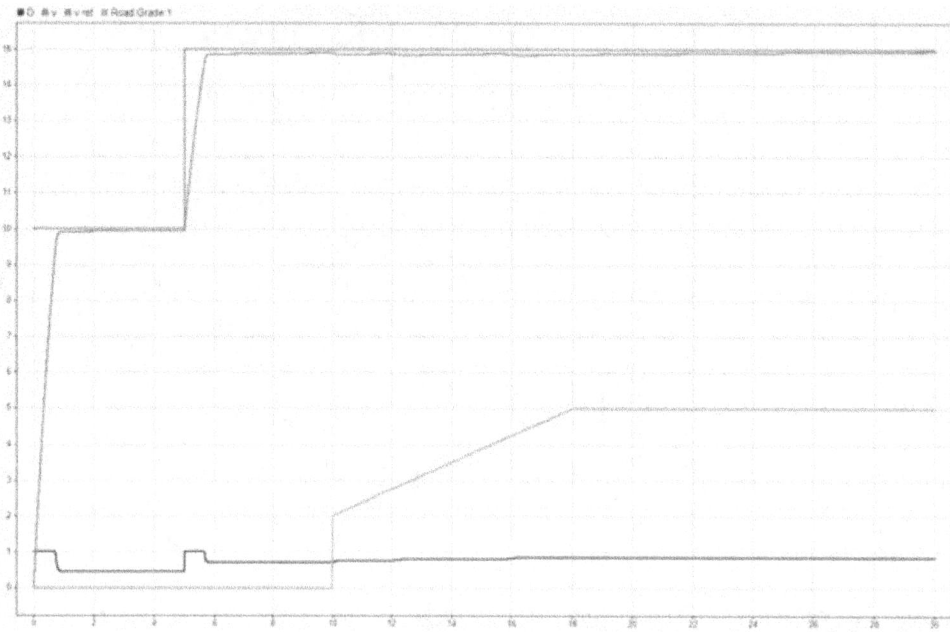

Figure 25. Source vs. modulated current (steady state)

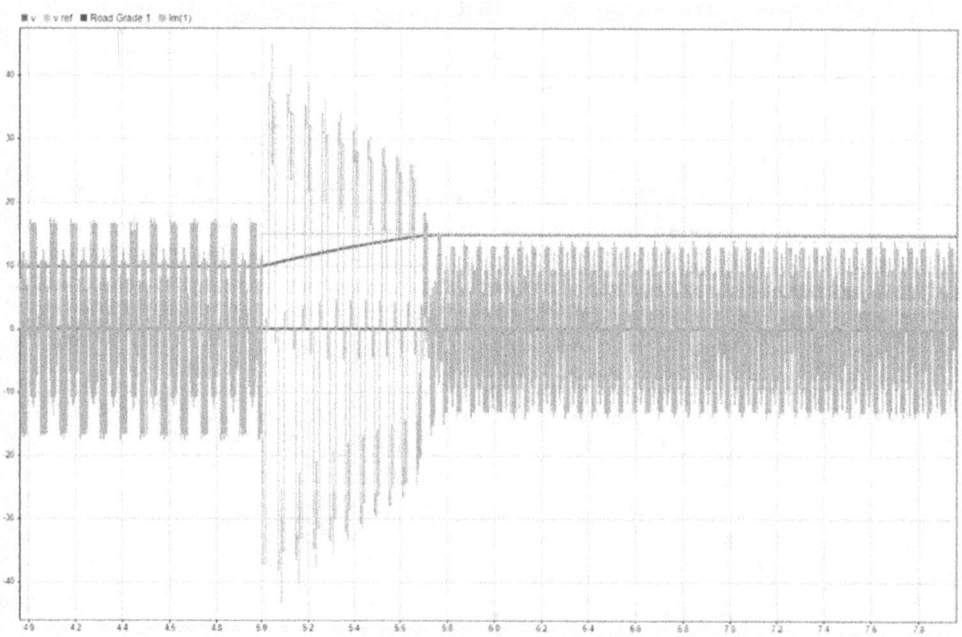

Figure 26. Input voltage of BLDC motor (steady state)

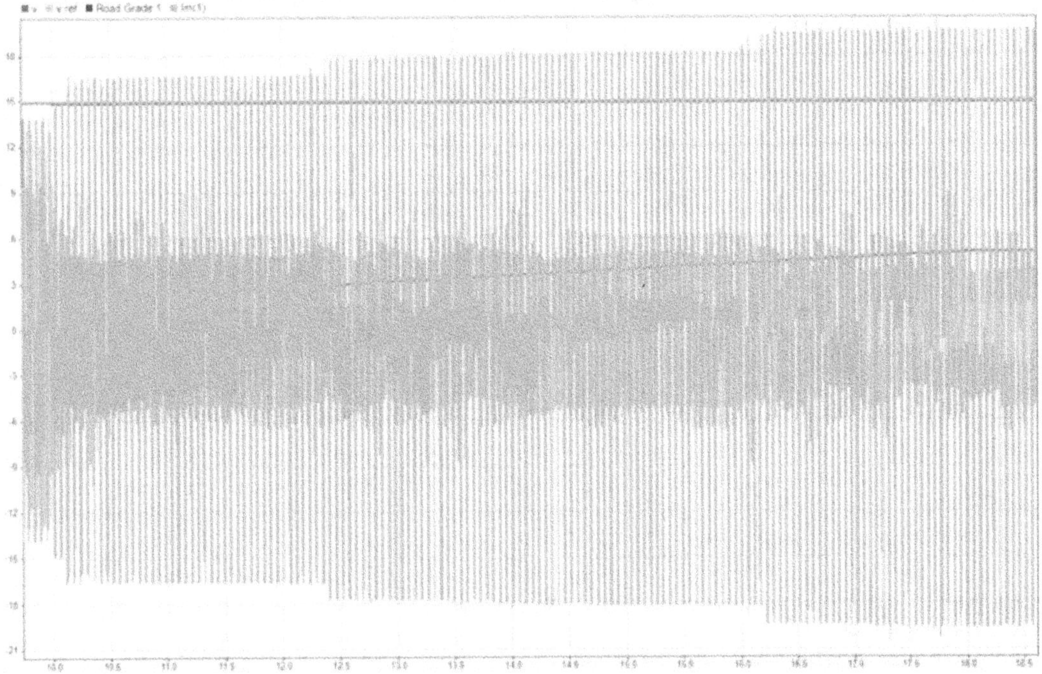

Figure 27. Input current of BLDC motor (steady state)

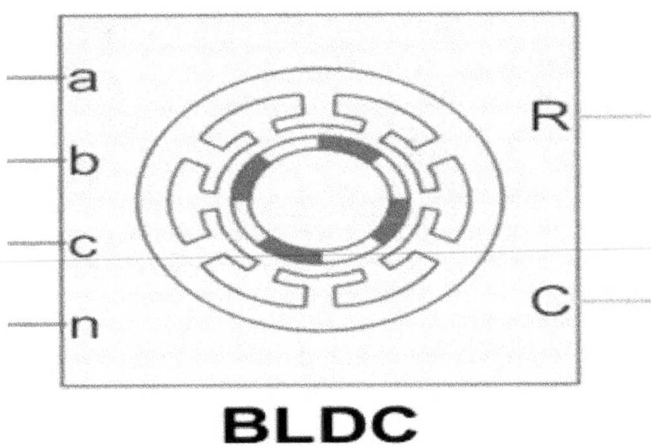

Figure 28. Duty cycle of buck converter

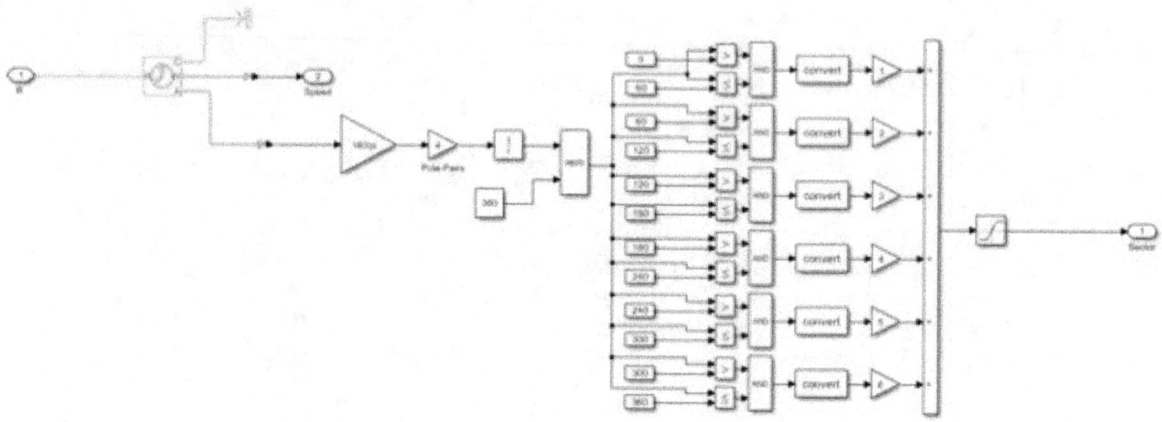

Figure 29. PID Control of eBike

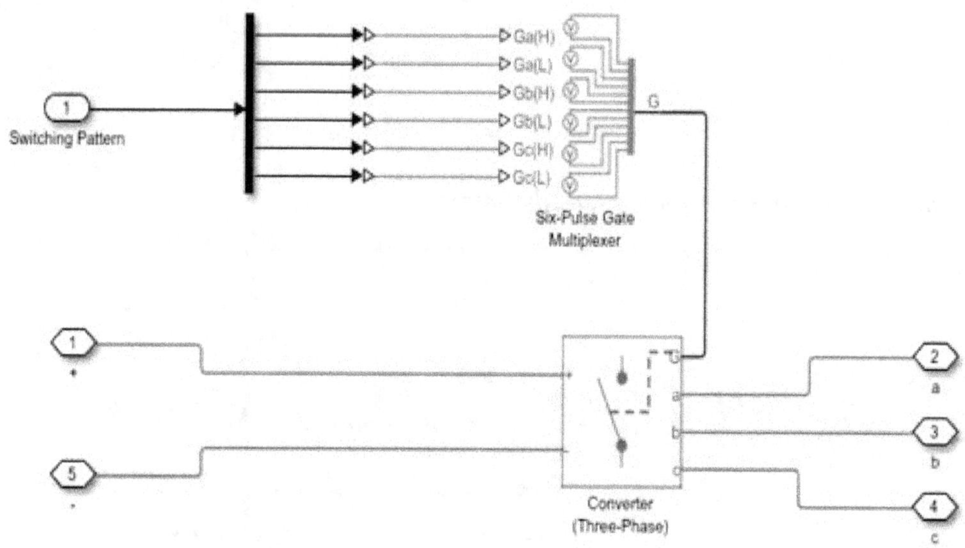

Figure 30. Speed of eBike at different speed and road grade

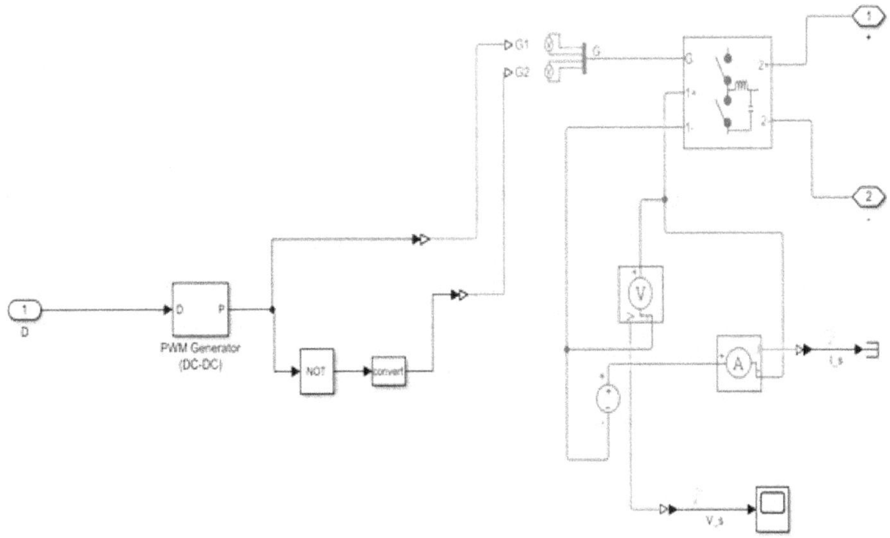

Figure 31. Current drawn by motor at different speed

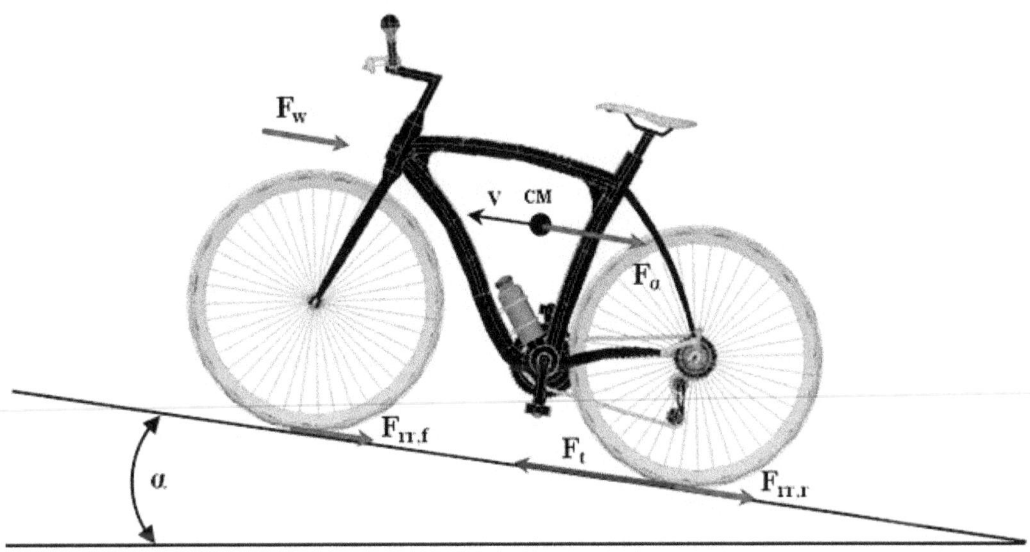

Figure 32. Current drawn by motor at different road grade

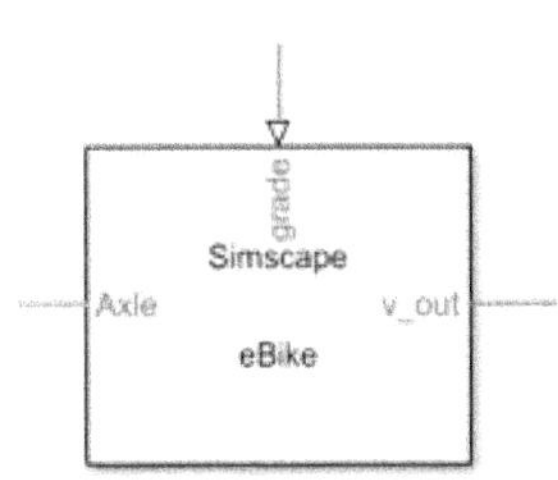

REFERENCES

Abagnale, C., Cardone, M., Iodice, P., Strano, S., Terzo, M., & Vorraro, G. (2015). Derivation and Validation of a Mathematical Model for a Novel Electric Bicycle. *Proceedings of the World Congress on Engineering*.

Babu, P. R., Ramprasath, S., & Paranthagan, B. (2012). Modeling and Dynamic Simulation of Permanent Magnet Brushless DC Motor (PMBLDCM) Drives. AIM 2012: Mobile Communication and Power Engineering, 556-564.

Hung, N. B., & Lim, O. (2019). A simulation and experimental study of dynamic performance and electric consumption of an electric bicycle. *Energy Procedia*, *158*, 2865–2871. doi:10.1016/j.egypro.2019.01.937

Mclean, R., Williamson, C., & Kattan, L. (2021). Simulation Modeling of Urban E-Scooter Mobility. *International Symposium on Modeling, Analysis, and Simulation of Computer and Telecommunication Systems*.

Rao, P. C., Obulesh, Y. P., & Babu, S. (2012). Mathematical modeling and BLDC motor with closed loop speed control using PID controller under various loading conditions. *Journal of Engineering and Applied Sciences*, *7*(10).

Sathiyan, S. P. (2018). Effective Design and Analysis of Pedelec e-Bike using MBD Approach. *International Journal of Mechanical Engineering and Technology*, *9*(13), 568–577.

Thejasree, G., & Maniyeri, R. (2019). E-bike System Modeling and Simulation. *IEEE International Conference on Intelligent Systems and Green Technology*.

Chapter 7
Enhancement of Suspension Performance of the Vehicle Model Using Sliding Mode Controllers

Yuvapriya T
School of Electrical Engineering, Vellore Institute of Technology, Vellore, India

Lakshmi P
College of Engineering, Anna University, Chennai, India

ABSTRACT

In commercial vehicles, seat suspension plays a pivotal role since it provides ride comfort and safety for the driver. This chapter presents the design of terminal-based sliding mode controller (TSMC) and fractional order plus terminal-based SMC (FOTSMC) for a full car with driver seat (FCD) vehicle active suspension system. The FCD model considered in this chapter is eight degree of freedom (DOF) linear vehicle model, which is simulated in MATLAB/Simulink environment. The vehicle model is excited under bumpy road and random road disturbance (ISO8608), which represents actual roads in real time. Time domain analysis is carried out for seat acceleration (SA) and the percentage of reduction of SA root mean square (RMS) values by the action of the controller. The performance criteria analysed are ride comfort analysis (ISO2631 standards), power spectral density (PSD) analysis, and stability analysis. These performance analysis and simulation results prove that FOTSMC gives better results when compared to TSMC for a FCD suspension system.

1. INTRODUCTION

In a real time, scenario application of controller design requires a driver for riding the car. Even though driverless cars have been invented but they are still not available for commercial use. The ride comfort analysis is very useful to analyse the vertical vibration of the Full Car Model (FCM) but it does not

DOI: 10.4018/978-1-6684-6631-5.ch007

include the dynamics of seat suspension for a driver. Therefore, it is necessary to include the seat suspension for a driver model for the analysis of the complete dynamics of FCM. Passive seat suspension, Semi-active seat suspension and Active seat suspension are the three types of seat suspension. In Passive seat Suspension System (PSS) the constant spring stiffness and damping coefficient of the vehicle model has a very low attenuating power. Similarly in semi-active seat suspension system, the damping force of the shock absorber is controlled by using variable damper. Whereas in active seat suspension the force is generated using actuator with respect to the road conditions.

The issues of nonlinearities of 8 DOF vehicle with driver model due to friction in the dampers are addressed in (Güçlü, 2003). The objective of this proposed PID control scheme is to arrest the vibration in the seat section and body of the vehicle. The PID control technique decreases vibration and linearizes the nonlinear model, improving ride comfort for the passenger. An optimal control technique for 8 DOF Active Seat Suspension (ASS) has been developed (Sadati et al., 2008). The road handling and ride comfort is considered as the performance index to show the effectiveness of the proposed method. To validate the proposed method, the stochastic inputs are simulated and the result confirms that the optimal kalman observer provides better performance index.

The Neural Network (NN) controller for nonlinear 8 DOF full vehicle with driver model to obtain desirable output has been designed (Guclu & Gulez, 2008). The time and frequency responses of the vehicle are plotted for with and without controllers. The proposed NN control technique provides zero steady state error and better travelling comfort to the passenger. The simulation results show that NN outperforms than uncontrolled system. An illustrated static and dynamic comfort of the human with seat model is shown in (Siefert et al., 2008). The virtual realisation of the human model is adopted to analyse the static and dynamic loading effects. The vibration analysis is successfully carried out in human seat model. The ride comfort analysis for tractors cabin through experimental measurements and finite element modelling is demonstrated in (Zehsaz et al., 2011). According to ISO standards two different road profiles are chosen and accelerometer sensor is attached in the vehicle for measurement. Due to vertical acceleration, tractors operator experiences the discomfort. (Pazooki et al., 2012) depicts the off road dynamics of the vehicle model and three dimensional tyre model. An ASS and PSS results are evaluated for driver's seat location in terms of PSD analysis. The suspension design increases human health and ride comfort of the passenger in the off road vehicle.

An adaptive FLC for ASS was introduced in (Soleymani et al., 2012). The ride comfort is examined for different traffic conditions which clearly portraits the importance of controllers in the vehicle model. The changing speed also influences the quality of ride comfort in the ASS. The designed controller not only enhances the travel comfort and suspension travel but also reduces the energy consumption in the vehicle. The utilization of adaptive soft computing techniques for 8 DOF FCM is shown in (Khan et al., 2014). Two different road profiles such as bump and pot hole are considered to establish the importance of semi active and active suspension systems. The controller parameters are adjusted based on online adaptation which minimizes the cost function. The robustness of the back stepping based FLC can be examined against passive and semi ASS using different performance indices. The intercity bus with ten DOF model is considered in (Sekulić et al., 2013) for vibration analysis. The excitation to the bus is generated in terms PSD and the vibration effects are examined as per 1997 ISO 2631-1 standard by using vertical acceleration (Duarte et al., 2018). The comfort of the driver assessed by means of different spring stiffness and shock absorber. (Sharma et al., 2016) introduces primary and secondary suspension for ten DOF vehicle model. The effect of primary and secondary suspension on ride comfort at 30Kmph is considered for study. The simulation of ten DOF with step signal is simulated in the simulink environment. The simulation results confirm that the secondary suspension improves the ride comfort over primary suspension. In (Adam & Jalil, 2017) agricultural tractor driver health criterion for performance analysis was considered. In this study, driver has been exposed to large magnitude of Whole-Body Vibration (WBV) which deteriorates the human health such as low back pain. The different parameters have been taken for analysis and the WBV is very high in the vertical direction.

The designed robust H_∞ controller for 8 DOF vehicle model is shown in (Rizvi et al., 2018). The mathematical model of vehicle is modelled by using Euler equation. The effect of the road disturbance which degrades the performances of vehicle and passenger comfort. The vibration of the vehicle is suppressed by robust controller. (Loiseau et al., 2018) proposes robust control technique based on tyre and road contact. The road noise attenuation is considerably reduced with active noise control. The simulation and experimental results are investigated to prove the feedback control scheme.

2. MATHEMATICAL MODELLING FOR FULL CAR WITH DRIVER MODEL

The FCD is considered as an 8 DOF vehicle model and it is shown in Fig. 1. The FCM model has 7 DOF and the seat suspension added one DOF in the vertical direction.

Figure 1. FCD model

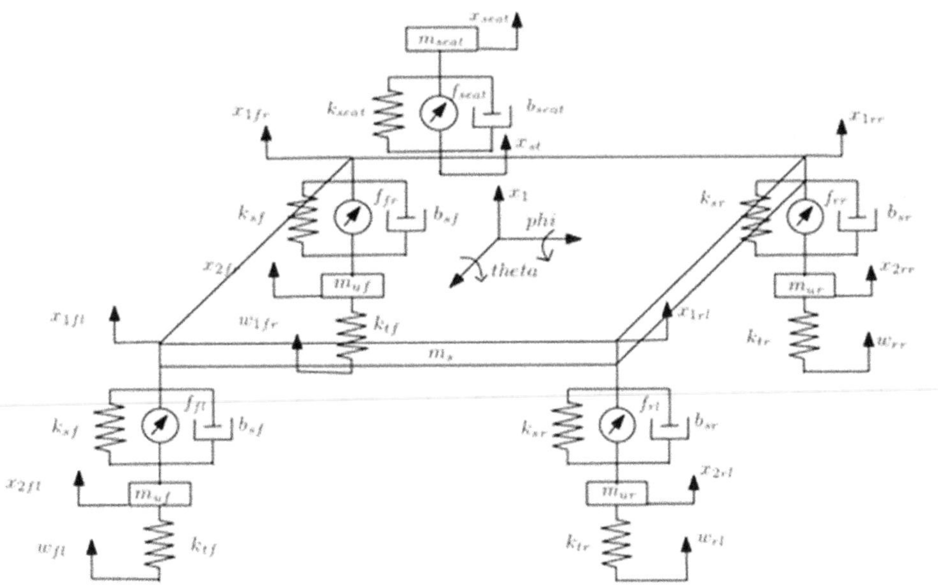

The vertical movement of sprung mass is given in Equation (1),

$$\ddot{X}_1 = \frac{1}{m_s}\Big[k_{sf}\left(-x_{1_{fl}} - x_{1_{fr}} + x_{2_{fl}} + x_{2_{fr}}\right) + k_{sr}\left(-x_{1_{rl}} - x_{1_{rr}} + x_{2_{rl}} + x_{2_{rr}}\right)$$
$$+b_{sf}\left(-\dot{x}_{1_{fl}} - \dot{x}_{1_{fr}} + \dot{x}_{2_{fl}} + \dot{x}_{2_{fr}}\right) + b_{sr}\left(-\dot{x}_{1_{rl}} - \dot{x}_{1_{rr}} + \dot{x}_{2_{rl}} + \dot{x}_{2_{rr}}\right) \quad (1)$$
$$+f_{fl} + f_{fr} + f_{rl} + f_{rr}\Big]$$

Pitch movement of sprung mass is

$$\bar{\theta} = \frac{1}{I_\theta}\Big[k_{sf}l_f\left(-x_{1_{fl}} - x_{1_{fr}} + x_{2_{fl}} + x_{2_{fr}}\right) - k_{sr}l_r\left(-x_{1_{rl}} - x_{1_{rr}} + x_{2_{rl}} + x_{2_{rr}}\right)$$
$$+b_{sf}l_f\left(-\dot{x}_{1_{fl}} - \dot{x}_{1_{fr}} + \dot{x}_{2_{fl}} + \dot{x}_{2_{fr}}\right) - b_{sr}l_r\left(-\dot{x}_{1_{rl}} - \dot{x}_{1_{rr}} + \dot{x}_{2_{rl}} + \dot{x}_{2_{rr}}\right) \quad (2)$$
$$-l_f f_{fl} - l_f f_{fr} + l_r f_{frl} + l_r f_{rr}\Big]$$

Roll movement of sprung mass is

$$\ddot{\Phi} = \frac{1}{I_\phi}\Big[k_{sf}t_f\left(-x_{1_{fl}} + x_{1_{fr}} + x_{2_{fl}} - x_{2_{fr}}\right) + k_{sr}t_r\left(-x_{1_{rl}} + x_{1_{rr}} + x_{2_{rl}} - x_{2_{rr}}\right)$$
$$+b_{sf}t_f\left(-\dot{x}_{1_{fl}} + \dot{x}_{1_{fr}} + \dot{x}_{2_{fl}} - \dot{x}_{2_{fr}}\right) + b_{sr}t_r\left(-\dot{x}_{1_{rl}} + \dot{x}_{1_{rr}} + \dot{x}_{2_{rl}} - \dot{x}_{2_{rr}}\right) \quad (3)$$
$$-t_f f_{fl} + t_2 f_{fr} - t_r f_{rl} + t_r f_{rr}\Big]$$

Vertical displacement of front left wheel is

$$m_{uf}\ddot{x}_{2_{fl}} = -\left[k_{sf}\left(-x_{1_{fl}} + x_{2_{fl}}\right) + k_{tf}\left(x_{2_{fl}} - w_{fl}\right) + b_{sf}\left(-\dot{x}_{1_{fl}} + \dot{x}_{2_{fl}}\right) - f_{fl}\right] \quad (4)$$

Vertical displacement of front right wheel is

$$m_{uf}\ddot{x}_{2_{fr}} = -\left[k_{sf}\left(-x_{1_{fr}} + x_{2_{fr}}\right) + k_{tf}\left(x_{2_{fr}} - w_{fr}\right) + b_{sf}\left(-\dot{x}_{1_{fr}} + \dot{x}_{2_{fr}}\right) - f_{fr}\right] \quad (5)$$

Vertical displacement of rear left wheel is

$$m_{ur}\ddot{x}_{2_{rl}} = -\left[k_{sr}\left(-x_{1_{rl}} + x_{2_{rl}}\right) + k_{tr}\left(x_{2_{rl}} - w_{rl}\right) + b_{sr}\left(-\dot{x}_{1_{rl}} + \dot{x}_{2_{rl}}\right) - f_{rl}\right] \quad (6)$$

Vertical displacement of rear right wheel is

$$m_{ur}\ddot{x}_{2_{rr}} = -\left[k_{sr}\left(-x_{1_{rr}} + x_{2_{rr}}\right) + k_{tr}\left(x_{2_{rr}} - w_{rr}\right) + b_{sr}\left(-\dot{x}_{1_{rr}} + \dot{x}_{2_{rr}}\right) - f_{rr}\right] \quad (7)$$

where

$$x_{1_{fl}} = x_1 + t_1\Phi + 1_f\theta \tag{8}$$

$$x_{1_{fr}} = x_1 - t_2\Phi + 1_f\theta \tag{9}$$

$$x_{1_{rl}} = x_1 + t_1\Phi - 1_r\theta \tag{10}$$

$$x_{1_{rr}} = x_1 - t_2\Phi - 1_r\theta \tag{11}$$

The FCD parameters descriptions are same as discussed in (Rizvi et al., 2018). The mass of the seat is taken as m_{seat}, coefficient of seat damper as b_{seat}, coefficient of seat stiffness as $k_{seat,}$ whereas x_{st} and x_{seat} are sprung and unsprung mass displacement of seat. The s_1 and s_2 are the distance of the seat from the centre of gravity along roll and pitch direction respectively. w_1 and w_2 are the distance of the left and right wheels. The actuator force given to the driver seat is f_{seat}. The FCD parameters (Rizvi et al., 2018) are used for simulation.

3. CONTROLLER DESIGN

SMC is a type of variable control design formed by discontinuous feedback control structure that switches the system state to remain in a specified space within the state space called as sliding surface. The ideal sliding motion is when the system dynamics is confined to the sliding surface (Bayar & Khaneghah, 2020; Chen et al., 2011; Chen & Huang, 2008; Cheng et al., 2010; Ovalle et al., 2022). This shows a reduced order dynamics when compared to original plant. The advancement in SMC design led to design of TSMC and FOTSMC (Efe, 2010).

The simulation results of SMCs for FCM proves that TSMC and FOTSMC have the ability to enhance the ride comfort of the passenger to a great extent rather than other SMCs and optimized Proportional Integrator Derivative (PID) controllers. Hence, the suspension deflection and suspension velocity of seat are chosen as state variables to design TSMC and FOTSMC. In this case the Seat Acceleration (SA) is considered to analyse the driver seat dynamics of FCD model. Four control forces are exerted to each wheel in FCM to limit vibration. The control force equation is added for the driver seat in addition to control forces for the FCM. The sliding mode equations and control forces are clearly discussed and taken from (Yuvapriya et al., 2020).

To design control force for seat (f_{seat}), the state variables are

Suspension deflection of seat $x_1 = x_{st} - x_{seat}$ (14)

Body velocity of seat $x_2 = \dot{x}_{st}$ (15)

The existence condition for sliding mode is possible when

$$\dot{V} = s\dot{s} < 0 \text{ i.e } \dot{s} = 0 \tag{16}$$

The subsequent seat control force is generated using TSMC by substituting the state variables in the sliding surface.

$$s = \mathrm{x}^{\frac{p}{q}}{}_{2} + \lambda \mathrm{x}_1 \tag{17}$$

where p and q are integers which satisfy the condition p <q < 2 p. After differentiation

$$\dot{s} = \dot{\mathrm{x}}_2 + \lambda^{\frac{q}{p}} \mathrm{x}_1^{\;2-\frac{p}{q}} \tag{18}$$

$$0 = \dot{\mathrm{x}}_2 + \lambda^{\frac{q}{p}} \mathrm{x}_1^{\;2-\frac{p}{q}} \tag{19}$$

To design the actuator force (f_a), the state variables (Equations (14) and (15)) are substituted in the Equation (19) and simplified.

$$\begin{aligned}
\mathrm{f}_{seat} = -\Bigg\{ & \frac{1}{\mathrm{m}_s}\Big[\mathrm{k}_{sf}\left(-\mathrm{x}_{1_{fl}} - \mathrm{x}_{1_{fr}} + \mathrm{x}_{2_{fl}} + \mathrm{x}_{2_{fr}}\right) + \mathrm{k}_{sr}\left(-\mathrm{x}_{1_{rl}} - \mathrm{x}_{1_{rr}} + \mathrm{x}_{2_{rl}} + \mathrm{x}_{2_{rr}}\right) + \mathrm{b}_{sf}\left(-\dot{\mathrm{x}}_{1_{fl}} - \dot{\mathrm{x}}_{1_{fr}} + \dot{\mathrm{x}}_{2_{fl}} + \dot{\mathrm{x}}_{2_{fr}}\right) \\
& + \mathrm{b}_{sr}\left(-\dot{\mathrm{x}}_{1_{rl}} - \dot{\mathrm{x}}_{1_{rr}} + \dot{\mathrm{x}}_{2_{rl}} + \dot{\mathrm{x}}_{2_{rr}}\right) + \mathrm{b}_{seat}\left(\dot{\mathrm{x}}_{seat} - \dot{\mathrm{x}}_{st}\right) + \mathrm{k}_{seat}\left(\mathrm{x}_{seat} - \mathrm{x}_{st}\right) + \mathrm{f}_{fl} + \mathrm{f}_{fr} + \mathrm{f}_{rl} + \mathrm{f}_{rr}\Big] \\
& + \frac{1}{\mathrm{I}_\theta}\Big[\mathrm{k}_{sf}\mathrm{l}_f\left(-\mathrm{x}_{1_{fl}} - \mathrm{x}_{1_{fr}} + \mathrm{x}_{2_{fl}} + \mathrm{x}_{2_{fr}}\right) - \mathrm{k}_{sr}\mathrm{l}_r\left(-\mathrm{x}_{1_{rl}} - \mathrm{x}_{1_{rr}} + \mathrm{x}_{2_{rl}} + \mathrm{x}_{2_{rr}}\right) + \mathrm{b}_{sf}\mathrm{l}_f\left(-\dot{\mathrm{x}}_{1_{fl}} - \dot{\mathrm{x}}_{1_{fr}} + \dot{\mathrm{x}}_{2_{fl}} + \dot{\mathrm{x}}_{2_{fr}}\right) \\
& - \mathrm{b}_{sr}\mathrm{l}_r\left(-\dot{\mathrm{x}}_{1_{rl}} - \dot{\mathrm{x}}_{1_{rr}} + \dot{\mathrm{x}}_{2_{rl}} + \dot{\mathrm{x}}_{2_{rr}}\right) + \mathrm{b}_{seat}\mathrm{s}_2\left(\dot{\mathrm{x}}_{seat} - \dot{\mathrm{x}}_{st}\right) + \mathrm{k}_{seat}\mathrm{s}_2\left(\mathrm{x}_{seat} - \mathrm{x}_{st}\right) - \mathrm{l}_f\mathrm{f}_{fl} - \mathrm{l}_f\mathrm{f}_{fr} + \mathrm{l}_r\mathrm{f}_{frl} + \mathrm{l}_r\mathrm{f}_{rr}\Big] \\
& + \frac{1}{\mathrm{I}_\phi}\Big[\mathrm{k}_{sf}\mathrm{w}_2\left(-\mathrm{x}_{1_{fl}} + \mathrm{x}_{1_{fr}} + \mathrm{x}_{2_{fl}} - \mathrm{x}_{2_{fr}}\right) + \mathrm{k}_{sr}\mathrm{w}_1\left(-\mathrm{x}_{1_{rl}} + \mathrm{x}_{1_{rr}} + \mathrm{x}_{2_{rl}} - \mathrm{x}_{2_{rr}}\right) + \mathrm{b}_{sf}\mathrm{w}_2\left(-\dot{\mathrm{x}}_{1_{fl}} + \dot{\mathrm{x}}_{1_{fr}} + \dot{\mathrm{x}}_{2_{fl}} - \dot{\mathrm{x}}_{2_{fr}}\right) \\
& - \mathrm{b}_{sr}\mathrm{w}_1\left(-\dot{\mathrm{x}}_{1_{rl}} + \dot{\mathrm{x}}_{1_{rr}} + \dot{\mathrm{x}}_{2_{rl}} - \dot{\mathrm{x}}_{2_{rr}}\right) + \mathrm{b}_{seat}\mathrm{s}_1\left(\dot{\mathrm{x}}_{seat} - \dot{\mathrm{x}}_{st}\right) + \mathrm{k}_{seat}\mathrm{s}_1\left(\mathrm{x}_{seat} - \mathrm{x}_{st}\right) + \mathrm{k}_{seat}\mathrm{s}_1\left(\mathrm{x}_{seat} - \mathrm{x}_{st}\right) \\
& + \mathrm{s}_1\mathrm{f}_{seat} - \mathrm{w}_1\mathrm{f}_{fl} + \mathrm{w}_2\mathrm{f}_{fr} - \mathrm{w}_1\mathrm{f}_{rl} + \mathrm{w}_2\mathrm{f}_{rr}\Big]\Bigg\} - \lambda\frac{\mathrm{q}}{\mathrm{p}}\mathrm{m}_{seat}\left(\dot{\mathrm{x}}_{st} - \dot{\mathrm{x}}_{seat}\right)^{\left(2-\frac{\mathrm{p}}{\mathrm{q}}\right)} - \mathrm{ksign(s)}
\end{aligned} \tag{20}$$

Similarly, the control force of seat is designed by using FOTSMC as

$$
\begin{aligned}
f_{seat} = -\Bigg\{ & \frac{1}{m_s}\Big[k_{sf}\left(-x_{1_{fl}} - x_{1_{fr}} + x_{2_{fl}} + x_{2_{fr}}\right) + k_{sr}\left(-x_{1_{rl}} - x_{1_{rr}} + x_{2_{rl}} + x_{2_{rr}}\right) + b_{sf}\left(-\dot{x}_{1_{fl}} - \dot{x}_{1_{fr}} + \dot{x}_{2_{fl}} + \dot{x}_{2_{fr}}\right) \\
& + b_{sr}\left(-\dot{x}_{1_{rl}} - \dot{x}_{1_{rr}} + \dot{x}_{2_{rl}} + \dot{x}_{2_{rr}}\right) + b_{seat}\left(\dot{x}_{seat} - \dot{x}_{st}\right) + k_{seat}\left(x_{seat} - x_{st}\right) + f_{fl} + f_{fr} + f_{rl} + f_{rr}\Big] \\
& + \frac{1}{I_\theta}\Big[k_{sf} l_f\left(-x_{1_{fl}} - x_{1_{fr}} + x_{2_{fl}} + x_{2_{fr}}\right) - k_{sr} l_r\left(-x_{1_{rl}} - x_{1_{rr}} + x_{2_{rl}} + x_{2_{rr}}\right) + b_{sf} l_f\left(-\dot{x}_{1_{fl}} - \dot{x}_{1_{fr}} + \dot{x}_{2_{fl}} + \dot{x}_{2_{fr}}\right) \\
& - b_{sr} l_r\left(-\dot{x}_{1_{rl}} - \dot{x}_{1_{rr}} + \dot{x}_{2_{rl}} + \dot{x}_{2_{rr}}\right) + b_{seat} s_2\left(\dot{x}_{seat} - \dot{x}_{st}\right) + k_{seat} s_2\left(x_{seat} - x_{st}\right) - l_f f_{fl} - l_f f_{fr} + l_r f_{frl} + l_r f_{rr}\Big] \\
& + \frac{1}{I_\phi}\Big[k_{sf} w_2\left(-x_{1_{fl}} + x_{1_{fr}} + x_{2_{fl}} - x_{2_{fr}}\right) + k_{sr} w_1\left(-x_{1_{rl}} + x_{1_{rr}} + x_{2_{rl}} - x_{2_{rr}}\right) + b_{sf} w_2\left(-\dot{x}_{1_{fl}} + \dot{x}_{1_{fr}} + \dot{x}_{2_{fl}} - \dot{x}_{2_{fr}}\right) \\
& - b_{sr} w_1\left(-\dot{x}_{1_{rl}} + \dot{x}_{1_{rr}} + \dot{x}_{2_{rl}} - \dot{x}_{2_{rr}}\right) + b_{seat} s_1\left(\dot{x}_{seat} - \dot{x}_{st}\right) + k_{seat} s_1\left(x_{seat} - x_{st}\right) + s_1 f_{seat} \\
& - w_1 f_{fl} + w_2 f_{fr} - w_1 f_{rl} + w_2 f_{rr}\Big]\Bigg\} - \lambda \frac{q}{p}\left(D^{1-\alpha}\right)^{\frac{p}{q}} m_{seat}\left(\dot{x}_{st} - \dot{x}_{seat}\right)^{\left(2-\frac{p}{q}\right)} - k\,\mathrm{sign}(s)
\end{aligned} \tag{21}
$$

4. SIMULATION RESULTS AND DISCUSSION

The simulation results for the FCD model from the TSMC and FOTSMC are examined in this section. The MATLAB/Simulink environment simulates the FCD model and the controllers. Trial and error are used to choose and use the SMCs parameters for simulation. The chosen parameters are $\alpha = 0.8$, $\lambda = 0.8$, $K = 1$, p=21, q=24. The performances of the FCD model under Bump Input (BI) (Rizvi et al., 2018) and Random Input (RI) (Yuvapriya et al., 2018) road profiles are investigated for PSS, TSMC and FOTSMC. Initially, the FCD runs at an average speed of 30 km/h for a time period of 10 seconds.

4.1 Performance Analysis of TSMC and FOTSMC with Bump Input (BI)

In order to evaluate the performances of designed controllers for FCD model, the Seat Acceleration (SA) of the driver is considered for quantitative and qualitative analysis rather than body acceleration. The time response of SA without controller i.e., Passive Suspension System (PSS) is illustrated in Fig. 2 and the time responses of SA for TSMC and FOTSMC with BI is shown in Fig. 3. The oscillation of SA for FCD is reduced and it suppresses the vibration to zero level with ASS. Fig. 4 depicts the control effort of TSMC and FOTSMC. The time response of BA indicates that the FOTSMC performs better than TSMC and PSS. The designed controller takes necessary control action to control the vibration with BI. These forces counteract road disruption and diminish vehicle vibration.

Figure 2. Time response of SA without controller-FCD- BI

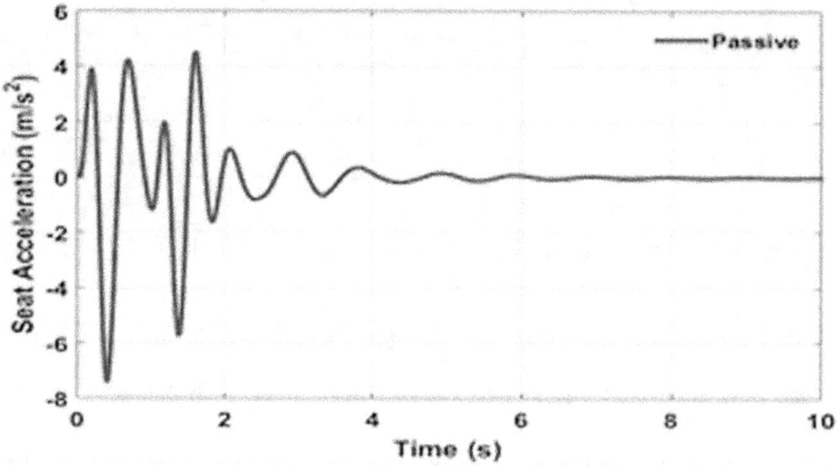

The performance indices RMS, Frequency Weighted based RMS (FWRMS), and Vibration Dose Value (VDV) of SA are selected for numerical analysis in order to evaluate the vibration suppression of SMCs. These are provided in Table 1 along with the numerical analysis results. In comparison to TSMC, FOTSMC has a 46% reduction in RMS values of SA.

4.2 Performance Analysis of TSMC and FOTSMC with Random Input (RI)

Investigating the real road as the FCD model's disruption is crucial to assessing the performances of the TSMC and FOTSMC. According to ISO 8608, a random road input is created. Class E, RI road profile

Figure 3. Time response of SA with controllers-FCD-BI

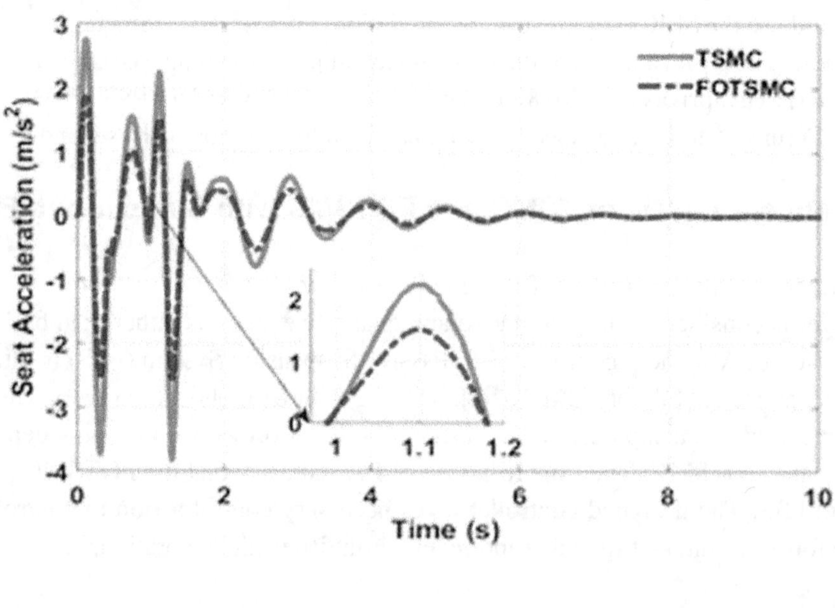

Table 1. Performance Indices of SMCs-BI

Controllers	RMS of SA (m/s^2)	% of Reduction	FWRMS of SA (m/s^2)	Level of Discomfort	VDV of SA ($m/s^{1.75}$)
Passive	1.37	-	1.46	VU	4.38
TSMC	0.71	46%	0.335	LU	2.41
FOTSMC	**0.475**	**65%**	**0.125**	**NU**	**1.63**

is used to assess the seat suspension performance. The time response of SA without controller and time responses of designed SMCs with RI are illustrated in Fig. 5 and Fig. 6. The figure shows that employing FOTSMC as opposed to TSMC and PSS reduces the oscillation of SA with RI. According to Fig. 7, FOTSMC produced a greater control action than TSMC.

The RMS values of SA for the designed TSMC and FOTSMC are shown in Table 2. For RI, the RMS values of SA are used to figure out the percentage of substantial reduction. Hence the percentage of reductions for FOTSMC with RI is decreased to 65.4%. The comparative analysis of RMS values shows significant reduction with FOTSMC rather than TSMC and PSS.

4.3 Ride Comfort Analysis

The FWRMS and VDV of BA are used to measure the passenger's ride pleasure in accordance with ISO-2631 in order to confirm the passenger comfort of the specified FOTSMC approach (Burgess-Limerick, 2016). As can be observed from Table 1, the driving pleasure with PSS is "Very Uncomfortable," while for FOTSMC with BI, it has improved to "Not Uncomfortable" (NU). Additionally, the PSS calculation for RI (Table 2) yields the result "Uncomfortable" (U). Further it is enhanced to 'Little Uncomfortable'

Figure 4. Control forces of seat produced by FCD-BI

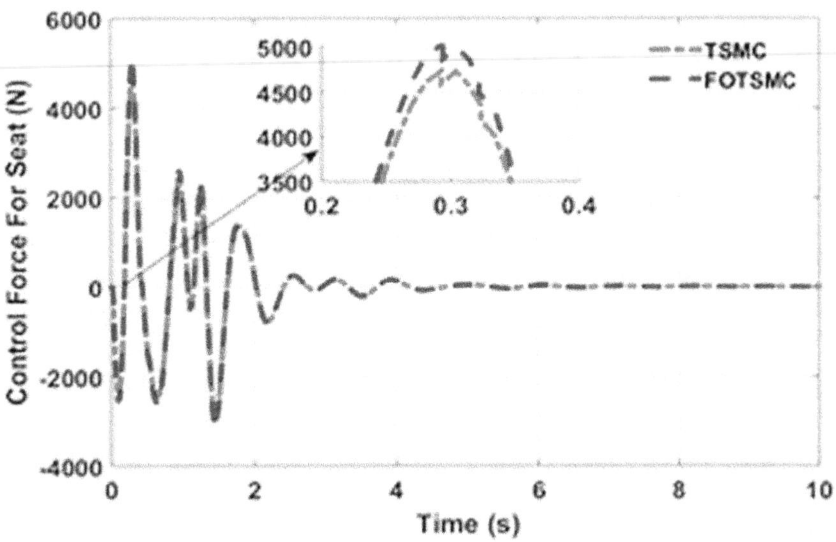

Figure 5. Time response of SA without controller-RI

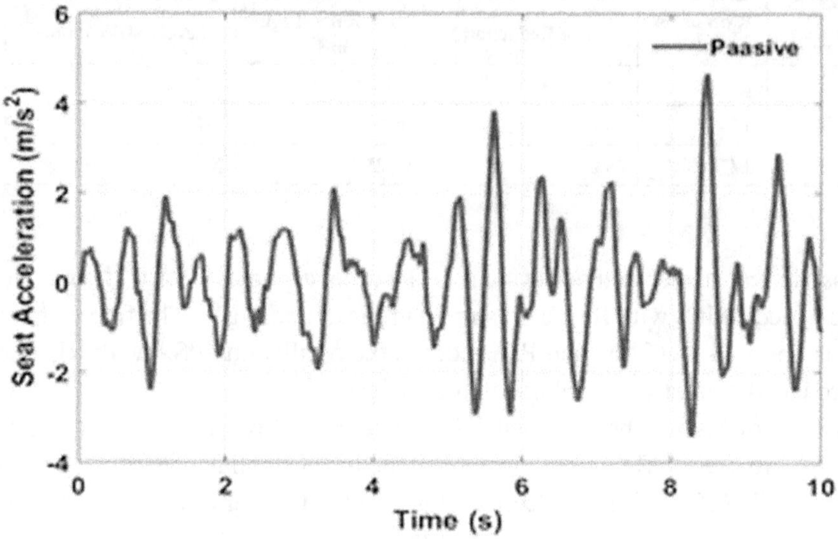

(LU) with FOTSMC. When comparing to other TSMC and PSS, the performance indices of VDV for BI and RI have significantly dropped with FOTSMC.

4.4 Power Spectral Density Analysis

The Power Spectral Density (PSD) response is plotted for BI (Fig. 8) and RI (Fig. 9) road profiles. In Fig 8, the magnitude of PSD with FOTSMC is around 1 m^2/s^4 Hz for BI road profile and it is gradually

Figure 6. Time response of SA with controllers-RI

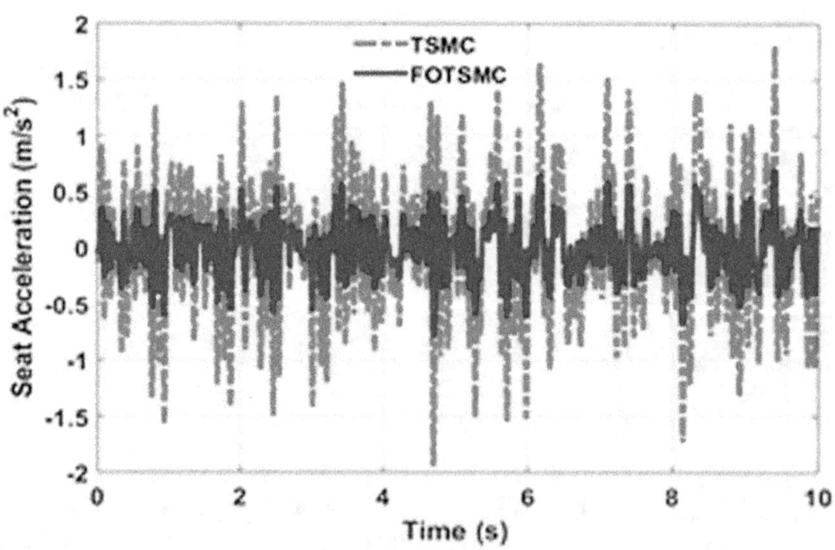

Figure 7. Control forces of seat produced by FCD-RI

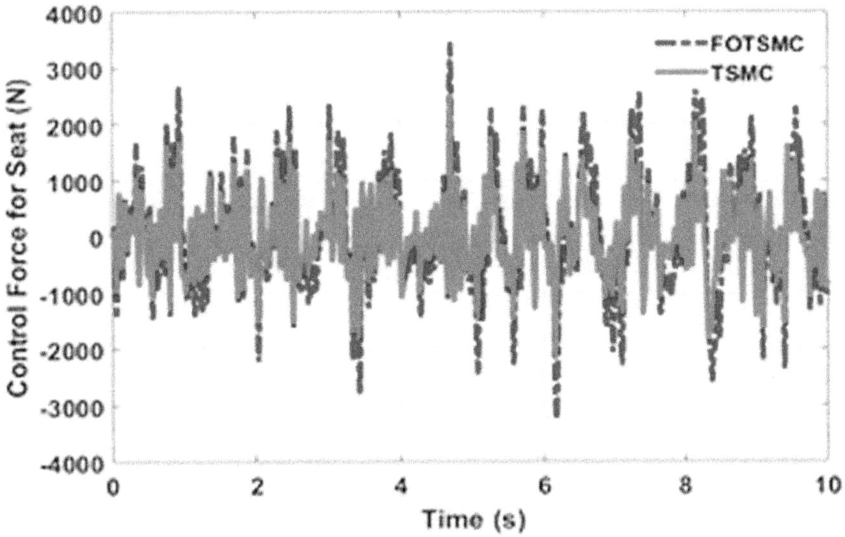

reduced when the frequency lies between 4-8 Hz. Similarly, the magnitude of PSD with FOTSMC is around 0.01 m^2/s^4 Hz (Fig. 9) for RI. The Magnitude of FOTSMC is comparatively lesser than TSMC and PSS for BI and RI road profile.

The speed of the FCD is chosen as 30 kmph and then the road profiles are simulated as per speed conditions. It is assumed as the low speed of the FCD is from 10 kmph and the high speed is up to 100 kmph.

Figure 8. PSD of SA-FCD-BI

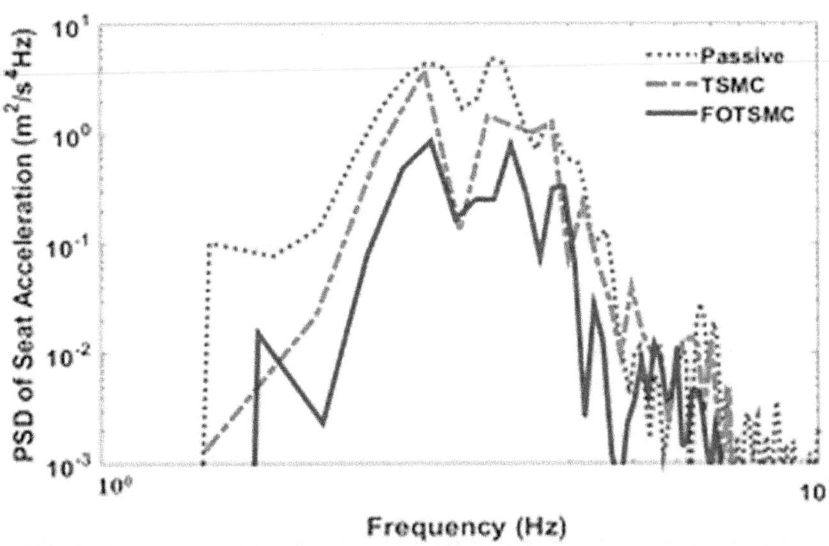

Figure 9. PSD of SA-FCD-RI Speed Analysis

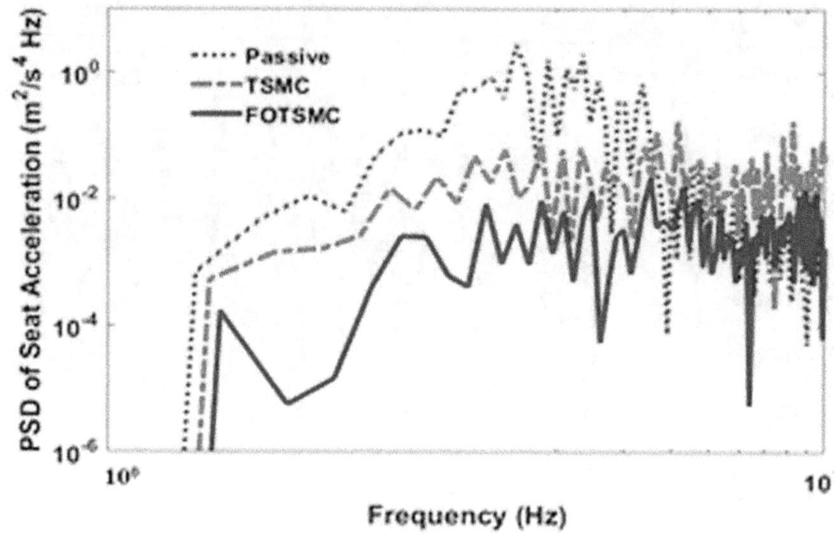

The percentages of reduction in RMS values are calculated and the responses are plotted for two different road conditions. From the Fig. 10, RMS values of SA are significantly reduced with FOTSMC from 10 to 40 kmph compared to other TSMC with BI. After 40 kmph, designed SMCs drastic reduction in RMS values of SA from 40 to 60 kmph. Then there is a gradual reduction in the RMS values of the designed TSMC and FOTSMC from 60 to 100 kmph.

Similarly, the reduction in RMS values of SA is computed and plotted for RI (Fig. 11) road profile. It is clearly depicted in Fig. 10 & Fig. 11 the performances of designed SMCs are slowly decaying when the vehicle travels at high speed.

Figure 10. Speed Analysis of SMCs-FCD-BI

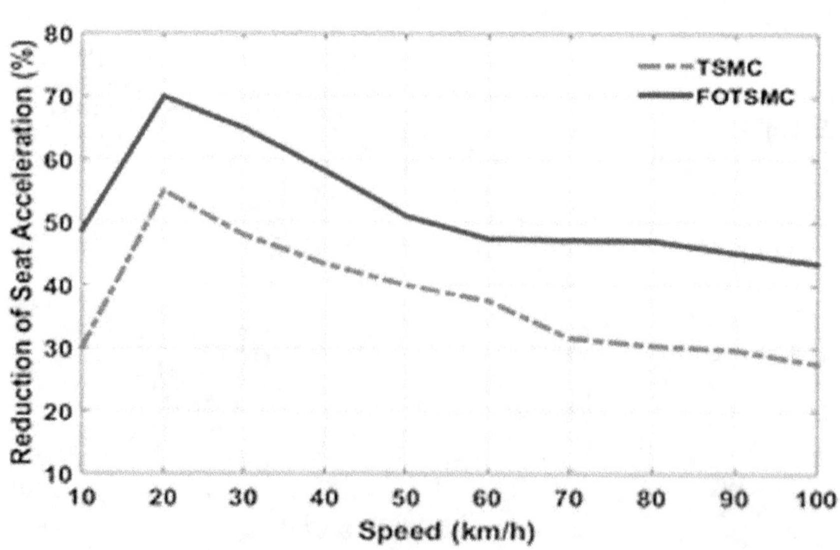

Figure 11. Speed Analysis of SMCs-FCD-RI

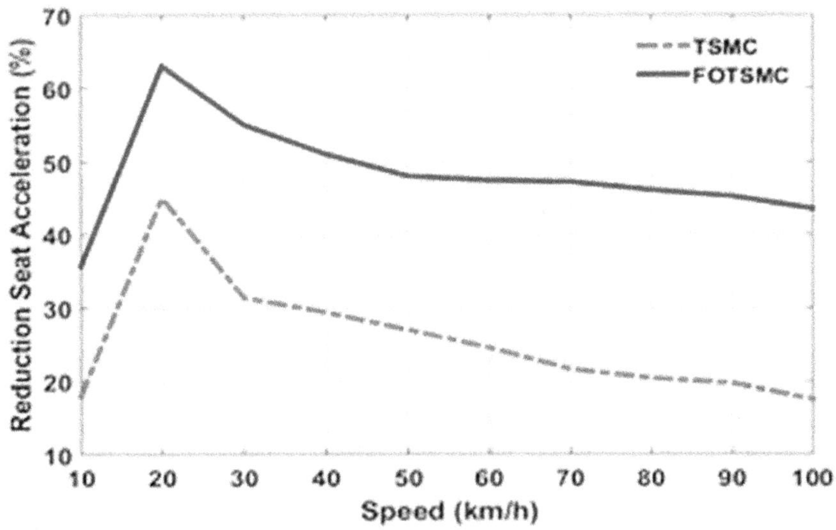

4.5 Stability Analysis

For BI, state variables are shown on the phase planes of designed TSMC (Fig. 12) and FOTSMC (Fig. 13). The trajectories of designed SMCs with BI are converging to origin.

The stability analysis with BI confirms that designed SMCs are stable for FCD. The trajectories of designed SMCs with BI are converging to origin. The stability analysis with BI confirms that designed SMCs are asymptotically stable (Slotine & Li, 1991) for FCD. Hence the FOTSMC is considered for remaining road profiles to confirm the stability of FCD. Similarly, the trajectories of FOTSMC with RI (Fig. 14) converges nor diverges.

5. CONCLUSION

TSMC and FOTSMC are designed and simulated for FCD model using MATLAB/Simulink. Using BI and RI road profiles, the designed controllers' performance is evaluated. When comparison with TSMC and PSS, FOTSMC exhibits a significant reduction of RMS values and oscillation. The ride comfort analysis and speed analysis are carried out to validate the potency of the designed controllers. According

Table 2. Performance Indices of SMCs-RI

Controllers	RMS of SA (m/s^2)	% of Reduction	FWRMS of SA (m/s^2)	Level of Discomfort	VDV of SA ($m/s^{1.75}$)
Passive	1.312	-	1.21	U	2.95
TSMC	0.9	48.4%	0.73	FU	0.798
FOTSMC	**0.59**	**65.4%**	**0.51**	LU	**0.3**

to the qualitative and quantitative investigation, FOTSMC improves ride comfort and reduces vibration for BI to zero. Further, the stability for SMCs with BI is stable and marginally stable for RI.

Figure 12. Phase Plane of TSMC-FCD-BI

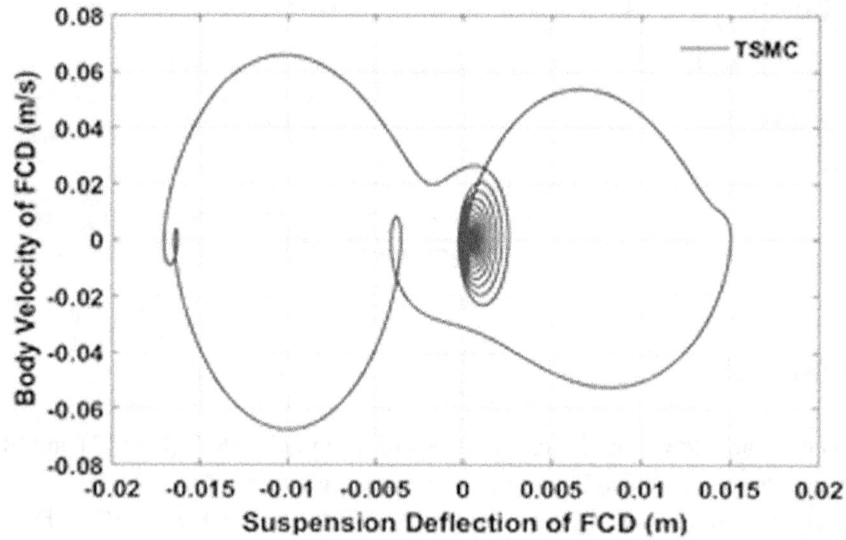

Figure 13. Phase Plane of FOTSMC-FCD-BI

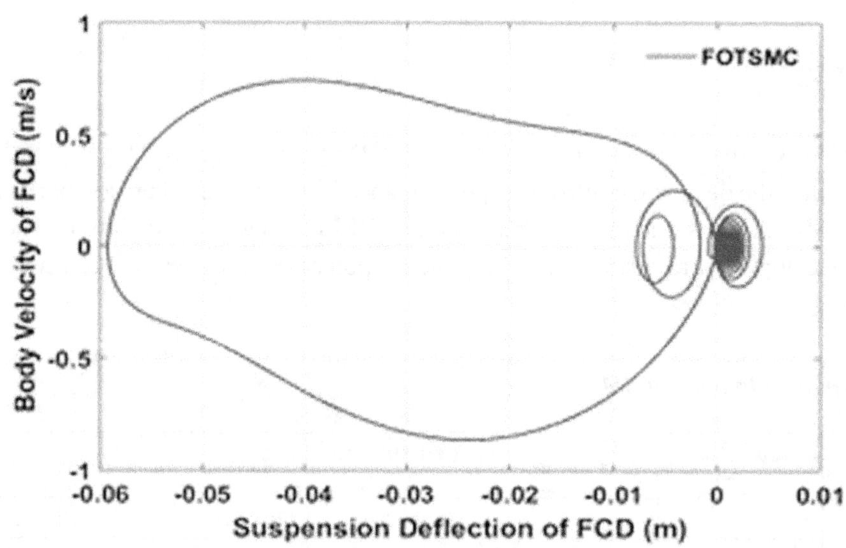

Figure 14. Phase Plane of FOTSMC-FCD-RI

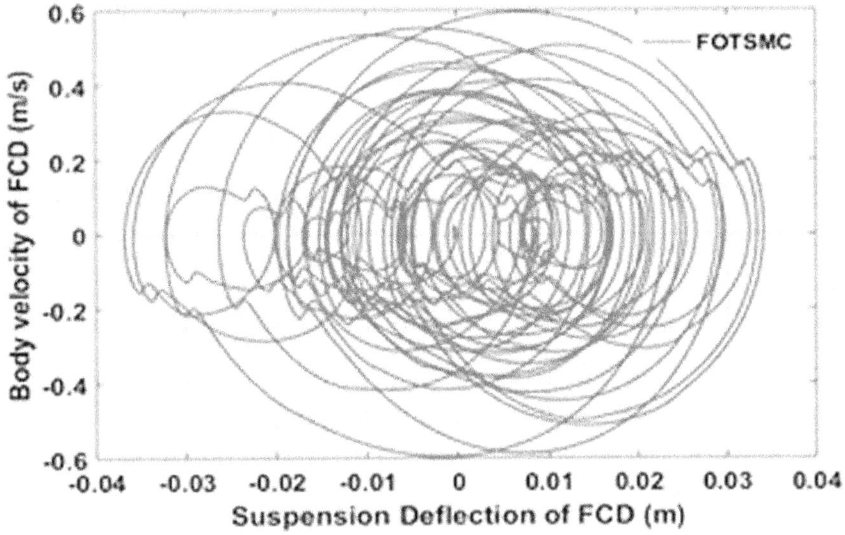

FUNDING

This research received no specific grant from any funding agency in the public, commercial, or not-for-profit sectors.

REFERENCES

Adam, S. A., & Jalil, N. A. (2017). Vertical suspension seat transmissibility and SEAT values for seated person exposed to whole-body vibration in agricultural tractor preliminary study. *Procedia Engineering*, *170*, 435–442. doi:10.1016/j.proeng.2017.03.070

Bayar, K., & Khaneghah, F. S. (2020). Optimal sliding mode control method for active suspension control. *IFAC-PapersOnLine*, *53*(2), 14285–14291. doi:10.1016/j.ifacol.2020.12.1178

Burgess-Limerick, R. (2016). Measuring and managing workplace whole-body vibration exposures. *Acoustics Australia*, *44*(1), 129–135. doi:10.100740857-015-0032-5

Chen, B. C., Shiu, Y. H., & Hsieh, F. C. (2011). Sliding-mode control for semi-active suspension with actuator dynamics. *Vehicle System Dynamics*, *49*(1-2), 277–290. doi:10.1080/00423111003602376

Chen, H. Y., & Huang, S. J. (2008). A new model-free adaptive sliding controller for active suspension system. *International Journal of Systems Science*, *39*(1), 57–69. doi:10.1080/00207720701669453

Cheng, C. P., Chao, C. H., & Li, T. H. S. (2010). Design of observer-based fuzzy sliding-mode control for an active suspension system with full-car model. *IEEE International Conference on Systems, Man and Cybernetics*, 1939-1944. 10.1109/ICSMC.2010.5642261

Duarte, M. L., de Araújo, P. A., Horta, F. C., Del Vecchio, S., & de Carvalho, L. A. (2018). Correlation between weighted acceleration, vibration dose value and exposure time on whole body vibration comfort levels evaluation. *Safety Science*, *103*, 218–224. doi:10.1016/j.ssci.2017.11.008

Efe, M. Ö. (2010). Fractional order sliding mode control with reaching law approach. *Turkish Journal of Electrical Engineering and Computer Sciences*, *18*(5), 731–748. doi:10.3906/elk-0906-3

Güçlü. (2003). Active control of seat vibrations of a vehicle model using various suspension alternatives. *Turkish Journal of Engineering and Environmental Sciences*, *27*(6), 361–374.

Guclu, R., & Gulez, K. (2008). Neural network control of seat vibrations of a non-linear full vehicle model using PMSM. *Mathematical and Computer Modelling*, *47*(11-12), 1356–1371. doi:10.1016/j.mcm.2007.08.013

Khan, L., Qamar, S., & Khan, M. U. (2014). Comparative analysis of adaptive NeuroFuzzy control techniques for full car active suspension system. *Arabian Journal for Science and Engineering*, *39*(3), 2045–2069. doi:10.100713369-013-0729-4

Loiseau, P., Chevrel, P., Yagoubi, M., & Duffal, J. M. (2018). Robust active noise control in a car cabin: Evaluation of achievable performances with a feedback control scheme. *Control Engineering Practice*, *81*, 172–182. doi:10.1016/j.conengprac.2018.09.015

Ovalle, L., Ríos, H., & Ahmed, H. (2022). Robust control for an active suspension system via continuous sliding-mode controllers. *Engineering Science and Technology, an International Journal, 28*, 101026.

Pazooki, A., Rakheja, S., & Cao, D. (2012). Modeling and validation of off-road vehicle ride dynamics. *Mechanical Systems and Signal Processing*, *28*, 679–695. doi:10.1016/j.ymssp.2011.11.006

Rizvi, S. M., Abid, M., Khan, A. Q., Satti, S. G., & Latif, J. (2018). H¥ control of 8 degrees of freedom vehicle active suspension system. *Journal of King Saud University-Engineering Sciences*, *30*(2), 161–169. doi:10.1016/j.jksues.2016.02.004

Sadati, H., Malekzadeh, S., & Ghasemi, M. (2008). Optimal Control of an 8-DOF Vehicle Active Suspension System Using Kalman Observer. *Shock and Vibration*, *15*(5), 493–503. doi:10.1155/2008/725264

Sekulić, D., Dedović, V., Rusov, S., Šalinić, S., & Obradović, A. (2013). Analysis of vibration effects on the comfort of intercity bus users by oscillatory model with ten degrees of freedom. *Applied Mathematical Modelling*, *37*(18-19), 8629–8644. doi:10.1016/j.apm.2013.03.060

Sharma, S. K., Pare, V., Chouksey, M., & Rawal, B. R. (2016). Numerical studies using full car model for combined primary and cabin suspension. *Procedia Technology*, *23*, 171–178. doi:10.1016/j.protcy.2016.03.014

Siefert, A., Pankoke, S., & Wölfel, H. P. (2008). Virtual optimisation of car passenger seats: Simulation of static and dynamic effects on drivers' seating comfort. *International Journal of Industrial Ergonomics*, *38*(5-6), 410–424. doi:10.1016/j.ergon.2007.08.016

Slotine, J. J. E., & Li, W. (1991). *Applied nonlinear control*. Prentice Hall.

Soleymani, M., Montazeri-Gh, M., & Amiryan, R. (2012). Adaptive fuzzy controller for vehicle active suspension system based on traffic conditions. *Scientia Iranica, 19*(3), 443–453. doi:10.1016/j.scient.2012.03.002

Yuvapriya, T., Lakshmi, P., & Rajendiran, S. (2018). Vibration suppression in full car active suspension system using fractional order sliding mode controller. *Journal of the Brazilian Society of Mechanical Sciences and Engineering, 40*(4), 1–1. doi:10.100740430-018-1138-0

Yuvapriya, T., Lakshmi, P., & Rajendiran, S. (2020). Vibration control and performance analysis of full car active suspension system using fractional order terminal sliding mode controller. *Archives of Control Sciences, 30*(2), 295–324.

Zehsaz, M., Sadeghi, M. H., Ettefagh, M. M., & Shams, F. (2011). Tractor cabin's passive suspension parameters optimization via experimental and numerical methods. *Journal of Terramechanics, 48*(6), 439–450. doi:10.1016/j.jterra.2011.09.005

Chapter 8
Active Suspension Control of Full Car Model Using Bat Optimized PID Controller

Yuvapriya T
School of Electrical Engineering, Vellore Institute of Technology, Vellore, India

Lakshmi P
College of Engineering, Anna University, Chennai, India

ABSTRACT

Long drives on bumpy roads and configuration issues like discomfort in seating arrangements have a harmful impact on the human body. The passengers experience severe health problems and stress-related issues. The full car model (FCM) with seven degrees of freedom (DOF) is considered for vibration control analysis. The aim of this work is to optimize the parameters of proportional integral and derivative (PID) controller by grey wolf optimization (GWO) and bat algorithm for betterment in the ride comfort of the passengers. A comparative analysis between the most commonly used PID controller and proposed optimized PID controller was executed over bump input (BI), ISO standard random input (RI), and sinusoidal input (SI) road profiles in MATLAB. Simulation results demonstrate that bat tuned PID (Bat-PID) controller enhances the ride comfort by decreasing the root mean square (RMS), frequency weighted RMS (FWRMS), and vibration dose values (VDV) of the body acceleration (BA) of the vehicle passing over BI, RI, and SI road profiles, which ensures the stability of the vehicle.

1. INTRODUCTION

The reduction of vibrations is a challenging task in the Full Car Model (FCM) of the vehicle. The researchers are focused on different types of control strategies in the FCM Active Suspension System (ASS) to analyse the vehicle dynamics and increase the ride comfort of the passengers. Active suspensions offer the potential of being adaptable to the quality of the road surface, vehicle speed and comfort

DOI: 10.4018/978-1-6684-6631-5.ch008

requirements. The suspension system reduces the transmission of oscillations to the vehicle body from road surface disturbances.

Rajagopal, K., & Ponnusamy, L. (2014) discussed PSO and Biogeography-Based Optimization (BBO) to obtain optimal PID controller parameters for Quarter Car Model (QCM). The time domain and frequency domain analysis are considered to examine the effectiveness of the Genetic Algorithm (GA) based Fuzzy Logic Controller (FLC). The ride comfort and stability of the nonlinear model is enhanced by GA based FLC when compared to PID controller and PSS. Nagarkar, M P *et al.* (2018) introduced multi objective optimization for nonlinear quarter car model to detach the vehicle body from external disturbances. The trajectory optimization of FCM (Soh, M *et al.* 2018) and the damper structure optimization using Particle Swarm Optimization (PSO) (Wang, M & Xie, N 2018) have been discussed in earlier studies. Wang, W. et al. (2015) took into account the cultural algorithm to optimise the Fuzzy PID controller to lower the BA, which increases the passenger's comfort throughout the ride. Jahromi, AF & Zabi hollah, A (2010) designed Linear Quadratic Regulator (LQR) and FLC to reduce the vibration level of passenger car. The overshoot and steady state error are decreased to minimum value with LQR in comparison to FLC. Prabakar, RS *et al.* (2013) considered magneto-rheological damper as a semi active device to replace actuator. The performances of body acceleration and road handling are utilised to evaluate the potency of LQR. To examine the passenger's comfort during the ride, additional frequency responses of sprung mass acceleration are illustrated. The simulation (Yuvapriya T and Lakshmi, P 2020) and experimental QCM is investigated with sliding mode controllers (Yuvapriya T *et al.* 2020). The requirements are simultaneously fulfilled and the controller's parameters are adjusted via evolutionary multi-objective optimization. (Reynoso-Meza *et al.* 2014). The ride quality of the vehicle is improved by Zhao, J *et al.* (2019) using semi- ASS. The quarter car model utilises air spring and hydraulic damper. The cuckoo search algorithm optimises PID parameters which improves the passenger comfort using semi-ASS. The simulation and experimental analysis illustrate that the optimised PID controller outperforms the conventional PID controller for different driving conditions.

Mirjalili, S *et al.* (2014) proposed Grey Wolf Optimization (GWO) for engineering problems. This algorithm is tested on 29 benchmark functions and the comparative study is made with different optimization techniques. The convergence speed and memory are key factors of GWO and the results of engineering problems also evident for GWO. The fifth order model of generator is utilized as equivalent model by Zhou, J *et al.* (2016). The parameters identification of equivalent model is carried out with GWO. The simulation and experiment analysis has been done to show the accuracy of the generator model through proper identification of system parameters. The frequency stability and performances of the controller are improved by Zhang, Y *et al.* (2017). The GWO is introduced to optimize control parameters of pump turbine and the simulation study proves the effectiveness of novel GWO technique in comparison with traditional ones. Sun, J *et al.* (2018) presented Pareto optimisation technique to control velocity of the vehicle. The weighted RMS values are considered as performance index to ensure the driving safety. The driving and comfort performance are utilised for optimisation and it is transformed as multi objective optimisation problem. This optimization technique offers superior performance in terms of driving safety and ride comfort. Pedro, J O *et al.* (2018) addressed the nonlinear control technique with dynamic Neural Network (NN) to rectify the intrinsic issues of performance measures in nonlinear FCM. In this approach PSO is used to train the NN and parameters computation. The frequency and time domain analysis of suspension deflection clearly portraits the robustness of the optimized NN approach.

Prassad, GS & Mohan, MK (2019) proposed LQR tuned PSO for ASS. The road handling and comfort analysis are carried out with ISO-8608 random road terrain in the frequency range of 5 to 20Hz.

The LQR based PSO reduces overshoot and unwanted oscillations. Taha, IB & Elattar, EE (2018) introduced improved GWO to minimise generation cost, voltage and transmission power loss. To enhance the operation of power system, the reactive power resources problems are solved by improved GWO. The economic dispatch problem is resolved by Liang, H *et al.* (2018) using meta heuristic algorithm. This Bat algorithm is utilised to avoid the premature convergence and improve the global searching ability. The proposed Bat algorithm with five benchmark functions is tested by convergence speed and efficiency. Arora, U &Lodhi, MEA (2016) designed PID controller using Bat optimization approach for three different plants. The optimal PID controller parameters are obtained through Bat algorithm which reduces search space and increases the convergence speed compared to other meta heuristic algorithms. The fuzzy based Bat algorithm for magnetic ball suspension system is developed by Talbi, N (2019). The main goal of introducing control algorithm is to control the ball position in the system. The fuzzy rule and membership functions are optimized by Bat algorithm which overcomes the difficulty of forming rule base and membership functions.

In comparison to evolutionary programming, the GWO and Bat algorithms use far fewer operators to adapt and store information about all iterations to deliver the optimum solution. These significant factors led to the use of the GWO and Bat algorithms in this study's FCM selection to optimise the standard PID parameters and improve the rider's comfort.

The organization of the paper is as follows: Section 2 describes the mathematical modelling of the FCM. Section 3 elaborates the optimized PID controllers. Section 4 demonstrates the simulation results and comparative analysis of optimized PID controllers. Finally, conclusion is discussed in Section 5.

2. MODELLING OF THE FULL CAR

Figure 1 depicts the schematic diagram of the FCM with seven Degrees of Freedom (DOF) (Gaspar, Pet al. 2003). It is easy to understand how the system works owing to the dynamic model, which illustrates the relationship between road disturbance and outcome. The suspension system is represented as a linear model to obtain state variables of the FCM through mathematical modelling.

The car's body and wheels move vertically to indicate the state variables. To develop the linear model, it is assumed that the roll and pitch angles are modest. The front and rear unsprung masses are indicated by m_{uf} and m_{ur}, respectively, whereas the sprung mass is denoted by m_s. An actuator, a damper, and a spring make up an ASS, which produces forces in FCM. K_{sf} and K_{sr}, as well as K_{tf} and K_{tr}, respectively, stand for the front and rear suspension stiffness and the front and rear tyre stiffness. The front and rear suspension damping are denoted by b_{sf} and b_{sr}. The front and rear tread are denoted by t_f and t_r. The pitch and roll moment of inertia are I_θ and I_φ, respectively. The parameters of FCM (Darus, R& Sam, Y 2009) are used for simulation as given in Table.1.

Front and back wheels, respectively, have l_f and l_r as their centres of gravity. The front and rear displacement of the sprung mass on the left and right side are denoted by x_{1fl}, x_{1rl} and x_{1fr}, x_{1rr}, respectively. The front and rear displacement of the unsprung mass on the left and right side are denoted by x_{2fl}, x_{2rl}, x_{2fr} and x_{2rr}, respectively.

The front left, rear left, front right and rear right disturbances are represented in the FCM by the letters w_{fl}, w_{rl}, w_{fr} and w_{rr}. The front and rear left wheel produce forces of f_{fl} and f_{rl}. Similarly, front and rear right wheel wheel produce forces of f_{fr} and f_{rr}, respectively. The sprung mass has freedom to move in the vertical, pitch, and roll directions, while the FCM has 7 DOF. The sprung mass's pitch angle is

Figure 1. Full Car Model

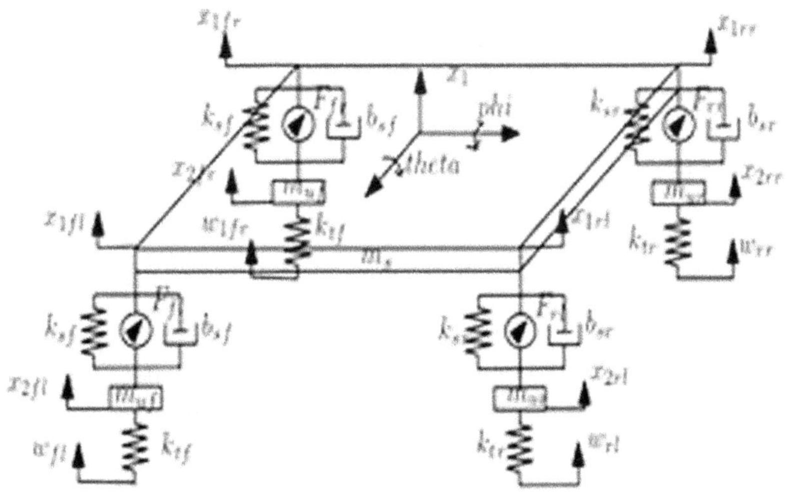

Θ, roll angle is φ, and vertical displacement is x_1. Each unsprung mass, such as x_{2fl}, x_{2rl}, x_{2fr} and x_{2rr}, has freedom of movement in the vertical direction.

The vertical movement of sprung mass is given in Equation (1),

$$\ddot{x}_1 = \frac{1}{m_s}\Big[k_{sf}\left(-x_{1_{fl}} - x_{1_{fr}} + x_{2_{fl}} + x_{2_{fr}}\right) + k_{sr}\left(-x_{1_{rl}} - x_{1_{rr}} + x_{2_{rl}} + x_{2_{rr}}\right) + b_{sf}\left(-\dot{x}_{1_{fl}} - \dot{x}_{1_{fr}} + \dot{x}_{2_{fl}} + \dot{x}_{2_{fr}}\right) + b_{sr}\left(-\dot{x}_{1_{rl}} - \dot{x}_{1_{rr}} + \dot{x}_{2_{rl}} + \dot{x}_{2_{rr}}\right) + f_{fl} + f_{fr} + f_{rl} + f_{rr}\Big] \quad (1)$$

Table 1. FCM Parameters

PARAMETERS	VALUE
m_s	1500 kg
m_{uf}, m_{ur}	59 kg
K_{sf}	35000 N/m
K_{sr}	38000 N/m
K_{tf}, K_{tr}	190000N/m
I_Θ	2160 kg/m²
I_φ	460 kg/m²
b_{sf}	1000 N/m
b_{sr}	1100 N/m
t_f	0.505 m
t_r	0.557 m
l_f	1.4 m
l_r	1.7 m

Pitch movement of sprung mass is

$$\ddot{\theta} = \frac{1}{I_\theta}\Big[k_{sf}l_f\left(-x_{1_{fl}} - x_{1_{fr}} + x_{2_{fl}} + x_{2_{fr}}\right) + k_{sr}l_r\left(-x_{1_{rl}} - x_{1_{rr}} + x_{2_{rl}} + x_{2_{rr}}\right) \\ + b_{sf}l_f\left(-\dot{x}_{1_{fl}} - \dot{x}_{1_{fr}} + \dot{x}_{2_{fl}} + \dot{x}_{2_{fr}}\right) + b_{sr}l_r\left(-\dot{x}_{1_{rl}} - \dot{x}_{1_{rr}} + \dot{x}_{2_{rl}} + \dot{x}_{2_{rr}}\right) - l_f f_{fl} - l_f f_{fr} + l_r f_{rl} + l_r f_{rr}\Big] \tag{2}$$

Roll movement of sprung mass is

$$\ddot{\Phi} = \frac{1}{I_\phi}\Big[k_{sf}t_f\left(-x_{1_{fl}} + x_{1_{fr}} + x_{2_{fl}} - x_{2_{fr}}\right) + k_{sr}t_r\left(-x_{1_{rl}} + x_{1_{rr}} + x_{2_{rl}} - x_{2_{rr}}\right) \\ + b_{sf}t_f\left(-\dot{x}_{1_{fl}} + \dot{x}_{1_{fr}} + \dot{x}_{2_{fl}} - \dot{x}_{2_{fr}}\right) + b_{sr}t_r\left(-\dot{x}_{1_{rl}} + \dot{x}_{1_{rr}} + \dot{x}_{2_{rl}} + \dot{x}_{2_{rr}}\right) - t_f f_{fl} + t_f f_{fr} - t_r f_{rl} + t_r f_{rr}\Big] \tag{3}$$

Vertical displacement of front left wheel is

$$m_{uf}\ddot{x}_{2_{fl}} = -\left[k_{sf}\left(-x_{1_{fl}} + x_{2_{fl}}\right) + f_{tf}\left(x_{2_{fl}} - w_{fl}\right) + b_{sf}\left(-\dot{x}_{1_{fl}} + \dot{x}_{2_{fl}}\right) - f_{fl}\right] \tag{4}$$

Vertical displacement of front right wheel is

$$m_{uf}\ddot{x}_{2_{fr}} = -\left[k_{sf}\left(-x_{1_{fr}} + x_{2_{fr}}\right) + k_{tf}\left(x_{2_{fr}} - w_{fr}\right) + b_{sf}\left(-\dot{x}_{1_{fr}} + \dot{x}_{2_{fr}}\right) - f_{fr}\right] \tag{5}$$

Vertical displacement of rear left wheel is

$$m_{ur}\ddot{x}_{2_{rl}} = -\left[k_{sr}\left(-x_{1_{rl}} + x_{2_{rl}}\right) + k_{tr}\left(x_{2_{rl}} - w_{rl}\right) + b_{sr}\left(-\dot{x}_{1_{rl}} + \dot{x}_{2_{rl}}\right) - f_{rl}\right] \tag{6}$$

Vertical displacement of rear right wheel is

$$m_{ur}\ddot{x}_{2_{rr}} = -\left[k_{sr}\left(-x_{1_{rr}} + x_{2_{rr}}\right) + k_{tr}\left(x_{2_{rr}} - w_{rr}\right) + b_{sr}\left(-\dot{x}_{1_{rr}} + \dot{x}_{2_{rr}}\right) - f_{rr}\right] \tag{7}$$

where

$$x_{1_{fl}} = x_1 + t_f\Phi + l_f\theta \tag{8}$$

$$x_{1_{fr}} = x_1 - t_f\Phi + l_f\theta \tag{9}$$

$$x_{1_{rl}} = x_1 + t_r\Phi - l_r\theta \tag{10}$$

$$x_{1_{rr}} = x_1 - t_r\Phi - l_r\theta \tag{11}$$

3. PID CONTROLLER

The basic form of Proportional Integral Derivative (PID) controller for each wheel of FCM is designed and simulated in the MATLAB/Simulink environment. The control forces of each wheel are derived as given below

Control force for front left and front right wheel are

$$f_{fl} = K_{p1}e_1(t) + K_{i1}\int e_1(t)dt + K_{d1}\frac{de_1(t)}{dt} \tag{12}$$

$$f_{fr} = K_{p2}e_2(t) + K_{i2}\int e_2(t)dt + K_{d2}\frac{de_2(t)}{dt} \tag{13}$$

Control force for rear left and rear right wheel are

$$f_{rl} = K_{p3}e_3(t) + K_{i3}\int e_3(t)dt + K_{d3}\frac{de_3(t)}{dt} \tag{14}$$

$$f_{rr} = K_{p4}e_4(t) + K_{i4}\int e_4(t)dt + K_{d4}\frac{de_4(t)}{dt} \tag{15}$$

Where, [K_{p1}, $K_{p2,}$ $K_{p3,}$ K_{p4}]-Proportional gain for each wheel,

[K_{i1}, K_{i2}, K_{i3}, K_{i4}]-Integral gain for each wheel,
[K_{d1}, K_{d2}, K_{d3}, K_{d4}]-Derivative gain for each wheel,
[e_1(t), e_2(t), e_3(t), e_4(t)]-Error signal of each wheel,
[de_1(t)/dt, de_2(t)/dt, de_3(t)/dt, de_4(t)/dt]-Derivative of error for each wheel

The PID controller parameters are chosen by trial-and-error method. The objective of optimizing the PID parameters is to minimize the cumulative value of BA which is given by

$$\text{minimize}\, J = \sqrt{\frac{1}{T}\int_0^T \|\ddot{x}_1\|^2\, dt} \tag{16}$$

Where J is cost function, $\ddot{x}_1$ is body acceleration and T is total time period.

The PID controller parameters are shown in Table 2.

Table 2. PID Parameters-FCM

	PID_1		PID_2		PID_3		PID_4	
PID		11985		6812		8342		6942
GWO-PID	K_{p1}	16978	K_{p2}	7784	K_{p3}	9528	K_{p4}	7934
Bat-PID		18772		8213		11252		8250
PID		10.2		15.03		13.01		12.53
GWO-PID	K_{i1}	9.04	K_{i2}	12.01	K_{i3}	10.03	K_{i4}	8.53
Bat-PID		7.2		6.98		8.52		5.75
PID		0.04		0.053		0.074		0.037
GWO-PID	K_{d1}	0.025	K_{d2}	0.033	K_{d3}	0.0624	K_{d4}	0.017
Bat-PID		0.0115		0.0124		0.0364		0.0127

4. OPTIMIZATION TECHNIQUES

To increase travel experience and limit the peak value of the ASS, the PID controller parameters are adjusted in this work using the Optimization approach and Grey Wolf Optimization (GWO) methods. (Yuvapriya T and Lakshmi, P 2020). Due to these significant factors, the GWO and Bat optimization methods (Mangaiyarkarasi, P *et al.* 2019) are used for the selected FCM to tune the traditional PID settings that improve the passenger's ride comfort.

4.1 Grey Wolf Optimization (GWO)

The Grey wolf is a member of the Canidae family (Canis lupus). Apex predators are thought to be those at the top of the food chain (Grey wolves). Leaders, often known as alphas. The alpha is in charge of making decisions like where to sleep, where to hunt, when to wake up, etc. (Mech, L. D. 1999). Beta wolves are the next species in the food chain for grey wolves. When making decisions or engaging in other pack activities, the beta wolves support the alpha. Omega, the lowest level of the food chain, serves as the scapegoat. Group hunting is the next level of grey wolves' social structure (Muro, C *et al.* 2011). The mathematical illustration of the wolves' surrounding habit (Mirjalili, S *et al.* 2014) is given by,

$$\vec{X}(t+1) = \vec{X}_p - \vec{A}.\vec{D} \tag{17}$$

$$\vec{D} = \left|\vec{C}.\vec{X}_p - \vec{X}(t)\right| \tag{18}$$

Where, $\vec{X}(t+1)$ indicates where the wolf will go next, $\vec{A}$ and $\vec{C}$ are coefficient vector and $\vec{D}$ is a vector which is given in Equation (18). The random value of $\vec{A}$ is in range of [-2*a*, 2*a*].

$$\vec{A} = 2\vec{a}.\vec{r}_1 - \vec{a} \tag{19}$$

$$\vec{C} = 2.\vec{r}_2 \tag{20}$$

The range of the random variable $\vec{C}$ is [0, 2]. Over iterations, components drop linearly from 2 to 0 to determine the position of the wolf ($\vec{X}_i$ and $\vec{a}$). The wolves can travel to any position between the points to the random vectors ($\vec{r}_1$ and $\vec{r}_2$). The top three answers are regarded as the best answers and the top wolves in the pack, respectively (Mirjalili, S *et al.* 2014). Updates to the position are as follows:

$$\vec{D}_\alpha = \left|\vec{C}_1.\vec{X}_\alpha - \vec{X}\right|, \vec{D}_\beta = \left|\vec{C}_2.\vec{X}_\beta - \vec{X}\right|, \vec{D}_\delta = \left|\vec{C}_3.\vec{X}_\delta - \vec{X}\right| \tag{21}$$

$$\vec{X}_1 = \vec{X}_\alpha - \vec{A}_1.(\vec{D}_\alpha), \vec{X}_2 = \vec{X}_\beta - \vec{A}_2.(\vec{D}_\beta), \vec{X}_3 = \vec{X}_\delta - \vec{A}_3.(\vec{D}_\delta) \tag{22}$$

$$\vec{X}(t+1) = \frac{\vec{X}_1 + \vec{X}_2 + \vec{X}_3}{3} \tag{23}$$

When the victim stops moving, the grey wolves attack it to end the hunt. The following steps explain the GWO algorithm: Updates to the position are as follows:

Step1: Initialize the vectors C, A, a well as the population size $\vec{X}_i$.
Step2: To create wolves at random based on pack size
Step3: In order to determine each search agent's fitness value, use Equation (18) and (20)
Step4: Equations (19) and (22) are used to find the best search agent $\vec{X}_\alpha$, second best agent $\vec{X}_\beta$, and third best agent $\vec{X}_\delta$.
Step5: Using Equation (23), update the hunt agent's location
Step6: To determine each hunt's fitness value
Step7: To alter search agents' value $\vec{X}_\alpha, \vec{X}_\beta$, and $\vec{X}_\delta$
Step8: to halt after the optimization has gone through all possible iterations. Then, if step3 is reached, print the best value otherwise.

In this study, grey wolves' hunting behaviour is mathematically modelled to create and imitate GWO. Iterative calculations are used to calculate the solutions. The location is then changed based on the answer after comparing the current one to the prior one. The best outcomes are then shown. By reducing HDV vibration, the PID parameters are improved for the improvement of travel pleasure.

4.2 Bat Algorithm

The Bat algorithm is a meta-heuristic-based algorithm that was inspired by the ecological hunting habits of bats. (Arora, U & Lodhi, MEA 2016). The following equations update the bats' position and velocity for each time step t (Messaoudi, I & Kamel, N 2019).

$$f_i = f_{\min} + (f_{\max} - f_{\min}).\rho \tag{24}$$

$$v_i^j(t) = v_i^j(t+1) + [x_{gbest}^i - x_i^j(t-1)].f_i \tag{25}$$

$$x_i^j = x_i^j(t-1) + v_i^j(t) \tag{26}$$

First bat population is created, and each bat's position x_i is determined by its velocity (v_i), pulse rate (r_i), volume (A_i), and frequency (f_i). Here, $\rho \epsilon$ [0, 1] stands for a random number and x_{gbest}^i provides the most effective solutions at the moment. One solution is chosen for the local search portion out of all the finest selections currently available. Using random-walk, a fresh solution is generated locally for each bat, and the average bat loudness is provided by A_{mean}^{old}.

$$x_i^{new} = x_i^{old} + \sigma.A_{mean}^{old} \tag{27}$$

The following steps provides a description of the Bat algorithm.

Step1: Initializing frequency (f_i), bat population (x_i), pulse rate (r_i), loudness (A_i), and velocity (v_i)
Step2: To compute each bat's fitness value
Step3: Equation (24) should be used to alter the frequency, and Equations (25) and (26) should be used to update the velocity and location to construct a new solution
Step4: To create a fresh solution based on the best one. Create a local solution as one of the best solutions based on pulse rate.
Step5: To calculate each new bat's fitness value.
Step6: Accept new solutions if their fitness values are lower than those of the existing solutions and their random numbers are lower than the present loudness.
Step7: To decrease the loudness and quicken the pulse rate. Next, arrange the bats to determine the current best solution.
Step8: Print the best solution and its fitness value once the maximum number of generations has been reached. Otherwise, return to step 3.

The solutions are initially computed using the Bat parameters and the objective function. A number of current solutions are introduced by altering these parameters, which then reveal the prey and its location. The best outcomes are ultimately investigated. As a result, the Bat algorithm adjusts the PID settings to improve the suspension system's performance.

Figure 2. Road Profile-Bump Input

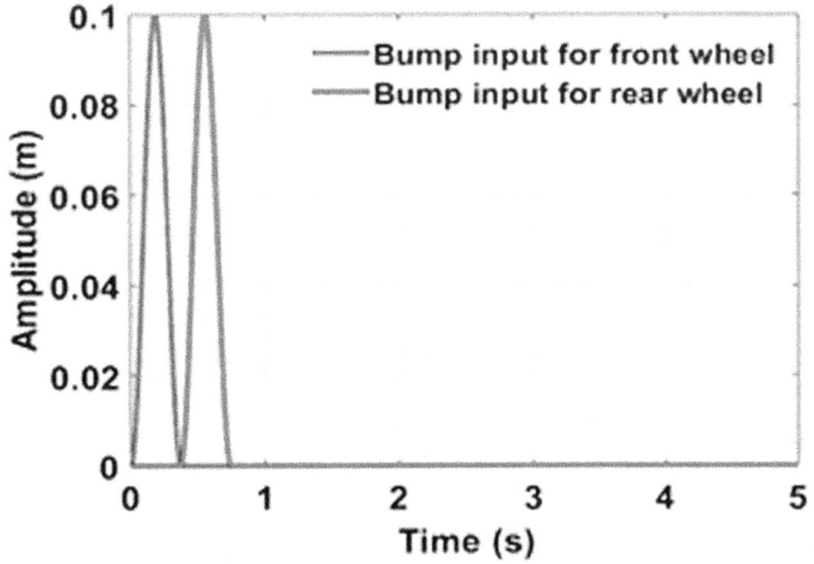

5. SIMULATION RESULTS AND DISCUSSION

In this work, the controllers for FCM with Bump Input (BI), Random Input (RI) and Sinusoidal Input (SI) road profiles are designed and simulated in the MATLAB/Simulink environment. The performances of the optimized PID controllers are compared to conventional PID controller and PSS for three different road profiles.

5.1 Performance Analysis of Optimized PID Controllers with BI

The main motive of this work is to enhance the ride comfort of the FCM. Since the ride quality of the vehicle is directly associated to the Body Acceleration (BA) experienced in the vehicle, therefore, an ultimate aim is to reduce the BA by means of optimization techniques. The optimized PID controller is tested with BI (Figure.2) road profile (Yuvapriya T *et al.* 2018) and their effects on FCM is also analysed. The delay between the front and rear wheel is 0.37 sec. Table.3 lists the parameters taken into account for the GWO and Bat algorithms.

Plotting the fitness function (Figure.3) reveals the rate of convergence and total number of iterations taken into account for the two algorithms. As seen in Figure.3, the Bat optimization converges after 68 iterations whereas the GWO converges after 82 iterations. Additionally, in comparison to GWO, the Bat algorithm provides a minimal fitness value.

Table.4 presents the statistical analysis of the FCM model using the GWO and Bat algorithms. The results show that the Bat algorithm outperforms the GWO-PID controller. Table 4 displays the optimised PID controller parameters.

The time responses of BA, Pitch Acceleration (PA) and Roll Acceleration (RA) with BI are illustrated in Figures.4-6. It has been observed from the Figures that the oscillations are gradually reduced with BI.

Table 3. GWO and Bat optimisation parameters-FCM

Parameter	Notation	GWO	Bat
Iterations	N	200	200
Dimension	D	12	12
Q_{min}	-	-	0
Q_{max}	-	-	2
Loudness	A_i		0.6
Pulse rate	r_i	-	0.7
Search agents	-	30	30
Random vectors	(r_1, r_2)	[0 1]	-

Similarly, controlled responses of BA, PA and RA are depicted in Figures.7-9. The clear observations has been made from these Figures that it reaches steady state output after 3 seconds.

Further the worst transients with PSS are greatly reduced by introducing GWO-PID controller and Bat-PID controller. Evidently, the Bat-PID controller is significantly decreases the oscillation when compared to GWO-PID controller, PID controller and PSS. Figures.8 and 9 depict the PA and RA of Bat-PID controller, GWO-PID controller and PID controller. Later, the oscillation of optimized PID is rapidly decreases compared to conventional PID controller. The control forces of front left, front right, rear left and rear right wheels under Bat-PID controller, GWO-PID controller and PID controller are illustrated in Figures.10-13, respectively. From these figures, the Bat-PID controller has taken more control effort to suppress vibration in the ASS compared to GWO-PID controller, PID controller and PSS.

For further assessing the enhancement of ride comfort of the vehicle, the RMS values of BA is computed for BI and it is shown in Table.5. The percentage of reduction in RMS values is computed with

Figure 3. Fitness function plot-FCM-BI

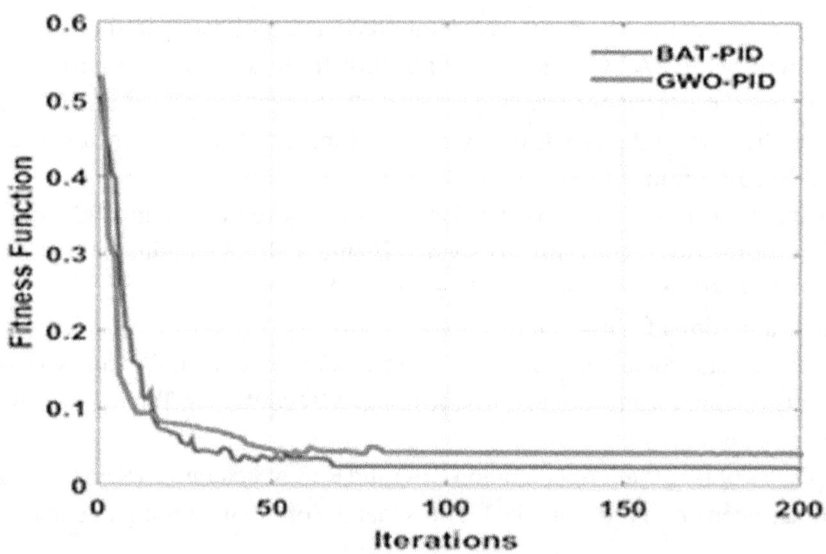

Figure 4. Time response of BA without controller-FCM-BI

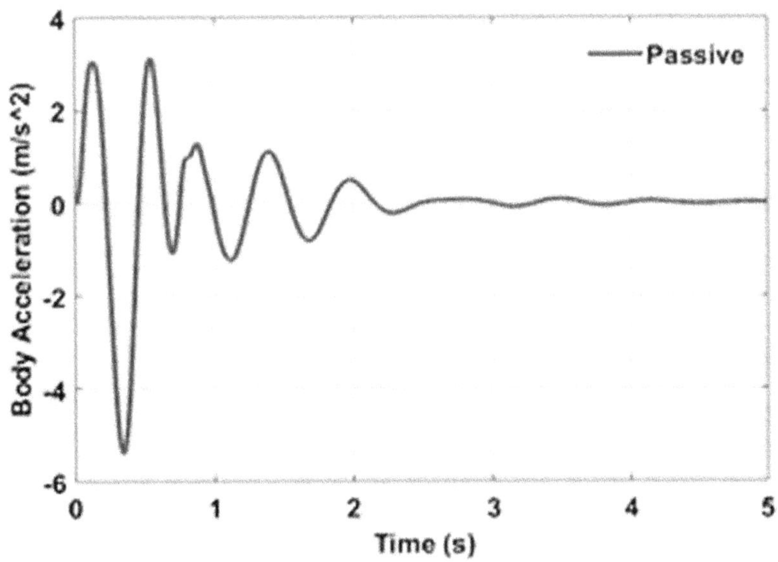

respect to PSS and shown in Figure.14. Clearly, it is observed from the Figure.14; the BA of Bat-PID controller is reduced by 51.3%, PA by 67.6% and RA by 54.7%. From the time domain analysis, Bat-PID controller has the potency to enhance the ride comfort in terms of reduction in RMS values, FWRMS and VDV of BA. Hence, the numerical and simulation results of Bat-PID controller offer significant improvement in vibration control over both GWO-PID controller and conventional PID controller.

5.2 Performance Analysis of Optimized PID Controllers with RI

The performances of the optimized PID controllers are tested with class E poor road surface (Yuvapriya T *et al.* 2020). Since the real road surface is not in uniform i.e. random in nature. The ISO 8608 class E RI road profile (shown in Figure.15) is chosen for the investigation of optimized PID controllers based on reduction in RMS values and oscillation. Hence the GWO and Bat optimized PID parameters used in the BI remains same for RI to prove efficacy of optimized PID controller.

The responses of BA, PA and RA of RI with PSS are illustrated in Figures.16-18. From these Figures, the oscillation level of PSS is not satisfactory to enhance the ride quality of the FCM.

Table 4. Statistical analysis of GWO and Bat algorithms-FCM

PARAMETER	GWO	Bat
Mean	0.0588	0.0470
Standard Deviation	0.0731	0.0581
Best Fit	0.0385	0.0225
Worst Fit	0.5285	0.5019

Figure 5. Time response of PA without controller-FCM-BI

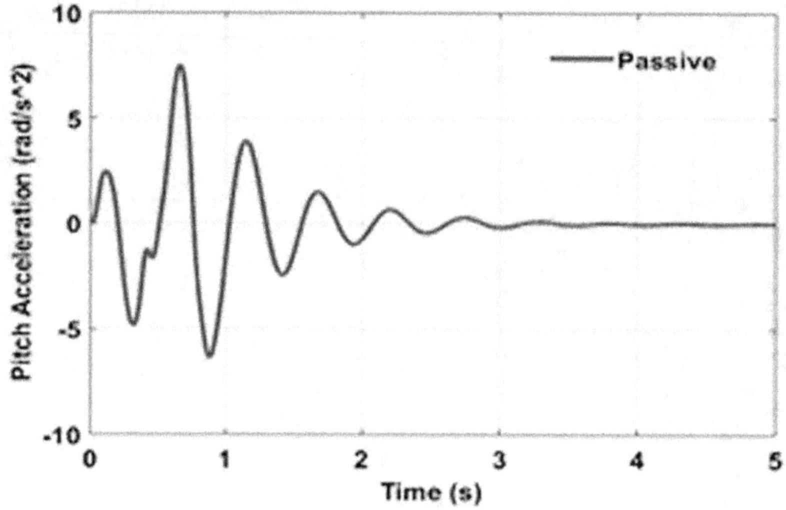

Figures.19-21 demonstrates the time trajectories of BA, PA and RA with optimized PID controller and conventional PID controller. Figure.19 shows BA of controlled responses of PID, GWO-PID controller and Bat-PID controller. It is noted that the oscillation level of BA is gradually decreased with Bat-PID controller compared to GWO-PID controller and PID controller. Also, PA (Figure.20) and RA (Figure.21) of Bat-PID controller shows little oscillation when compared to GWO-PID controller and PID controller.

Figure 6. Time response of RA without controller-FCM-BI

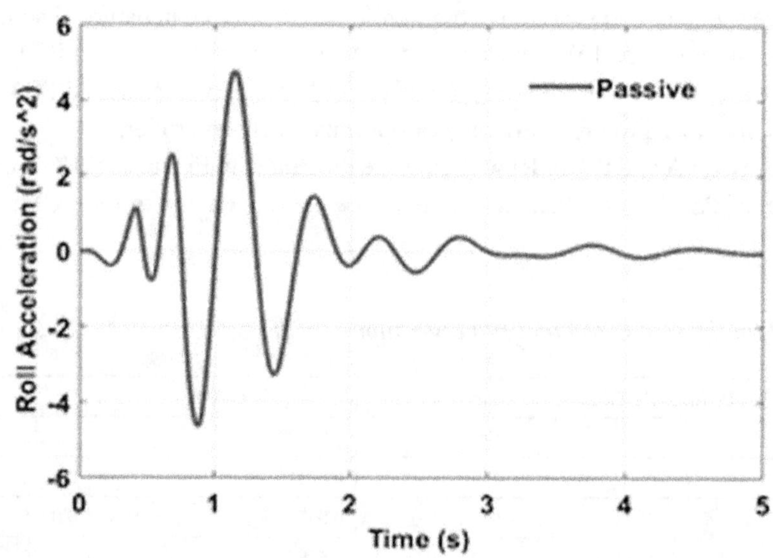

Figure 7. Time response of BA with controllers-FCM-BI

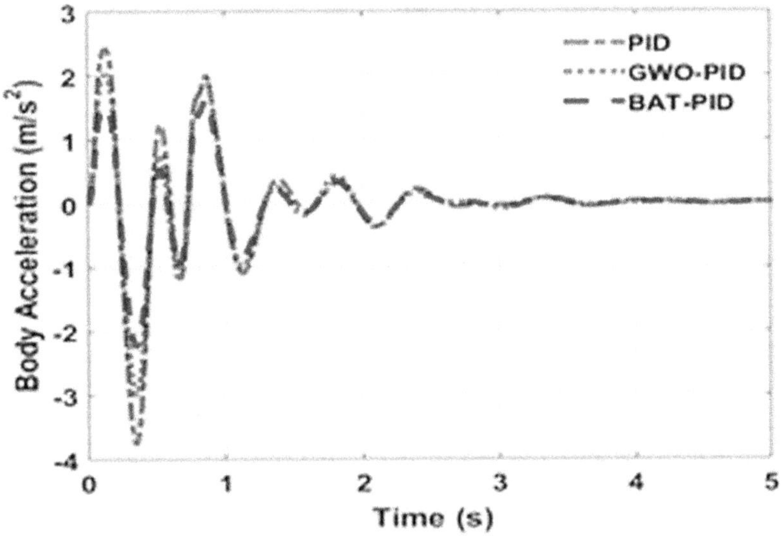

The control efforts taken by Bat-PID controller is comparatively high compared to GWO-PID controller and PID controller. The corresponding control forces for front left and right are shown in Figures.22 and 23. Similarly, the control forces generated for rear left and right are shown in Figures.24 and 25.

For further assessing the enhancement of ride comfort of the vehicle, the RMS values are computed for RI and it is depicted in Table.6. With relation to PSS, the percentage of reduced RMS values is calculated (Figure.26). Thus, the Bat-PID suppresses vibration to a minimum level and maximum reduction

Figure 8. Time response of PA with controllers-FCM-BI

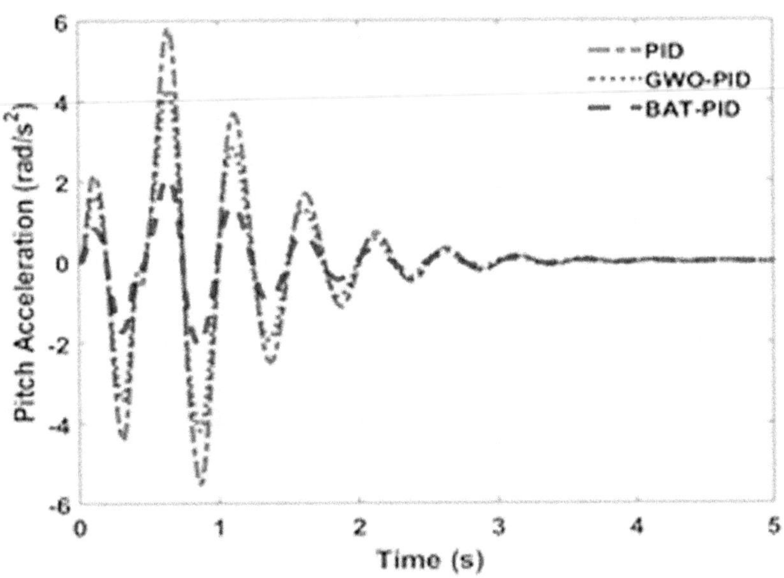

in RMS values. Hence, the numerical and simulation results of Bat-PID controller provide satisfactory vibration control over both GWO-PID controller and conventional PID controller.

Figure 9. Time response of RA with controllers-FCM-BI

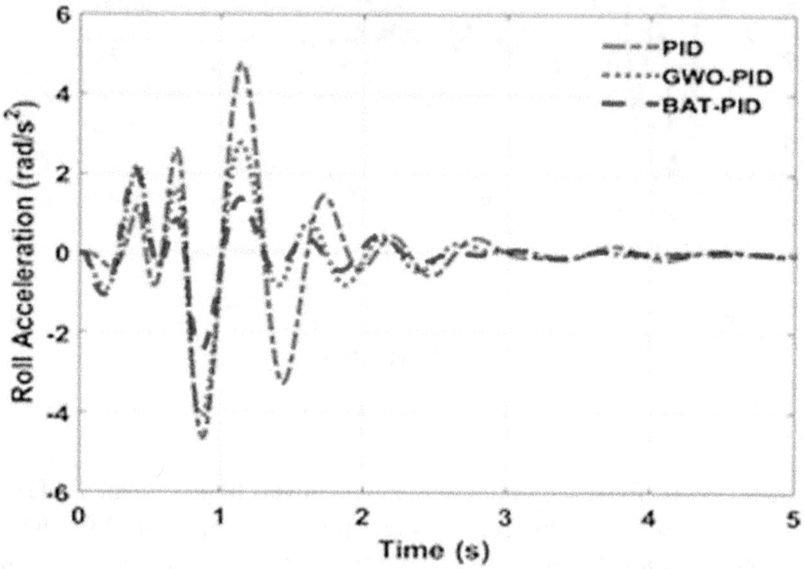

Figure 10. Control forces of front left wheel produced by FCM-BI

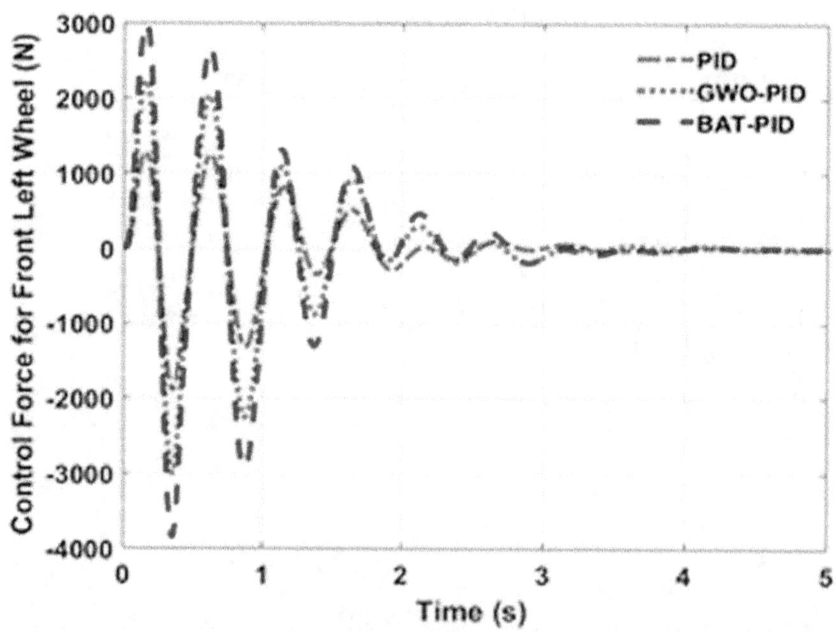

Figure 11. Control forces of front right wheel produced by FCM-BI

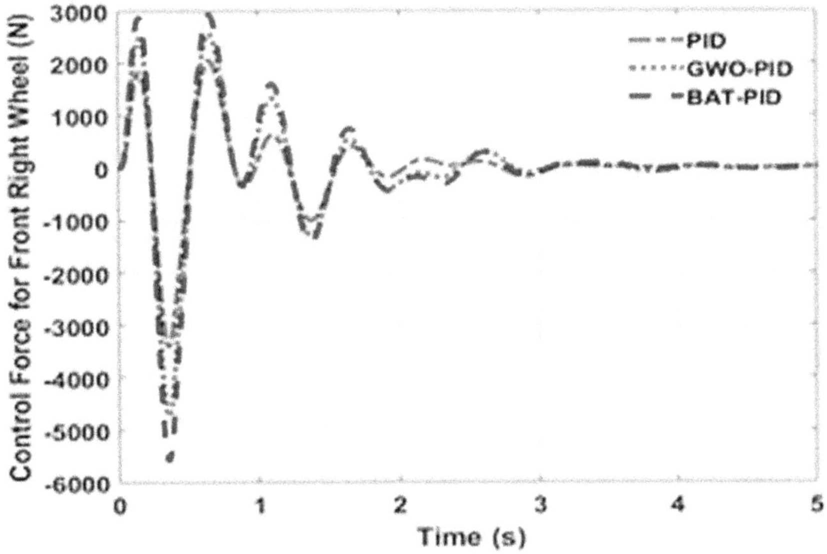

5.3 Performance Analysis of Optimized PID Controllers with SI

The performances of the optimized PID controllers are analysed with sinusoidal input road surface (shown in Figure.27). The GWO and Bat-PID controller parameters used in the BI remains same for SI to investigate the enrichment of optimized PID controllers.

Figure 12. Control forces of rear left wheel produced by FCM-BI

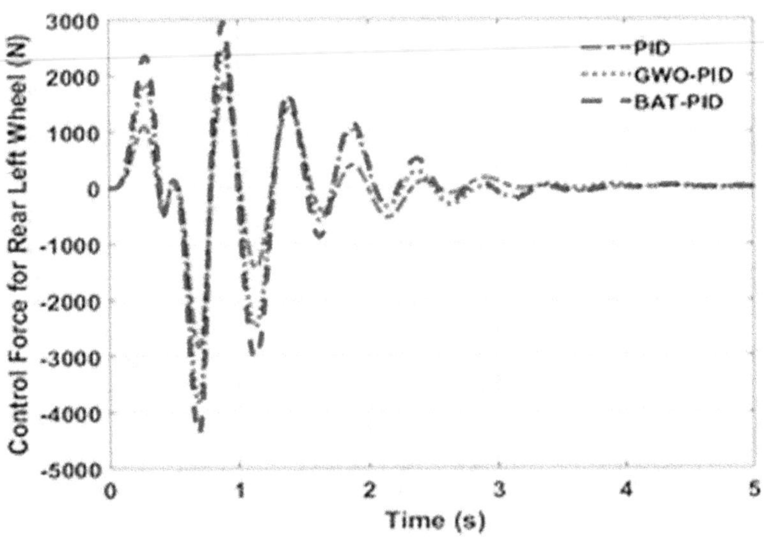

Figure 13. Control forces of rear right wheel produced by FCM-BI

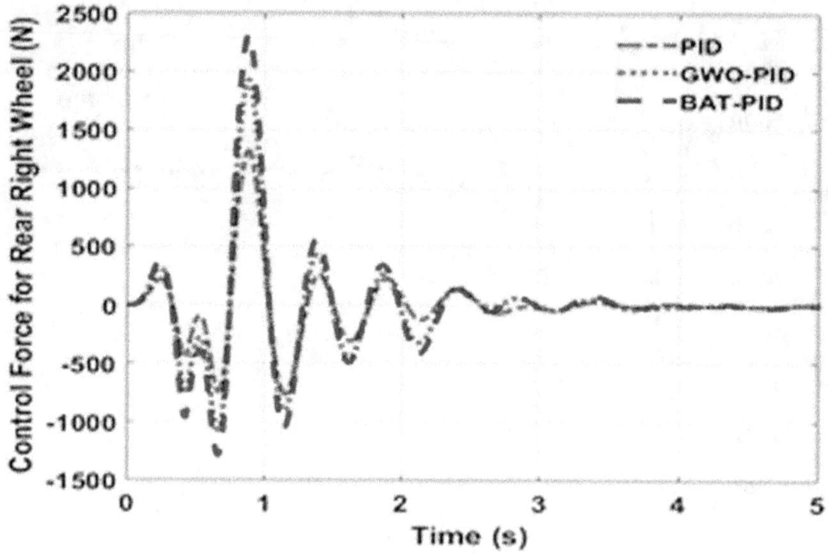

The responses of BA, PA and RA of SI with PSS are shown in Figures.28-30. From these Figures, the maximum amplitude of BA, PA and RA responses are not satisfactory to enhance the ride quality of the FCM.

Figure 14. Reduction in RMS values of BA-BI Road Profile

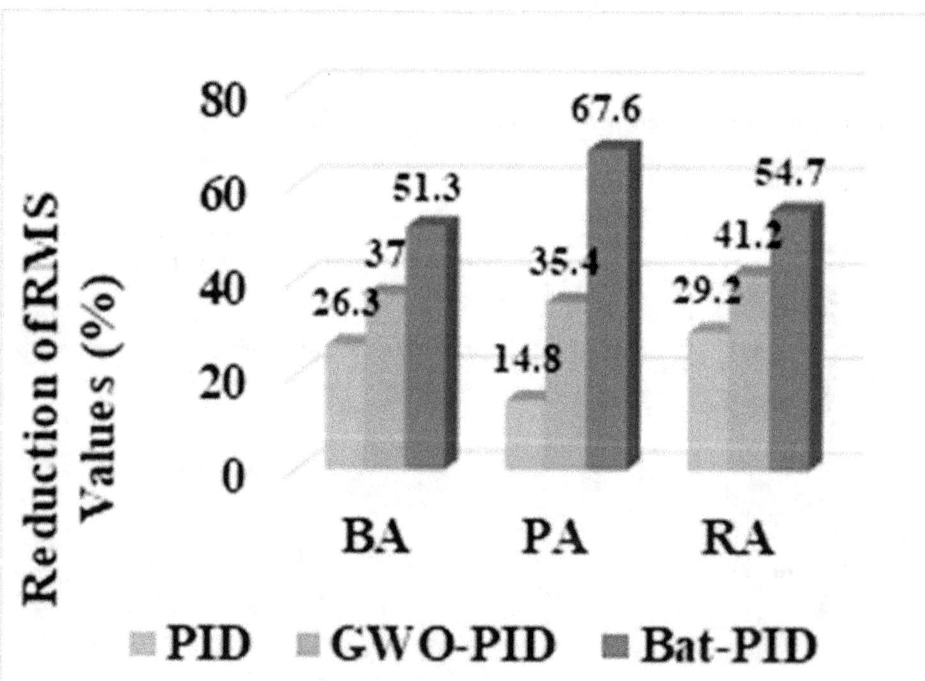

Table 5. Performance Indices of FCM-BI

Controllers	BA (m/s^2)	PA (rad/s^2)	RA (rad/s^2)	FWRMS of BA (m/s^2)	Level of Dis-comfort	VDV of BA ($m/s^{1.75}$)
Passive	1.09	1.91	1.33	0.602	FU	1.595
PID	0.803	1.626	0.942	0.433	LU	1.101
GWO-PID	0.69	1.234	0.782	0.391	LU	0.963
BAT-PID	**0.53**	**0.618**	**0.603**	**0.312**	**NU**	**0.768**

Figures.31-33 demonstrates the simulation results of BA, PA and RA with optimized PID and conventional PID controllers. Figure.31 shows BA of controlled responses of PID controller, GWO-PID controller and Bat-PID controller. It is noticed from the Figure that the amplitude of BA is gradually decreased with Bat-PID controller compared to GWO-PID controller and PID controller. Similarly, PA (Figure.32) and RA (Figure.33) of Bat-PID controller show gradual reduction when compared to GWO-PID controller and PID controller. The control effort generated by Bat optimised PID controller is comparatively higher than that of GWO-PID controller and PID controller. The corresponding control forces for front left and right are shown in Figures.34 and 35, respectively. Similarly the control forces of rear left and right are shown in Figures.36 and 37, respectively.

For further assessing the enrichment in ride comfort of the vehicle, the RMS values are computed for SI as shown in Table.7. The percentage of reduction of BA with the conventional PID controller, Bat-PID controller and GWO-PID controller are 13%, 24% and 41% (Figure.38). The Bat-PID controller provides significant reduction in RMS values. Hence, the numerical and simulation results of Bat-PID controller provide adequate vibration control over both GWO-PID controller and conventional PID controller.

Figure 15. Road Profile-Random Input

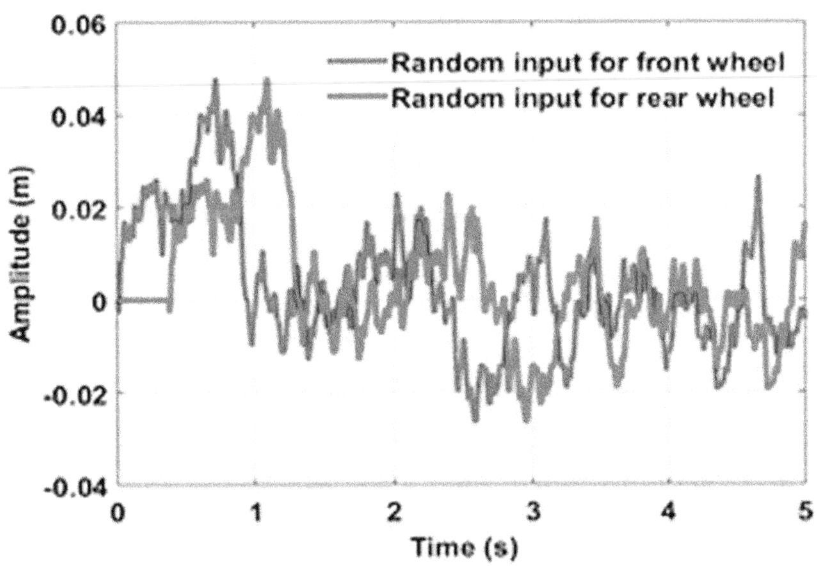

Figure 16. Time response of BA without controller-FCM-RI

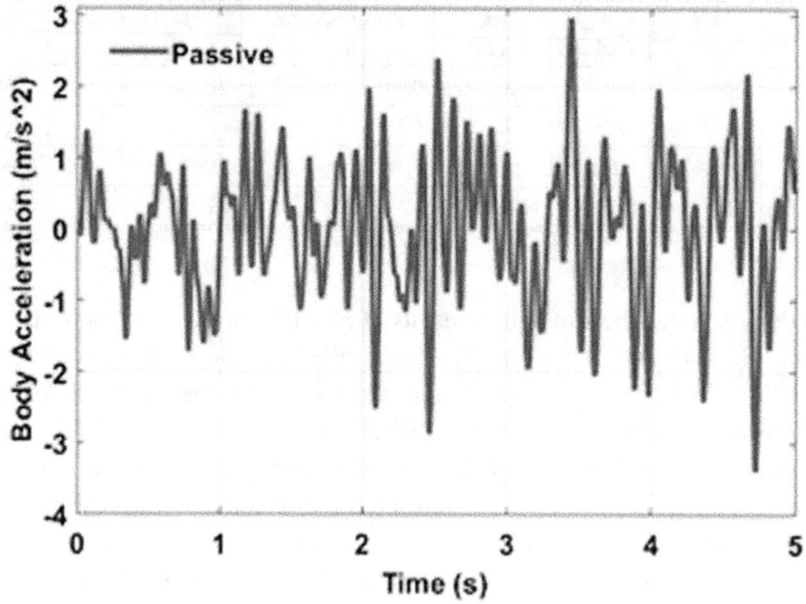

5.4 Ride Comfort Analysis

The FWRMS and VDV of BA based on ISO 2631 standard are computed to analyse the ride quality of the

Figure 17. Time response of PA without controller-FCM-RI

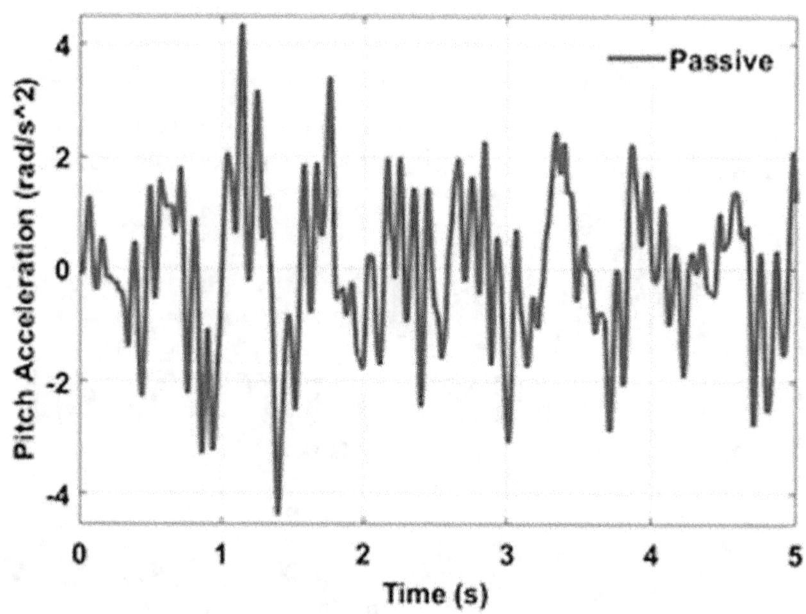

Figure 18. Time response of RA without controller-FCM-RI

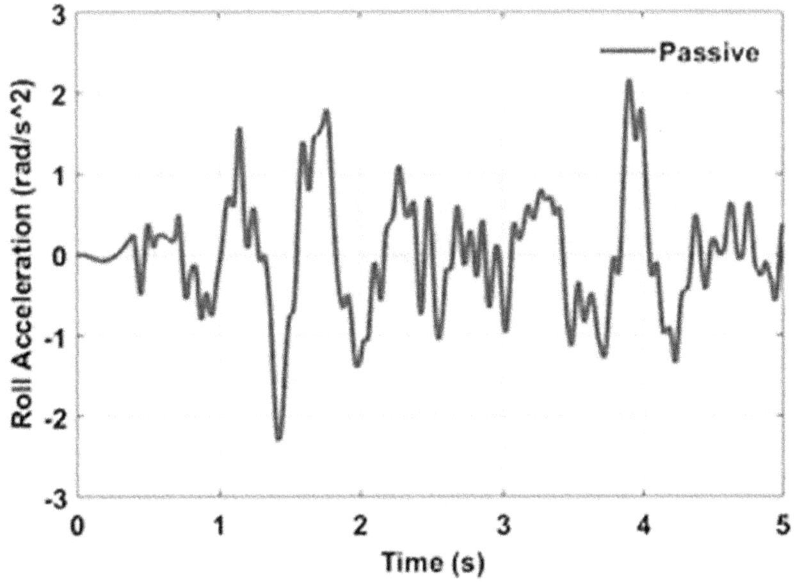

passenger which is shown in Tables.5-7. The health criterion of the passengers is ensured by ride comfort analysis with BI, RI and SI. The VDV of BA is gradually reduced with Bat-PID controller compared to GWO-PID controller, PID controller and PSS. The comfortable level of passenger is evaluated with FWRMS. It is clearly evident from the Tables.5 and 6, the FWRMS values of BA under PID and GWO-

Figure 19. Time response of BA with controllers-FCM-RI

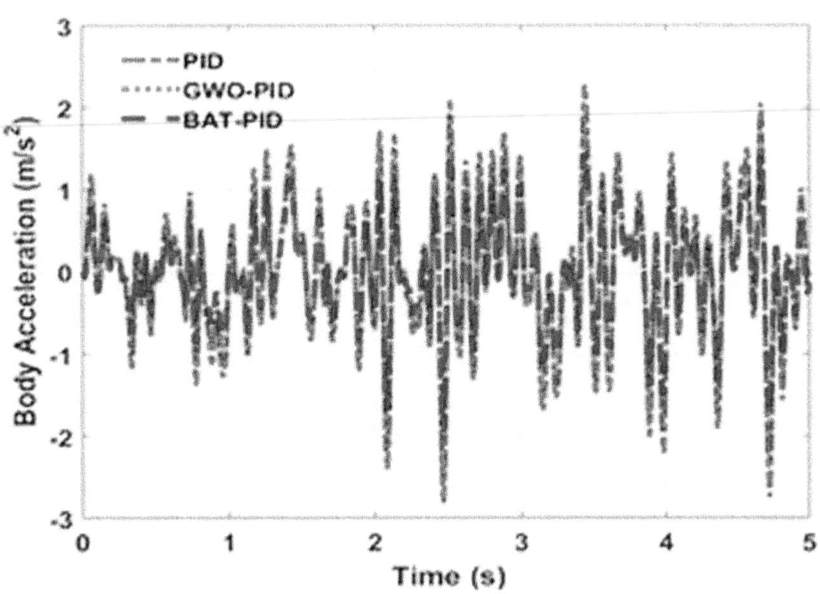

Figure 20. Time response of PA with controllers-FCM-RI

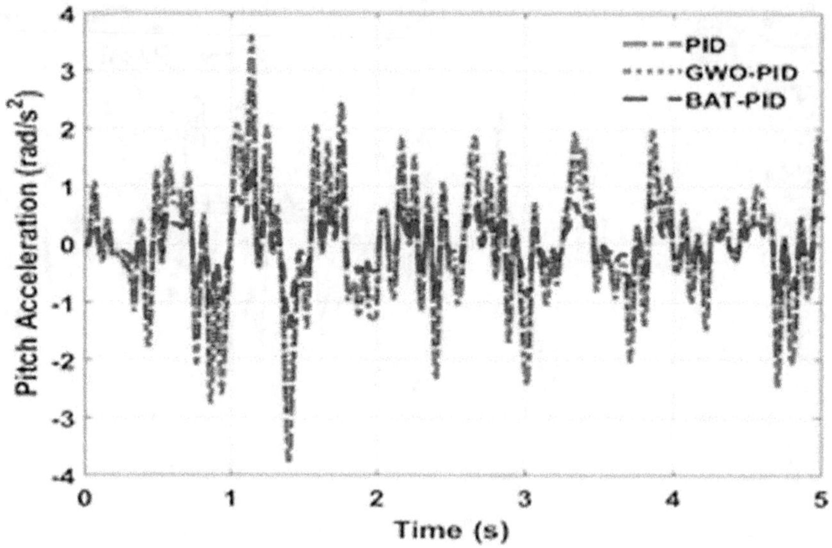

PID controller shows 'Little Uncomfortable' whereas Bat-PID controller offers 'Not Uncomfortable' for both BI and RI. Also, the analysis is carried out for SI which shows 'Extremely Uncomfortable' for PSS and optimized PID controllers. Hence, the level of comfort has been improved with Bat-PID controller for BI and RI when compared to GWO-PID controller, PID controller and PSS.

Figure 21.Time response of RA with controllers-FCM-RI

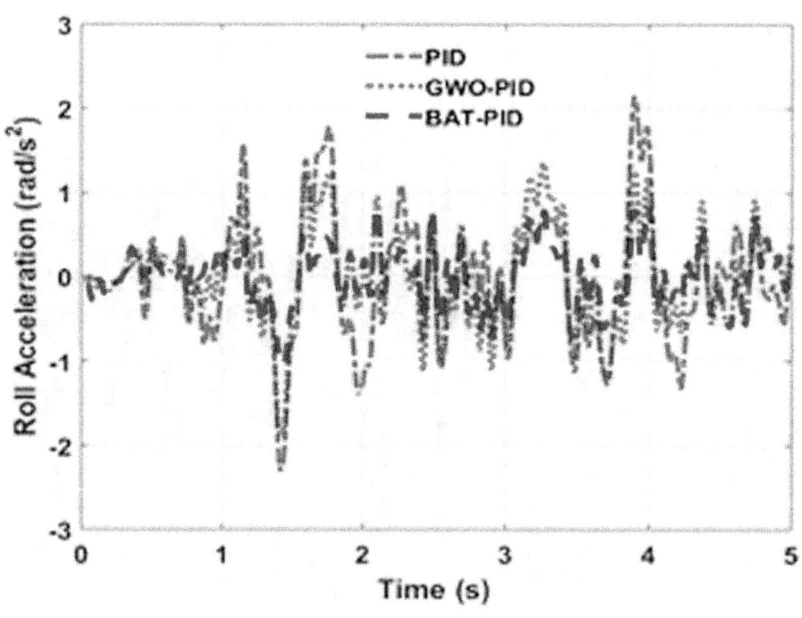

Figure 22. Control forces of front left wheel produced by FCM-RI

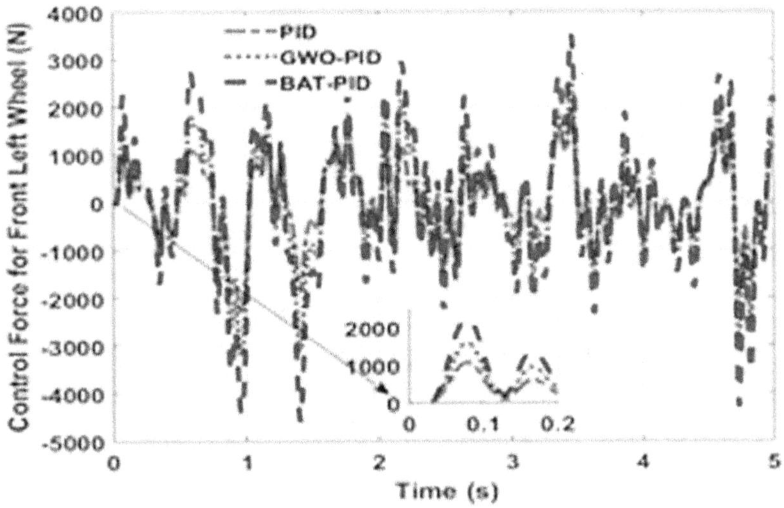

5.5 Power Spectral Density Analysis

For BI, RI, and SI, the Power Spectral Density (PSD) of BA is illustrated. Figure.39 shows that when compared to GWO-PID controller, PID controller, and PSS with BI, Bat-PID controller diminished the BA, which is in the human sensitive frequency range (4–8Hz). Similarly, the PSD analysis has been carried out for RI (Figure.40) and SI (Figure.41). The magnitude of BA for Bat-PID controller is considerably

Figure 23. Control forces of front right wheel produced by FCM-RI

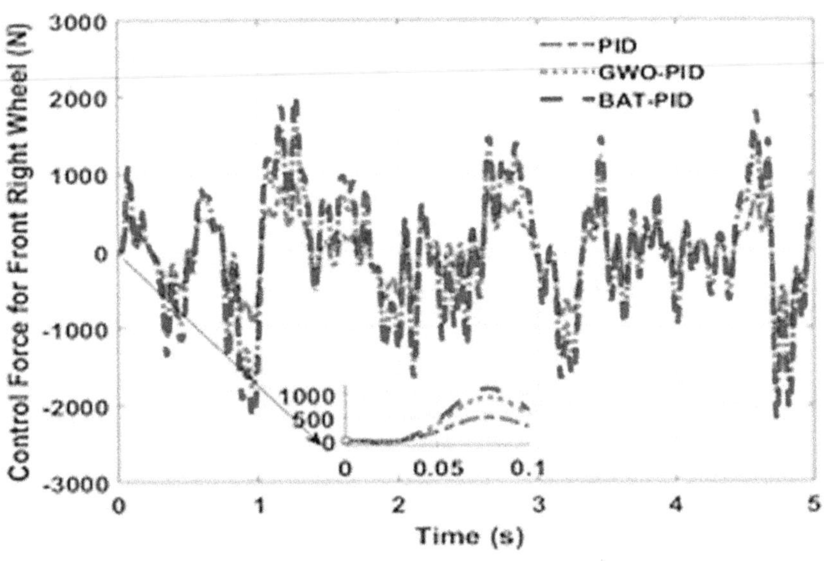

reduced in comparison with GWO-PID controller, PID controller and PSS. Hence the ride comfort and health criterion of the passenger is enhanced by Bat-PID controller.

Figure 24. Control forces of rear left wheel produced by FCM-RI

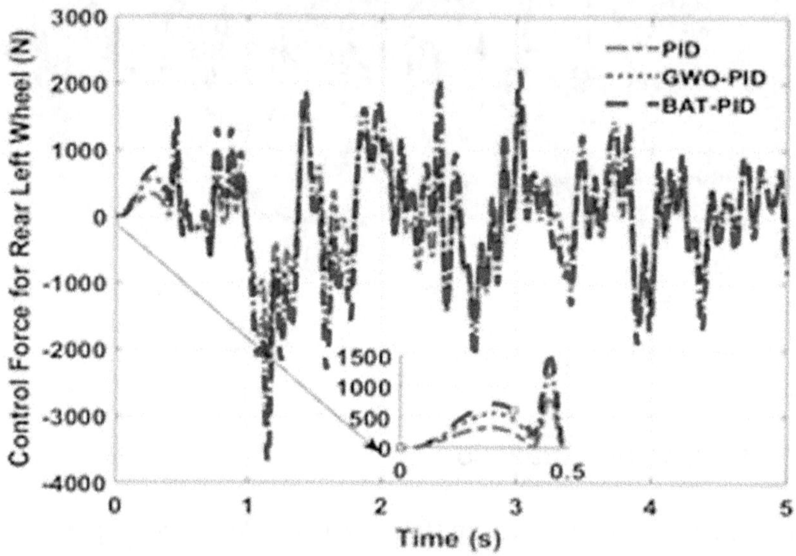

Figure 25. Control forces of rear right wheel produced by FCM-RI

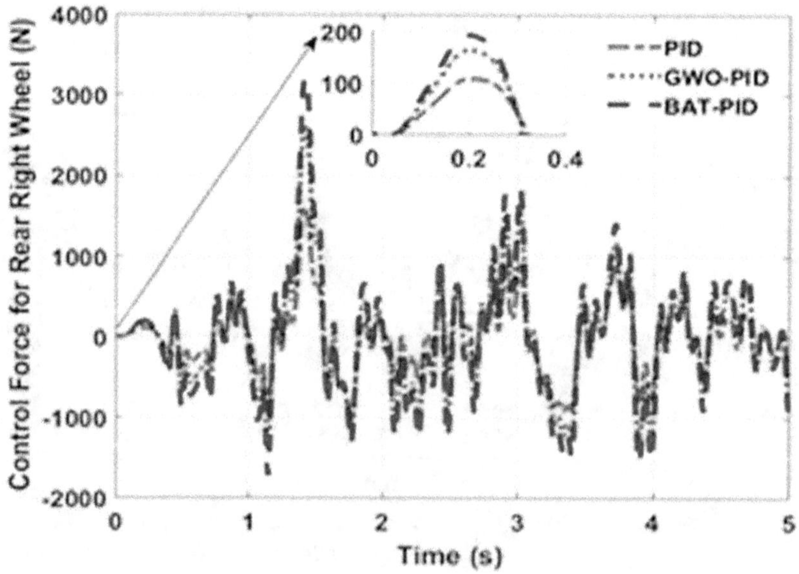

Table 6. Performance Indices of FCM-RI

Controllers	BA (m/s^2)	PA (rad/s^2)	RA (rad/s^2)	FWRMS of BA (m/s^2)	Level of Discomfort	VDV of BA ($m/s^{1.75}$)
Passive	0.99	1.35	0.73	0.546	FU	1.05
PID	0.86	1.1	0.584	0.431	LU	0.841
GWO-PID	0.75	0.83	0.513	0.369	LU	0.719
Bat-PID	**0.58**	**0.414**	**0.36**	**0.286**	**NU**	**0.558**

5.6 Speed Analysis

The percentages of reduction in RMS values are computed for every 10 kmph variations and the responses are plotted for three different road profiles (Figures.42-44). From the Figure.42, RMS values of BA are significantly reduced with Bat-PID controller from 10 to 40 kmph compared to other GWO-PID controller and PID controller with BI. After 40 kmph, the RMS values of BA are drastically decreased from 40 to 60 kmph while using optimised PID controllers. The RMS values of the optimised PID controllers are then gradually decreased from 60 to 100 kmph. Similarly, the decrease in BA's RMS values is calculated and plotted for RI (Figure.43) and SI (Figure.44) road profiles. It is clearly demonstrated in Figures.42-44, the performances of optimized PID controllers are slowly decaying when the vehicle travels at high speed.

5.7 Comparative Analysis

To evaluate the potency of the optimized PID controllers, the results are compared with existing LQR technique. The BI road profile (Figure.45) is simulated for 10kmph and the Q and R weight matrices are considered from (Darus, R& Sam, Y 2009).

Figure 26. Reduction in RMS values of BA-RI Road Profile

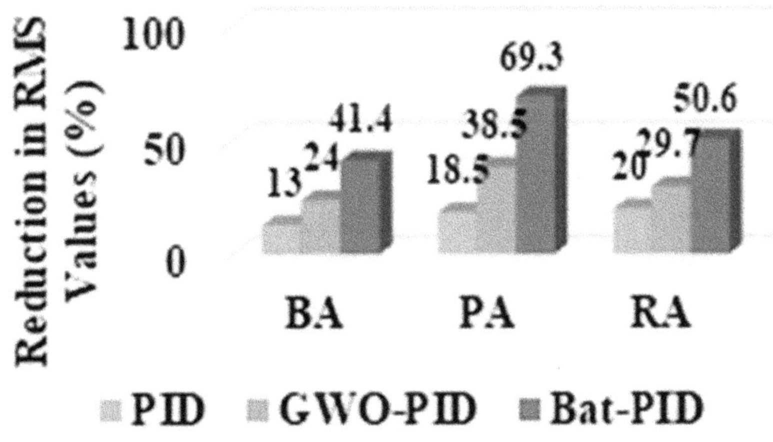

Figure 27. Road Profile-Sinusoidal Input

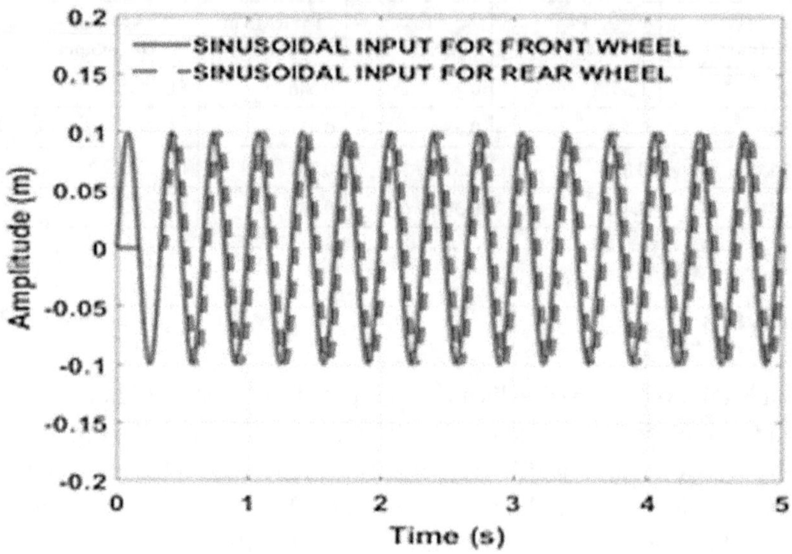

The simulation result of Body Displacement (BD) with LQR is considered for comparative analysis with optimized PID controllers. The time response of BD (Darus, R & Sam, Y 2009) is plotted in Figure.46. Further BA is taken into account to test the ride comfort of the passenger (Figure.47). It can be noted from the Figure.48 and 49, the oscillation of BD and BA are greatly reduced with optimized PID controllers in comparison with existing LQR technique. The corresponding control force of each wheel is illustrated in Figures.50-53.

Figure 28. Time response of BA without controller-FCM-SI

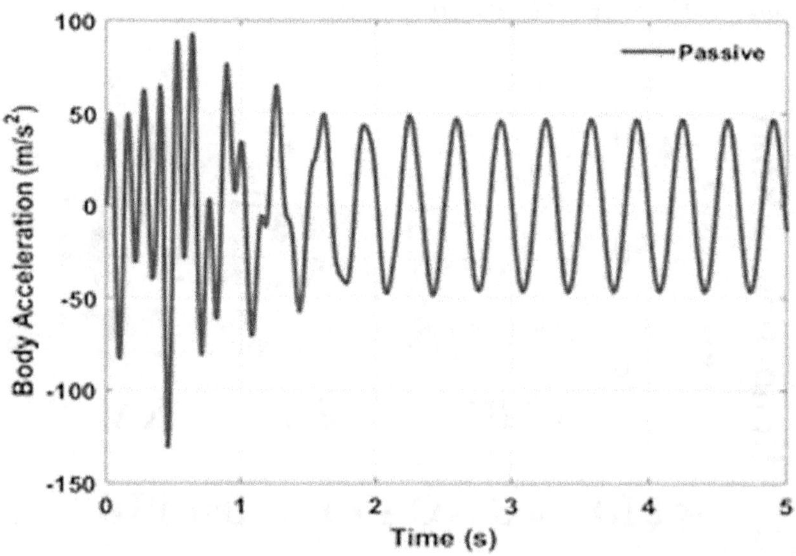

Figure 29. Time response of PA without controller-FCM-SI

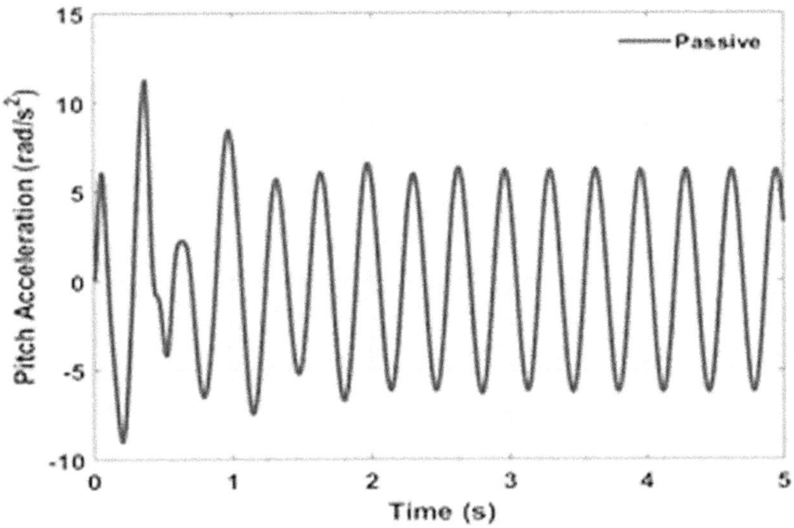

The RMS values of BD and BA are computed for passive, LQR, PID controller and optimized PID controllers (Table.8). The percentage of reduction with LQR is 43.4%, whereas PID controller is 62.2%, GWO-PID controller is 75% and Bat-PID controller is 86%. The RMS, FWRMS and VDV values of Bat-PID controller is significantly reduced in comparison with GWO-PID controller, LQR and PID controller.

The Bat-PID controller improves ride comfort from Uncomfortable to Not Uncomfortable level. The PSD of BA (Figure.54) is plotted for existing LQR and optimized PID controllers. Compared to the

Figure 30. Time response of RA without controller-FCM-SI

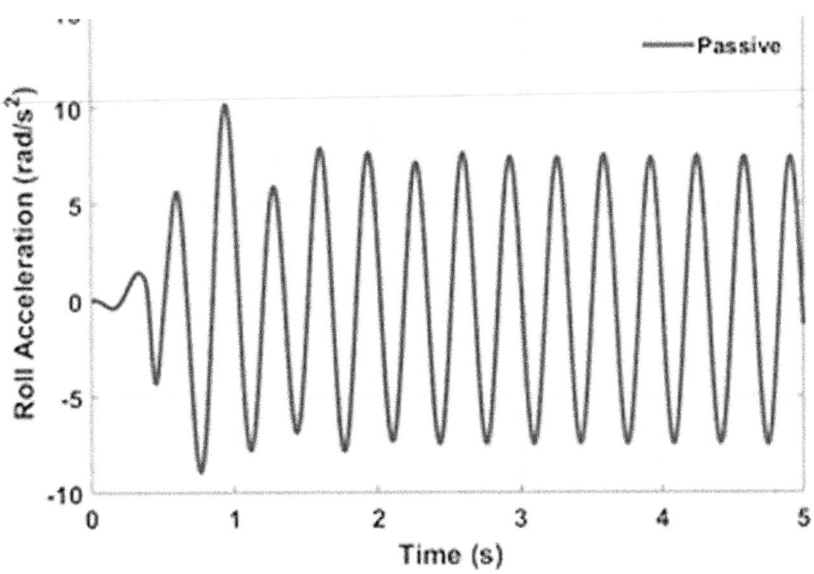

conventional LQR, PID, and GWO-PID controllers, the magnitude of BA is gradually reduced with the use of the Bat-PID controller

Figure 31. Time response of BA with controllers-FCM-SI

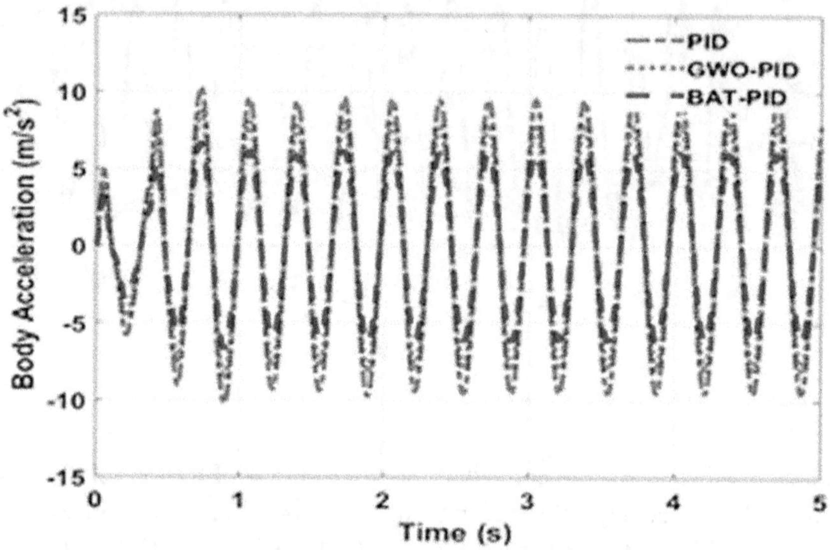

Figure 32. Time response of PA with controllers-FCM-SI

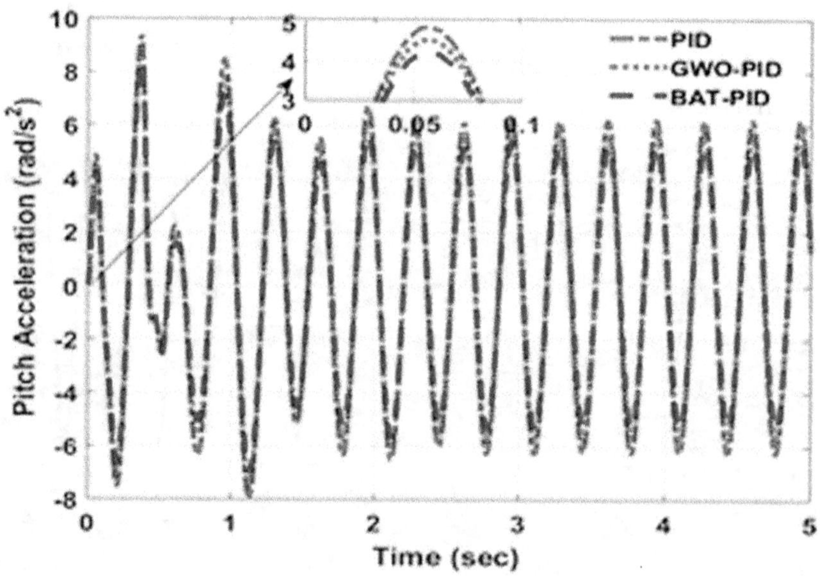

Figure 33. Time response of RA with controllers-FCM-SI

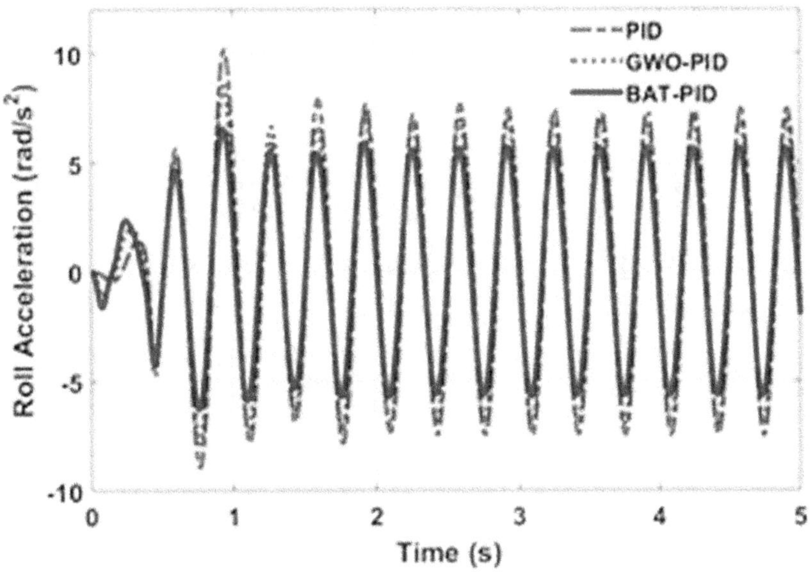

6. CONCLUSION

In this study, seven degrees of freedom FCM is considered for vibration control with and without control techniques. To improve the ride quality of the vehicle, the PID controller, Bat-PID controller, and GWO-PID controller are installed on an FCM. The simulation findings show that the failure to adapt to the erratic driving conditions caused the FCM with PSS to suddenly overshoot and oscillate. The suspen-

Figure 34. Control forces of front left wheel produced by FCM-SI

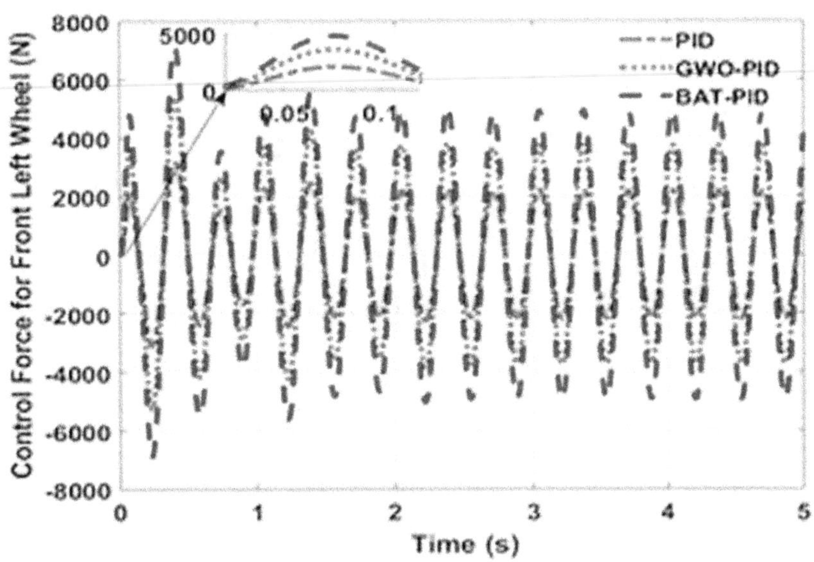

Figure 35. Control forces of front right wheel produced by FCM-SI

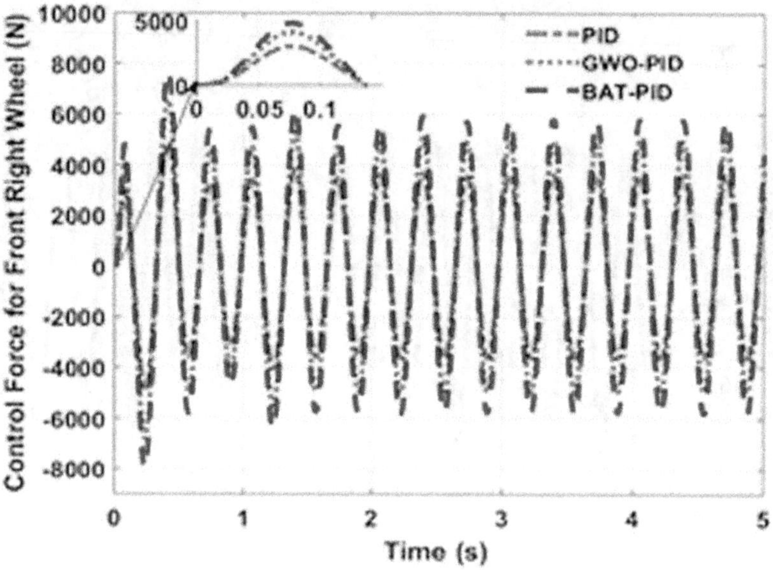

sion system's vibration and ongoing oscillation are reduced by the Bat-PID controller. The quantitative analysis from the Bat-PID controller that display the minimum RMS, FWRMS, and VDV values for the BA. Therefore, in terms of numerical and statistical analysis, the Bat-PID controller offers higher riding comfort. Consequently, it can be inferred from the simulation and numerical findings that the Bat-PID controller provides superior ride comfort (BI and RI) than the GWO-PID controller, the PID controller,

Figure 36. Control forces of rear left wheel produced by FCM-SI

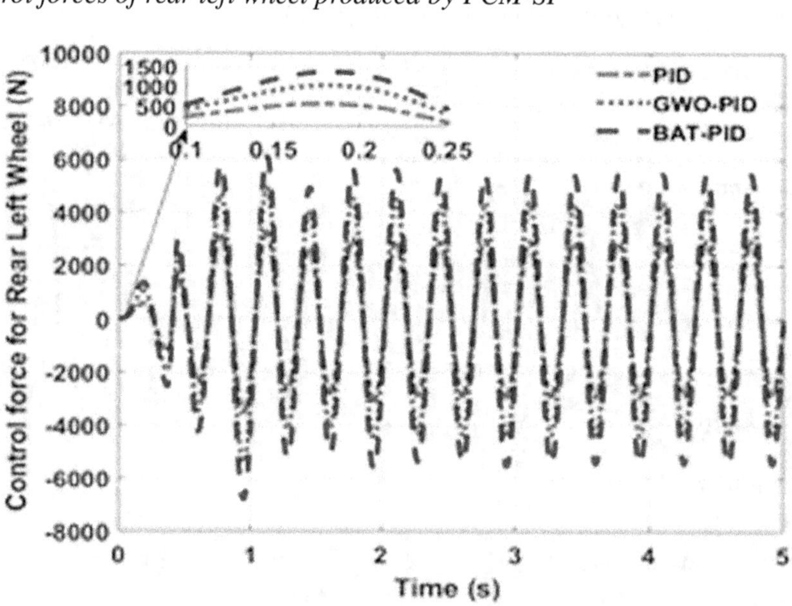

Figure 37. Control forces of rear right wheel produced by FCM-SI

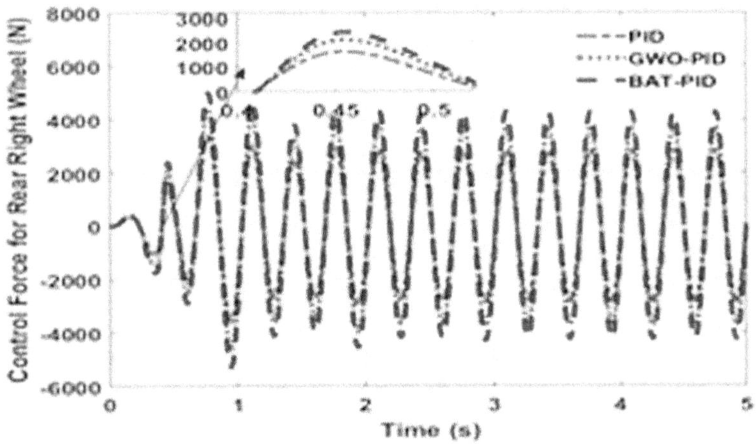

Table 7. Performance Indices of FCM-SI

Controllers	BA (m/s^2)	PA (rad/s^2)	RA (rad/s^2)	FWRMS of BA (m/s^2)	Level of Discomfort	VDV of BA ($m/s^{1.75}$)
Passive	8.82	4.81	5.08	6.22	Extremely Uncomfortable (EU)	10.54
PID	6.52	4.56	4.68	4.59	EU	7.79
GWO-PID	5.45	4.14	4.25	3.84	EU	6.52
Bat-PID	**4.162**	**3.82**	**3.84**	**2.94**	**EU**	**4.99**

Figure 38. Reduction in RMS values of BA-SI Road Profile

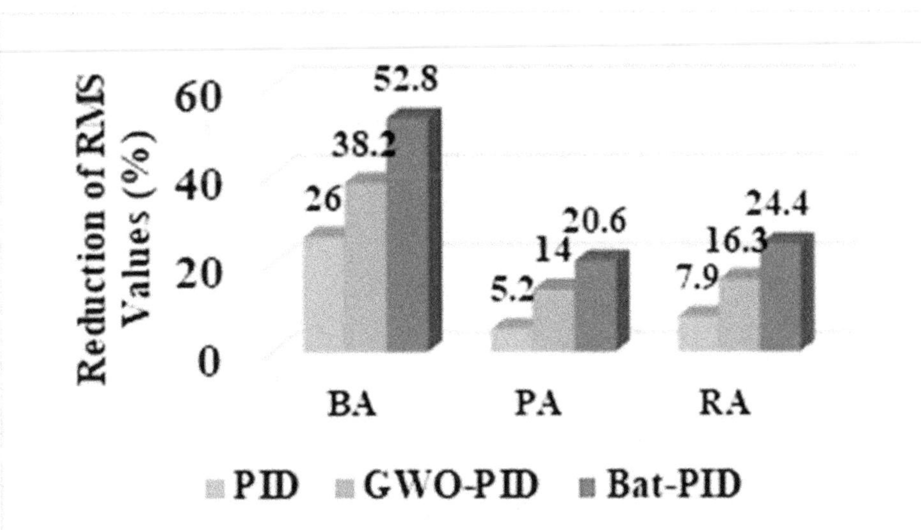

and the PSS. The riding comfort of standard PID controllers, PSS, and optimised PID controllers for SI road profiles all display "Extremely Uncomfortable." Future work may include adaptive and sliding mode controllers to improve the ride comfort of FCM with SI.

Figure 39. PSD of BA-FCM-BI

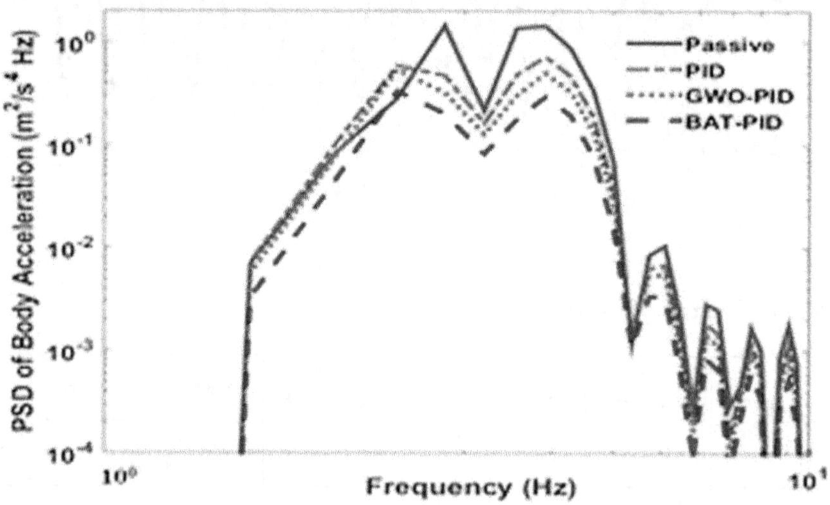

Figure 40. PSD of BA-FCM-RI

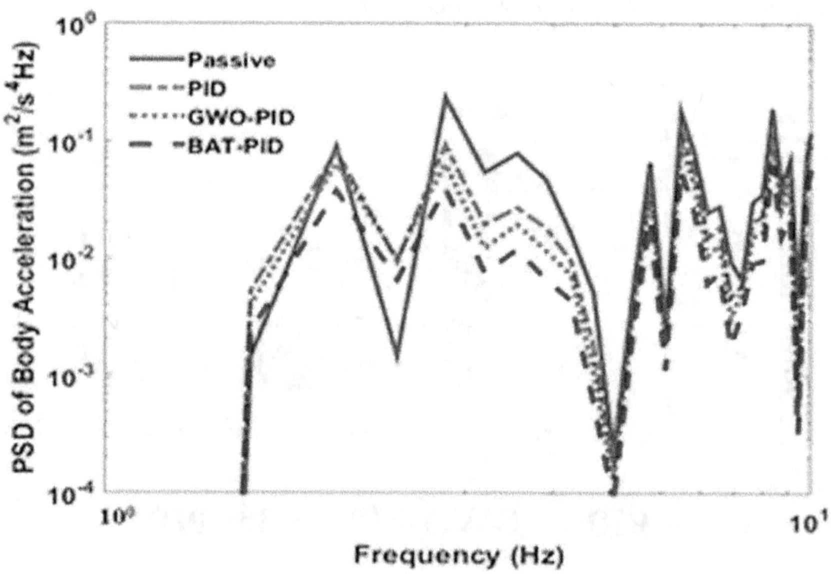

Figure 41. PSD of BA-FCM-SI

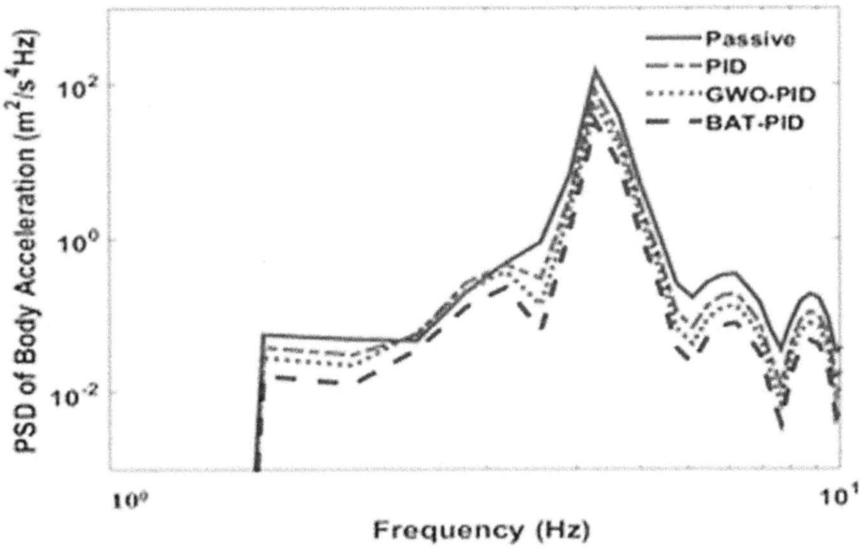

FUNDING

This research received no specific grant from any funding agency in the public, commercial, or not-for-profit sectors.

Figure 42. Speed analysis of optimized PID controllers-FCM-BI

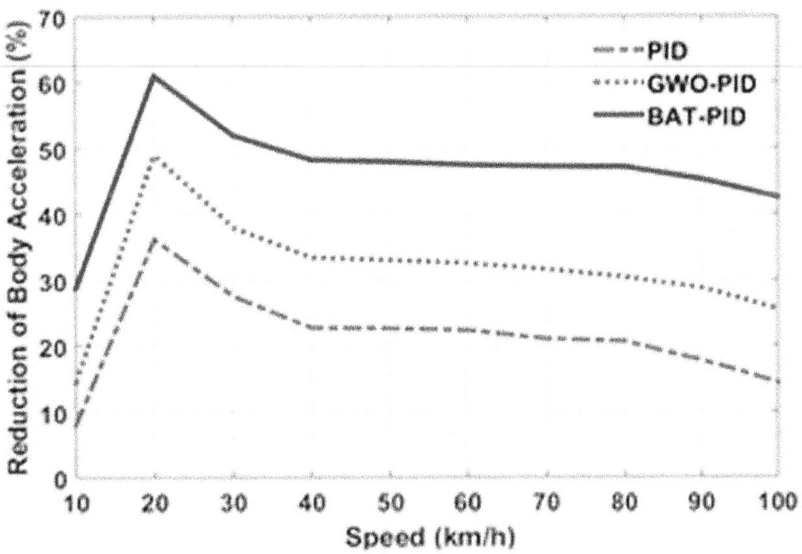

Figure 43. Speed analysis of optimized PID controllers-FCM-RI

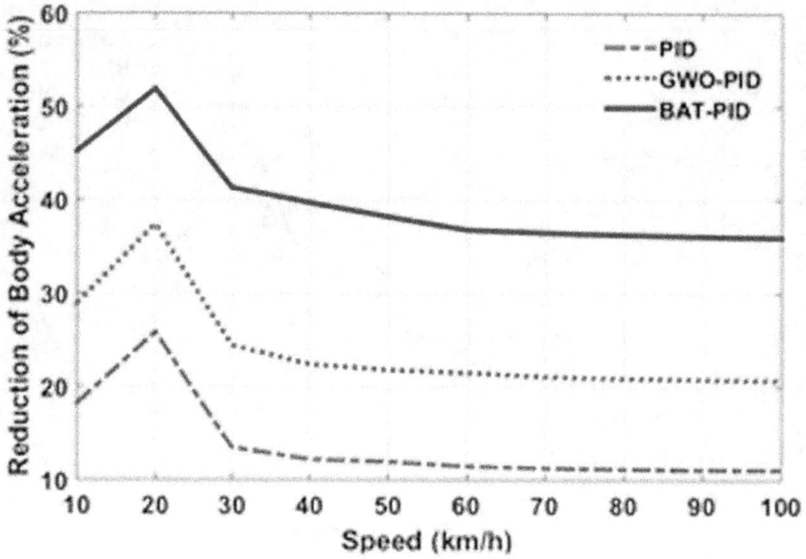

Figure 44. Analysis of optimized PID with different speed-FCM-SI

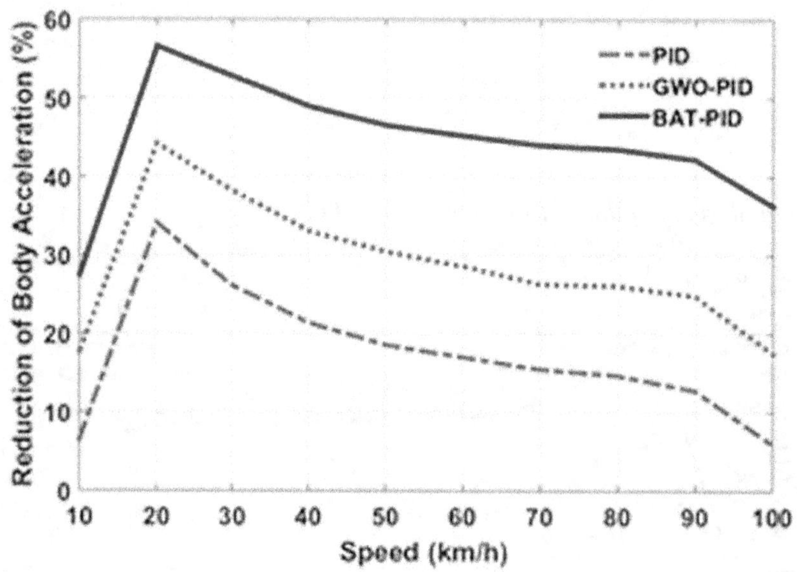

Figure 45. BI Road Profile (Darus, R& Sam, Y 2009)

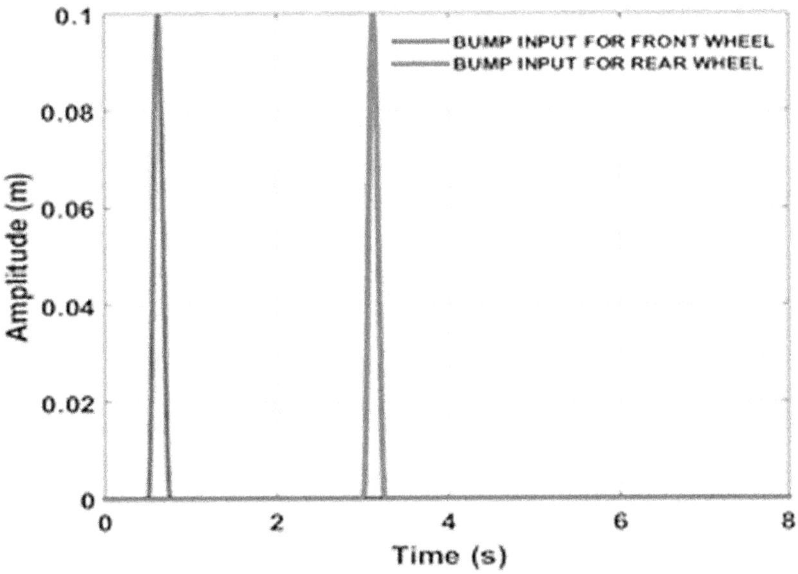

Figure 46. Time response of BD without controller-FCM-BI (Darus, R& Sam, Y 2009)

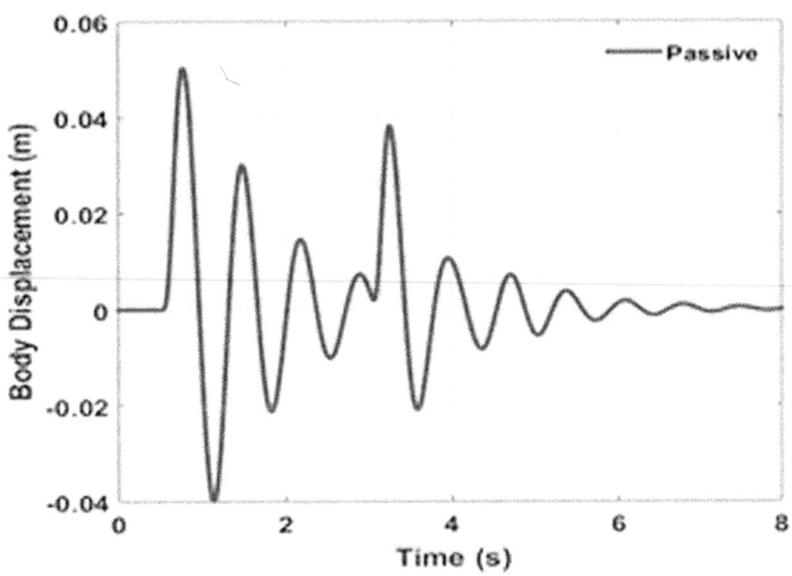

Figure 47. Time response of BA without controller-FCM-BI

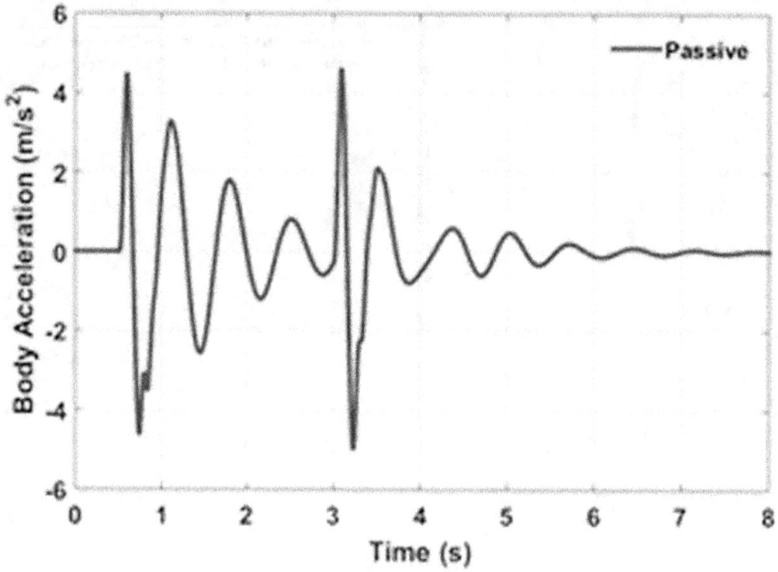

Figure 48. Time response of BD with controllers-FCM-BI (Darus, R & Sam, Y 2009)

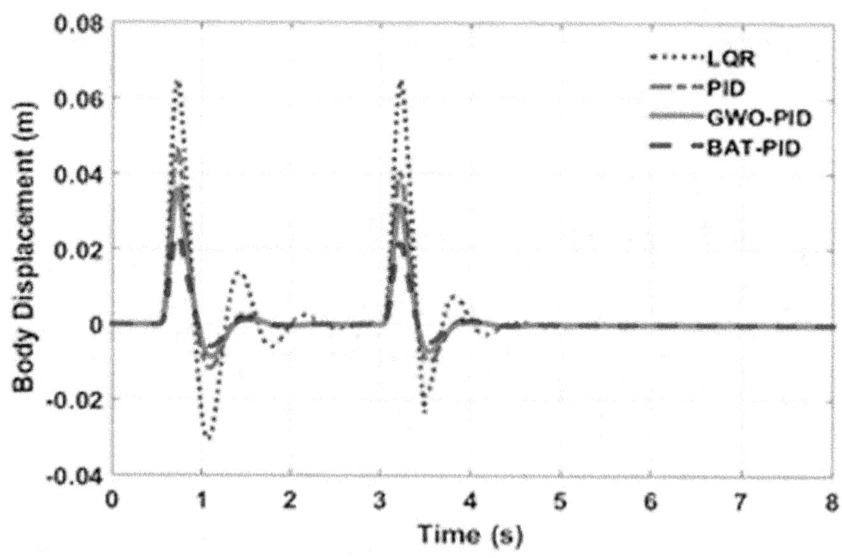

Figure 49. Time response of BA with controllers-FCM-BI

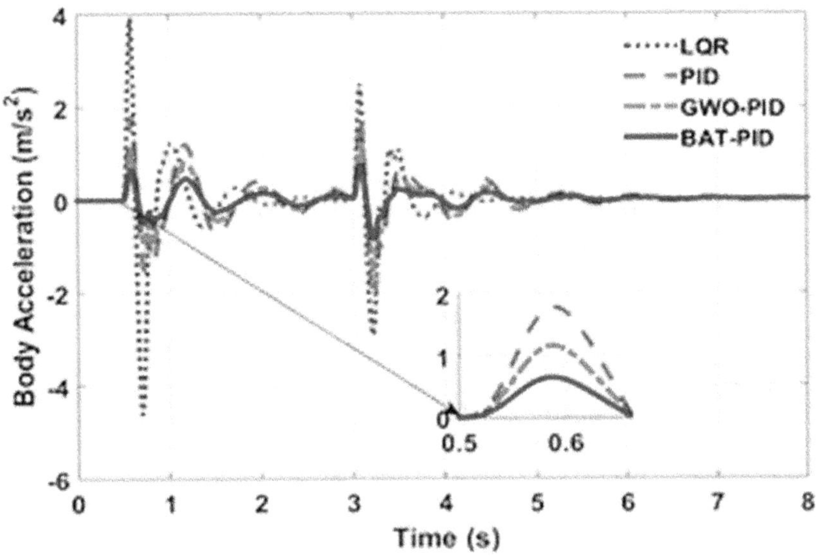

Figure 50. Control forces of front left wheel produced by FCM-BI

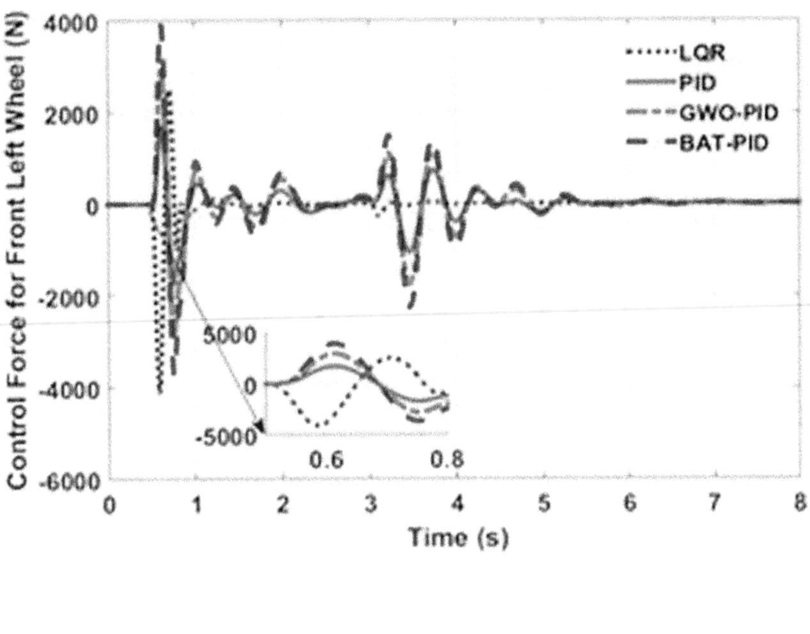

Figure 51. Control forces of front right wheel produced by FCM-BI

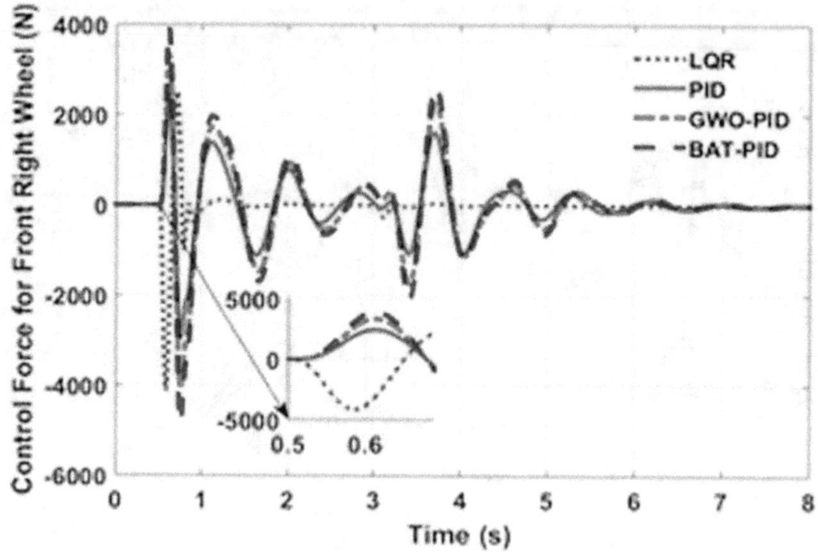

Figure 52. Control forces of rear left wheel produced by FCM-BI

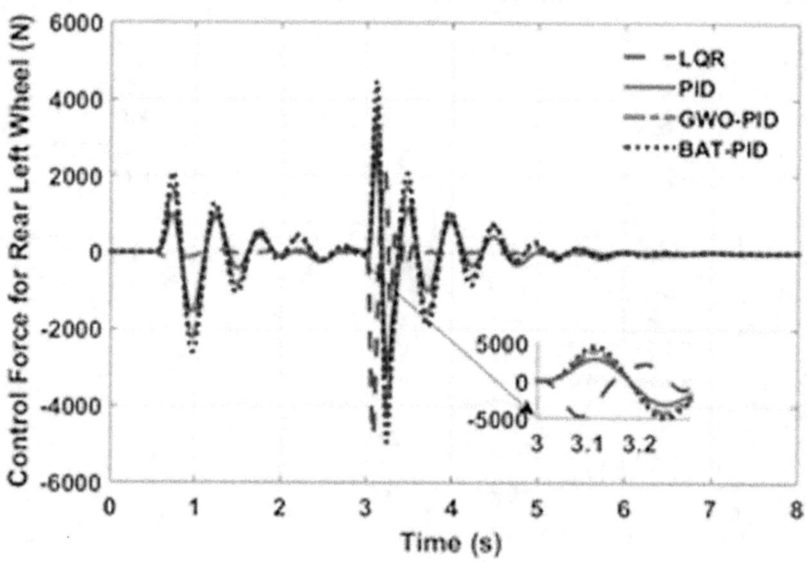

Figure 53. Control forces of rear right wheel produced by FCM-BI

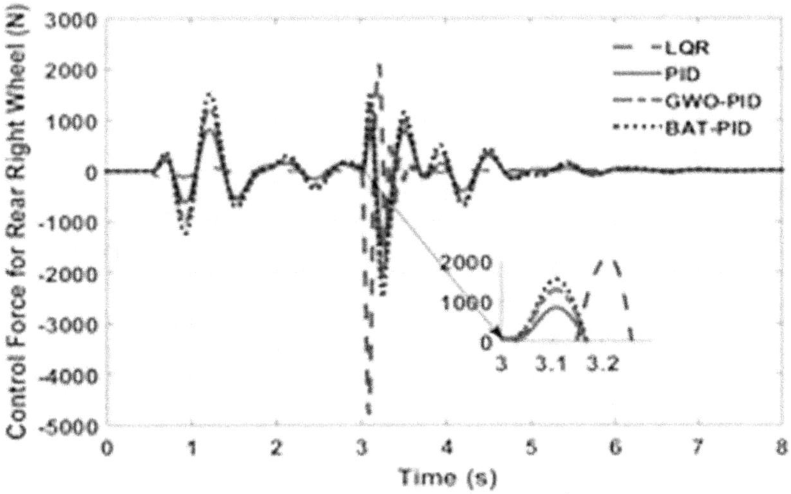

Table 8. Performance Indices of FCM-BI

Controllers	RMS Values of BD (m)	RMS of BA (m/s^2)	% of Reduction	FWRMS of BA (m/s^2)	Level of Discomfort	VDV of BA ($m/s^{1.75}$)
Passive	0.017	1.226	-	0.94	U	2.49
LQR	0.0138	0.694	43.4%	0.5	FU	1.589
PID	0.0085	0.463	62.2%	0.36	LU	1.01
GWO-PID	0.0067	0.31	75%	0.236	NU	0.68
Bat-PID	**0.00452**	**0.173**	86%	**0.134**	**NU**	**0.387**

Figure 54. PSD of BA-FCM-BI

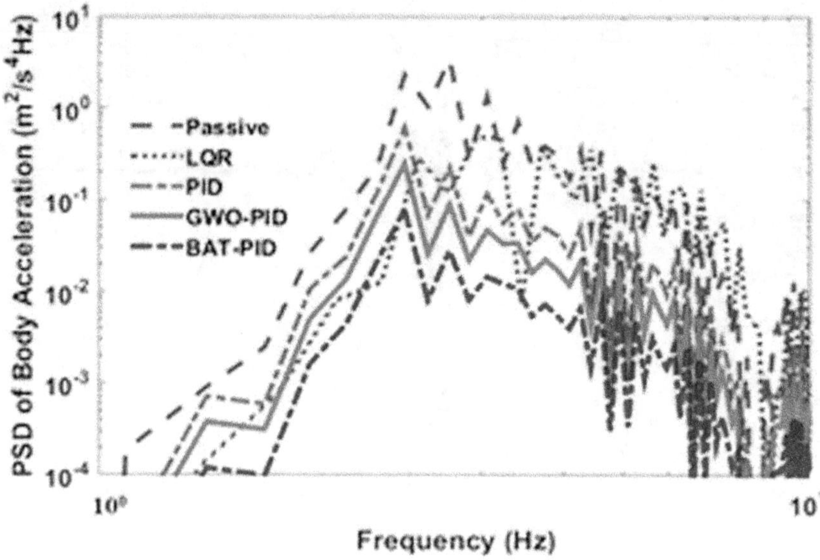

REFERENCES

Arora, U., Lodhi, M. E. A., & Saxena, T. K. (2016). PID parameter tuning using modified BAT Algorithm. *Journal of Automation and Control Engineering*, *4*(5), 347–352. doi:10.18178/joace.4.5.347-352

Darus, R., & Sam, Y. M. (2009). Modeling and control active suspension system for a full car model. *5th International Colloquium on Signal Processing & Its Applications*, 13-18. 10.1109/CSPA.2009.5069178

Gáspár, P., Szaszi, I., & Bokor, J. (2003). Design of robust controllers for active vehicle suspension using the mixed μ synthesis. *Vehicle System Dynamics*, *40*(4), 193–228. doi:10.1076/vesd.40.2.193.16541

Jahromi, A. F., & Zabihollah, A. (2010). Linear quadratic regulator and fuzzy controller application in full-car model of suspension system with magnetorheological shock absorber. In *Proceedings of 2010 IEEE/ASME International Conference on Mechatronic and Embedded Systems and Applications*. IEEE. 10.1109/MESA.2010.5552010

Liang, H., Liu, Y., Shen, Y., Li, F., & Man, Y. (2018). A hybrid bat algorithm for economic dispatch with random wind power. *IEEE Transactions on Power Systems*, *33*(5), 5052–5061. doi:10.1109/TPWRS.2018.2812711

Mangaiyarkarasi, P., Lakshmi, P., & Sasrika, V. (2019). Enhancement of vibration based piezoelectric energy harvester using hybrid optimization techniques. *Microsystem Technologies*, *25*(10), 3791–3800. doi:10.100700542-018-04291-1

Mech, L. D. (1999). Alpha status, dominance, and division of labor in wolf packs. *Canadian Journal of Zoology*, *77*(8), 1196–1203. doi:10.1139/z99-099

Messaoudi, I., & Kamel, N. (2019). A multi-objective bat algorithm for community detection on dynamic social networks. *Applied Intelligence*, *49*(6), 2119–2136. doi:10.100710489-018-1386-9

Muro, C., Escobedo, R., Spector, L., & Coppinger, R. P. (2011). Wolf-pack (Canis lupus) hunting strategies emerge from simple rules in computational simulations. *Behavioural Processes*, *88*(3), 192–197. doi:10.1016/j.beproc.2011.09.006 PMID:21963347

Nagarkar, M. P., Bhalerao, Y. J., Patil, G. J. V., & Patil, R. N. Z. (2018). GA-based multi-objective optimization of active nonlinear quarter car suspension system-PID and fuzzy logic control. *International Journal of Mechanical and Materials Engineering*, *13*(1), 1–20. doi:10.118640712-018-0096-8

Pedro, J. O., Dangor, M., Dahunsi, O. A., & Ali, M. M. (2018). Dynamic neural network-based feedback linearization control of full-car suspensions using PSO. *Applied Soft Computing*, *70*, 723–736. doi:10.1016/j.asoc.2018.06.002

Prabakar, R. S., Sujatha, C., & Narayanan, S. (2013). Response of a quarter car model with optimal magnetorheological damper parameters. *Journal of Sound and Vibration*, *332*(9), 2191–2206. doi:10.1016/j.jsv.2012.08.021

Prassad, G. S., & Mohan, M. K. (2019). A contemporary adaptive air suspension using LQR control for passenger vehicles. *ISA Transactions*, *93*, 244–254. doi:10.1016/j.isatra.2019.02.031 PMID:30837128

Rajagopal, K., & Ponnusamy, L. (2014). Biogeography-based optimization of PID tuning parameters for the vibration control of active suspension system. *Journal of Control Engineering and Applied Informatics*, *16*(1), 31–39.

Reynoso-Meza, G., Blasco, X., Sanchis, J., & Martínez, M. (2014). Controller tuning using evolutionary multi-objective optimisation:Current trends and applications. *Control Engineering Practice*, *28*, 58–73. doi:10.1016/j.conengprac.2014.03.003

Soh, M., Jang, H., Park, J., Sohn, Y., & Park, K. (2018). Development of Preview Active Suspension Control System and Performance Limit Analysis by Trajectory Optimization. *International Journal of Automotive Technology*, *19*(6), 1001–1012. doi:10.100712239-018-0097-x

Sun, J., Liu, Z., & Wu, J. (2018). An optimal driving strategy for the ride comfort performance of intelligent vehicles on uneven roads. *IFAC-PapersOnLine*, *51*(31), 984–989. doi:10.1016/j.ifacol.2018.10.052

Taha, I. B., & Elattar, E. E. (2018). Optimal reactive power resources sizing for power system operations enhancement based on improved grey wolf optimiser. *IET Generation, Transmission & Distribution*, *12*(14), 3421–3434. doi:10.1049/iet-gtd.2018.0053

Talbi, N. (2019). Design of fuzzy controller rule base using bat algorithm. *Energy Procedia*, *162*, 241–250. doi:10.1016/j.egypro.2019.04.026

Wang, M., & Xie, N. (2018). Structural optimization design of magnetic Shock absorber based on particle swarm optimization. *IOP Conference Series. Materials Science and Engineering*, *452*(2), 1–9. doi:10.1088/1757-899X/452/2/022012

Wang, W., Song, Y., Xue, Y., Jin, H., Hou, J., & Zhao, M. (2015). An optimal vibration control strategy for a vehicle's active suspension based on improved cultural algorithm. *Applied Soft Computing*, *28*, 167–174. doi:10.1016/j.asoc.2014.11.047

Yuvapriya, T. & Lakshmi, P. (2020). Numerical analysis and performance enhancement of active suspension system using bat optimization. *International Journal of Dynamics and Control*. doi:10.1007/s40435-020-00664-5

Yuvapriya, T., & Lakshmi, P. (2020). Simulation study on ride comfort enhancement for a heavy duty vehicle with fractional order sliding mode controller. *SN Appl. Sci.*, *2*(9), 1478. doi:10.100742452-020-03263-3

Yuvapriya, T., Lakshmi, P., & Rajendiran, S. (2018). Vibration suppression in full car active suspension system using fractional order sliding mode controller. *Journal of the Brazilian Society of Mechanical Sciences and Engineering*, *40*(4), 217. doi:10.100740430-018-1138-0

Yuvapriya, T., Lakshmi, P., & Rajendiran, S. (2020). Vibration control and performance analysis of full car active suspension system using fractional order terminal sliding mode controller. Archives of Control Sciences, 295-324.

Yuvapriya, T., Lakshmi, P., & Rajendiran, S. (2020). Design, Analysis and Experimental Validation of a Fractional Order Terminal Sliding Mode Controller on Active Suspension System. *Journal of the Institution of Electronics and Telecommunication Engineers*, 1–16.

Zhang, J., Deng, Y., Zhang, N., Zhang, B., Qi, H., & Zheng, M. (2019). Vibration performance analysis of a mining vehicle with bounce and pitch tuned hydraulically interconnected suspension. *Chinese Journal of Mechanical Engineering*, *32*(1), 1–17. doi:10.118610033-019-0315-0

Zhao, J., Wong, P. K., Xie, Z., Ma, X., & Hua, X. (2019). Design and control of an automotive variable hydraulic damper using cuckoo search optimized PID method. *International Journal of Automotive Technology*, *20*(1), 51–63. doi:10.100712239-019-0005-z

Zhou, J., Zhu, W., Zheng, Y., & Li, C. (2016). Precise equivalent model of small hydro generator cluster and its parameter identification using improved Grey Wolf optimiser. *IET Generation, Transmission & Distribution*, *10*(9), 2108–2117. doi:10.1049/iet-gtd.2015.1141

Section 3
Analytics in Electric Vehicles

Chapter 9
Intelligent Demand and Supply of Electrical Vehicles

B. V. A. N. S. S. Prabhakar Rao
https://orcid.org/0000-0002-7695-8827
Vellore Institute of Technology, Chennai, India

Rabindra Kumar Singh
Vellore Institute of Technology, Chennai, India

ABSTRACT

Today, it is impossible for a being to fathom living without a vehicle for transportation of any kind. All governments have put into place numerous regulations on the use of well-maintained vehicles to stop this, but there are always exceptions. The world is transitioning to an automated system that requires little human interaction as a result of the COVID-19 pandemic. This intelligent system involves identifying the predicting trends in the market analysis based on the expenses and revenues patterns of a particular group of people in a city or location. This the system needs to offer the insight into the direction for decision making of the market on demand of vehicle as per the supply with respect to the type of roads, season, and gender and age group patterns. Apart from this, the tax rules make a huge difference in this context. In some zones, many governments encourage by giving lots of subsidies or give tax relaxations to certain cc vehicles. In the future, based on the distance travelled, the system can impose tax fair or toll plaza fee.

SCOPE

Vehicle prices quickly increased as a result of the industry's lack of spare parts, space, people, and infrastructure as vehicle utilisation fell in the wake of the COVID-19 epidemic in 2020. In turn, the price drop sent a strong message to suppliers to reduce vehicle manufacturing. In contrast, higher oil prices in 2022 gave producers more motivation to increase output.

DOI: 10.4018/978-1-6684-6631-5.ch009

INTRODUCTION

Due to that huge traffic logjam; that leads to create lot of tension, stress and pollution. In terms of health and wealth this is no doubt pressing problem as per the current day needs. In addition to this, many people desire their own vehicles or to division with their loved ones, depending on accessibility and availability. Many times when people are travelling in their own vehicle the pollution they are generating and fuel consumed by the travellers and roots occupied by the their vehicles. In order to save nature in near future it's mandatory that there should be an alternative mechanism is required. With the help of computer vision technologies, hope to build a dependable, self-contained, and reasonably priced system in this text. However, our careless use of vehicles is harming society, whether consciously or unintentionally. Aside from that, will need to employ a different type of vehicle due to the increase in the price of petroleum products. Electric vehicles are in great demand in this situation. These are essential to everyone's daily lives. Since there are no clear rules to raise awareness of this issue, there is uncertainty among car owners regarding licencing, registration, and renewal of automobiles. However, as their significance in our lives increases, so does the amount of fraud and antisocial behaviour connected to these cars. An Intelligent System is required to forecast the need of the vehicle as per the requirement. Say some areas root is totally different than the normal or regular root. In some situations it's important that the decision making based on the number of traveller, weight of the luggage etc., will be treated as the parameters. Across the electrical vehicle manufacturing industry, there are plenty of patterns developed by many top level researchers with the latest techniques based on the technical analysis for trading and manufacturing system to forecast the price movements. This can be overcome with the use of Artificial Intelligence through the operative use of machine learning and deep learning techniques and strategies. The final objective of this intelligent framework can be achieved with the long run survival mode, welfare objectives, maximum sales revenue generation & utility of electric vehicles with profit maximization.

NEED OF ELECTRICAL VEHICLE DEMAND AND SUPPLY

Relationship between the required quantity of electric vehicle supply or service that manufacturers want to sell at different prices and the amount that consumers want to buy. It serves as the primary model for determining prices in economic theory. Since, no business will run with loss. The interaction of supply and demand in a market determines the price of different types of e-vehicle. The final price is known as the equilibrium price and signifies a compromise between the good's producers and customers. When a market is in equilibrium, the amount of a good that producer's supply and consumer's desire are equal.

When things are in equilibrium, there is no tendency for change. When there is no justification for the market price of the product to increase or decrease, the market is said to be in equilibrium. This happens at the price where the supply and demand are equal. The quantity that customers want to purchase and the amount that producers want to sell are identical at this price discussed with table 1.

- Battery Electric Vehicles (BEVs) are also known as All-Electric Vehicles (AEV)
- Hybrid Electric Vehicles (HEVs) are also known as series hybrid or parallel hybrid
- Plug-in Hybrid Electric Vehicles (PHEVs)
- Fuel Cell Electric Vehicles (FCEVs)

Table 1. Quantity of E-Vehicles Demand and Supplied

Quantity of E-Vehicles Required				Price	Quantity of E-Vehicles Supplied			
BEVs	**HEVs**	**PHEVs**	**FCEVs**	**₹**	**BEVs**	**HEVs**	**PHEVs**	**FCEVs**
1000	500	450	5000	4,00,000	1300	600	500	4500
2000	450	550	1500	55,000	2400	550	750	7500
3456	456	650	7500	35,000	3556	400	850	6700
4356	654	675	2000	54,000	4800	600	750	1800
4555	750	540	7200	23,000	4570	700	600	8200
7680	870	670	9800	75,000	7800	770	700	9910
7564	900	560	5000	4,00,000	7900	800	600	4500
5678	670	700	1500	55,000	6500	700	850	7500
4560	560	800	7500	35,000	6560	600	890	6700
3455	450	900	2000	54,000	6455	500	870	1800
2450	350	340	7200	23,000	5450	390	400	8200
4320	560	575	9800	75,000	7320	650	750	9910
3400	700	950	3500	45,000	6400	750	850	3450

At a price of ₹ 1,20,000 equilibrium is reached. E-vehicle in an equilibrium quantity of 8 When the price is over the equilibrium level of ₹ 1,20,000 there are more electric vehicles provided than there are demanded. At that pricing, businesses are unable to sell whatever they wish to. Due to the excess supply, there is pressure for the price to decrease. There is extra demand when the price is below equilibrium, which puts pressure on the price to increase. There is no pressure for the price to rise or fall except at the equilibrium price.

Role of Demand Estimation

The reliable forecasts about future demand would help an existing EV Industry in planning and scheduling of production, managing inventory, procurement of raw materials and other inputs, arrangement of finance etc. It also helps in reducing uncertainties in meeting market demand by avoiding underproduction and overproduction.

LITERATURE REVIEW

(Khan et al., 2019) proposed a fast DC charging station model. A grid connection was utilized to realize a DC bus through AC/DC converter. The decentralized control systems were designed for each vehicle and in order to achieve power transmission from AC to DC individual control were provided. Author also proposed power management system in order to achieve optimal power flow, which was integrated with solar PV systems to the charging station grid. In order to minimize the load on grid and conversion loss, the hybrid systems were used along EV batteries fleet.

(Rajagopalan et al., 2013) proposed fast charging characteristics along with advanced available technologies. It is very much evident that the numbers of EV are increasing very rapidly, since the fuel cost is rocketing very high and it is becoming unaffordable for common people. Therefore the demand of charging station will be very high in future. But merely increasing the number of charging station may serve the purpose, the charging station should be able to charge the EV as fast as possible like any other fuel station. Fast charging rate demands very high power charging station in infrastructure for supposing the charging station. Currently the establishment cost for the fast charging station is main hurdles. Hence, it requires sustainable and economic business model in order to match the demand and utilization.

(Mishra et.al. 2009) Proposed a fast acting dc-link voltage controller in order to compensate loads of ac and dc for three phases DSTATCOM (distribution static compensator). The DSTATCOM play a very vital role for compensating the rapidly varying unbalanced and non-liner loads. The DC-link voltage gets affected directly with any changes takes place in load. While a sudden rise in load would cause the dc-link voltage to drop below its reference value, a quick decrease in load would cause the opposite to occur. The dc-link voltage must vary while staying within the established parameters for DSTATCOM to function properly. Hence intelligent system is required to achieve the better control.

(Rigas et. al., 2014) published a survey on managing electric vehicles using artificial intelligence with respect of smart grid. They emphasize on to develop intelligent algorithm to minimize the charging cost using AI technologies to use grid system more effectively. The need of the Electric vehicles must be organized in such way that this should be able to avoid the peaks of power usage on the grid, because may leads in high electricity usage and in turn overload the local grid distribution.

(Ahmed, Moin, et al., 2021) proposed the adoption of electric vehicles and the role of AI in it. Currently the production of energy relies very heavily on fossil fuels like coal, oil and gas, which are unsustainable. These fuels very serious problem on climatic change and adverse effects on health. It is also understood that these fuels are withdrawn from limited resources, which is being depleted day by day. As a result the governments of developed and developing countries and the agencies have become vocal to use alternative and sustainable energy sources. Hence there will be a paradigm shift towards electrical vehicles. The optimal usage of energy can be achieved using artificial intelligence.

(Lee at al., 2020) proposed an analysis on the effects of AI on electric vehicle technology. The numbers of electric vehicles are increasing with great pace globally; this will majorly impact on reducing the greenhouse effect on gas emission on transportation. The emphasis was given on the classification using artificial intelligence algorithm for dynamic change pattern. In this study, they tries to demonstrate that the artificial intelligence play a major role in automation of electric vehicles.

BACKGROUND

This type of growth may be frequent in the gadget industry, where new consumer items with quick growth rates like the iPhone are not uncommon. Such a phenomenon is, however, incredibly uncommon in the world of transportation. Authors are not only talking about any nation here, where the majority of people may buy electric vehicles as their first mode of transportation; talking about a sizable population that is switching from their daily automobile commute to a new mode of transportation. If the news about the electric bicycle is true, I believe it will be the biggest mode of transportation since the excitement around personal automobiles at the turn of the past century.

Figure 1. Electric Vehicle Sales Trend in India

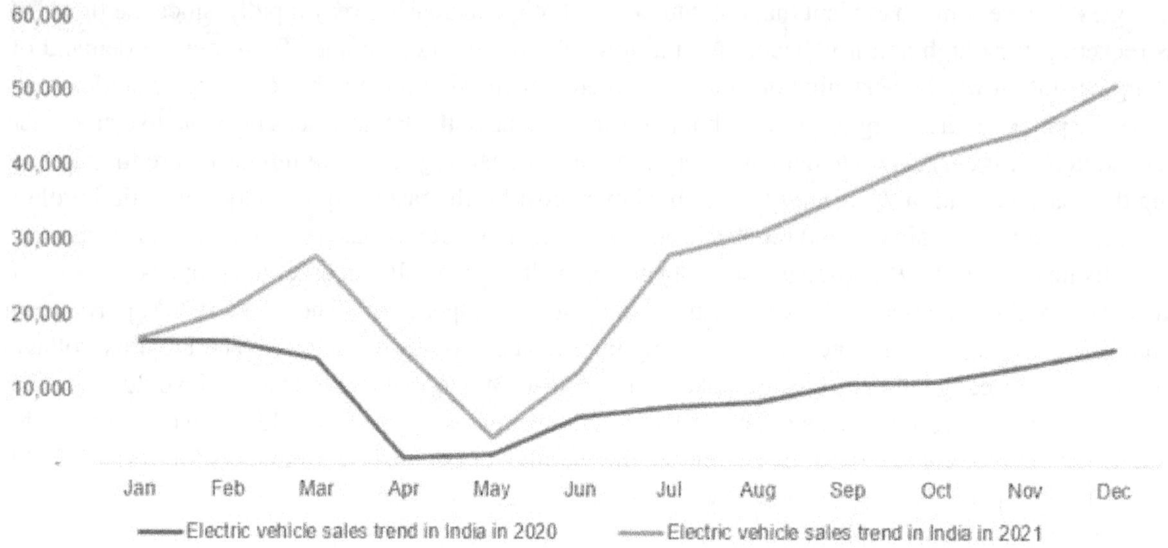

Figure 2. States -Wise-EV Sales Trend

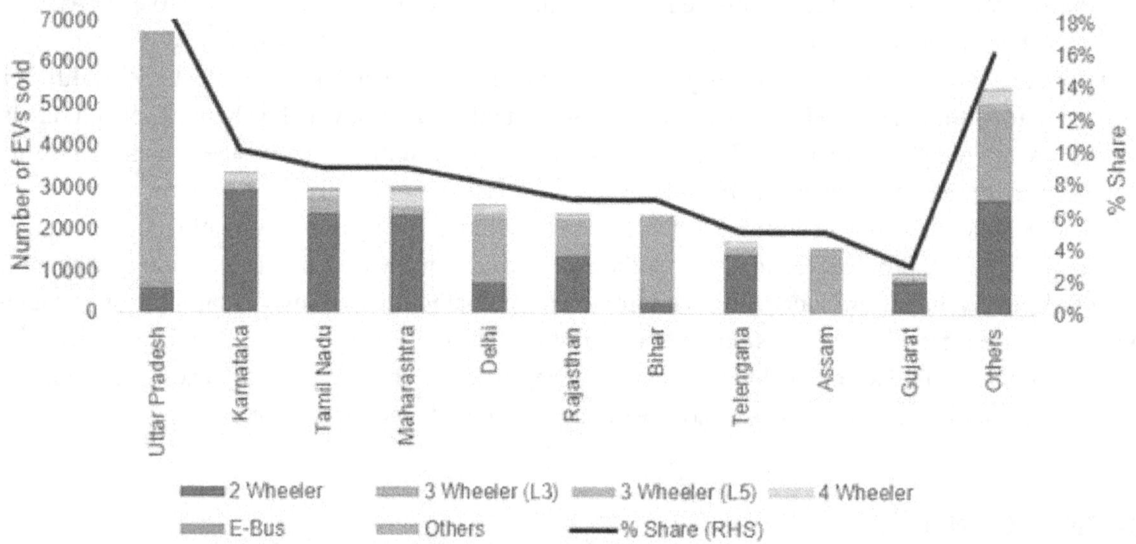

As per figure 1 and 2, almost all of the eighty million motorcycles sold worldwide each year now are powered by internal combustion engines. The expected surge in demand for electric bikes might more than treble the size of the two-wheeled market while also reducing demand for electric four-wheelers. According to the survey, electric bikes will account for 56% of the market for two-wheeled electric vehicles, followed by e-motorcycles at forty three per cent and e-scooters in a distant third place with less than one per cent as per table 2.

Table 2. Data Collection Parameters

S. No.	Demand	Supply	Advantages	Strategies
1 Based on Paper Survey	High	Low	Biased	Assessed based on the values
2 Based Online Survey	Low	Moderate	Un-biased	Assessed based on the IP Address
3 Based Interviews	Low	Moderate	Un-biased	Value based system
4 Based on Orders	Moderate	Low	Biased	Enhance the market strategy
5 Based on Sales	Moderate	Low	Biased	Change the market strategy
6 Based on Feedback	Low	Moderate	Un-biased	Improve the quality
7 Based on Demand	Low	Moderate	Biased	Strengthen the strategy
8 Based on Replacement	Moderate	High	Un-biased	Strengthen the quality
9 Based on Maintenance	Moderate	Moderate	Biased	Fix the resolution

APPROACH

Electric vehicle pricing is governed by the economic principle of supply and demand. It is assumed that, the unit price of a particular good or other traded good, such as labour or liquid financial assets, will fluctuate in a competitive market until it settles at a point where the quantity supplied and demanded at the current price are equal, establishing an economic equilibrium for price and quantity transacted. It offers the conceptual underpinning for modern economics.

A top-of-the-line electric bike, for instance, can create a sweeping 90 newton metres of torque, compared to the motor of an average bicycle, which provides 10 newton metres. If you are aware of this and are certain that 10 Newton metres would suffice for you, you can base your decision on this strategy.

METHODOLOGY

Some individuals believe that the popularity of e-bikes will increase as income levels rise and cities grow. It is obvious that there are other factors influencing the popularity of electric bikes in addition to wealth; their performance traits and inherent benefits make them a popular way of transportation for all income classes.

But can't ignore the fact that many nations are moving quickly to restructure their cities around the automobile. Similar to how they did in the West decades ago, houses are rising up in the suburbs today as per table 3. In this concept, the car is king because there is no widespread public transportation, making autos the only means of transportation between far-flung suburbs and business parks.

Similar to how the supply curve and marginal cost curve are related, the demand curve and marginal utility, measured in dollars, are as well. Consumers will be prepared to buy a specific quantity of an item at a specific price if the marginal value of further consumption is equal to the opportunity cost mandated by the price, that is, the marginal utility of other consumption possibilities. The demand schedule refers to a consumer's capacity and willingness to buy a particular good at a particular time.

The idea of a supply curve presupposes that businesses are perfect rivals with no impact on the market price by its very nature. This is due to the fact that, if a firm has market power—in contravention of the

Table 3. Vehicle types for Marginal Profit

S.No	Type of e-Vehicles	Demand	Marginal Profit
1	e-Ambulance	Moderate	High
2	e-Baby carriage	Moderate	High
3	e-Bicycle	High	Low
4	e-Boat	Moderate	Low
5	e-Bulldozer	Moderate	Low
6	e-Bus	Moderate	Low
7	e-Caravan	Moderate	Low
8	e-Carriage	High	Low
9	e-Cement mixer	High	Low
10	e-Delivery van	High	High
11	e-Dump truck	Moderate	Low
12	e-Motorcycle	Moderate	High
13	e-Mountain bike	Low	Low
14	e-Police	High	Low
15	e-Scooter	High	High
16	e-Taxi/cab	High	High
17	e-Tractor	High	High
18	e-Truck	High	High
19	e-Van	High	High

perfect competitor model—its choice of how much output to bring to market influences the market price. Each point on the supply curve answers the question, "If this firm is faced with this potential price, how much output will it sell?" As a result, the firm is not "fronted with" any particular pricing, necessitating the adoption of a more complex model, such as a monopoly, oligopoly, or differentiated-product model.

The concept of a demand curve, like the supply curve, necessitates that the buyer be a perfect competitor, that is, that the buyer have no influence over the market price. This is true because every point on the demand curve provides an answer to the question, "If buyers are faced with this potential price, how much of the product will they purchase?" However, if a buyer has market power (i.e., his purchases affect the price), he is not "faced with" any given price, and must use a more complex model.

IMPLEMENTATION FRAMEWORK

In order to prevent the above situation can forecast the demand and supply the vehicles as per the need so that it's advisable to establish showroom such a way that managing the things are absolutely smooth and fair enough.

Determinants of Price Elasticity of Demand

- Amount of Money Spent

- Brand loyalty
- Breadth of definition of a good
- Complementarity between Goods
- Consumer Income
- Duration of price change
- Existence of Substitutes
- Joint Demand
- Nature of Commodity
- Range of Prices
- Several Uses of Commodity
- Time and Elasticity
- Whether the Demand can be Postponed or not

Determinants of Elasticity of Supply

- Ease of Storage
- Effortlessness of Switching
- Length of the period of Production
- Mobility of factors
- Number of Producers
- Reaction of Costs
- Spare capacity
- The time frame of training

Economists make a distinction between the market supply curve and the supply curve of a specific firm. The market supply curve is the overall amount supplied by all firms; hence it represents the sum of the amounts given by all suppliers at each possible price (i.e., the supply curves of the individual firms are added horizontally). Discussed the parameters in detail in table 4.

Figure 3 shows the e-Vehicle sales for the recent years. Between the short-run and long-run supply curves, economists make a distinction. During the short run, one or more inputs are fixed (usually physical capital), and the number of businesses in an industry is also fixed (if it is a market supply curve). Long run refers to a time frame in which all inputs can be fully adjusted to any price adjustment and new firms can enter the market or current firms can leave it. Short-run supply curves are steeper than long-run ones (with quantity more sensitive to price, more elastic supply).

Figure 4 illustrates how a product's price P is decided by striking a balance between production at each price (supply S) and consumer demand at each price (demand D). The diagram depicts a rise in price (P) and quantity sold (Q) of the product as a result of a positive shift in demand from D1 to D2.

As per the above discussion finally can finalize on the common factors affecting supply include:

- Input costs, including labor costs
- The employed technology, productivity
- Future pricing predictions made by businesses
- Number of vendors

Table 4. Parameters Elasticity of Demand & Supply

S.No	Elasticity of Demand & Supply	E-Auto	C-Bicycle	E-Bike	E-Car	Auto	Bike	Car	Bicycle
Determinants of Price Elasticity of Demand									
1	Amount of Money Spent	Yes	Yes	Yes	No	Yes	No	May be	Yes
2	Brand loyalty	Yes	Yes	Yes	No	Not Now	No	Yes	Yes
3	Breadth of definition of a good	Yes	Yes	Yes	No	Not Now	No	Yes	Yes
4	Complementarity between Goods	No	Yes	Yes	No	No	No	Yes	Yes
5	Consumer Income	Yes	Yes	Yes	May be	Yes	Yes	Moderate	Yes
6	Duration of price change	Yes	Yes	Yes	No	Yes	Yes	May be	Yes
7	Existence of Substitutes	No	Yes	Later	Yes	Not Now	Yes	No	No
8	Joint Demand	Yes	Yes	Yes	No	Yes	No	Yes	Yes
9	Nature of Commodity	No	Yes	No	Yes	No	Yes	May be	No
10	Range of Prices	Yes	Yes	Yes	No	Yes	Yes	No	Yes
11	Several Uses of Commodity	No	Yes	Yes	No	Yes	No	Yes	Yes
12	Time and Elasticity	Yes	Yes	Yes	No	Yes	Yes	No	Yes
13	Whether the Demand can be Postponed or not	Yes	No	May be	No	Yes	No	May be	No
Determinants of Elasticity of Supply									
1	Ease of Storage	Yes	No	Yes	Yes	Yes	No	No	No
2	Effortlessness of Switching	Yes	Yes	Yes	May be	Yes	Yes	Moderate	Yes
3	Length of the period of Production	No	Yes	Yes	No	Yes	Yes	No	Yes
4	Mobility of factors	Yes	No	Yes	Yes	Yes	No	No	No
5	Number of Producers	Yes	Yes	Yes	May be	Yes	Yes	Moderate	Yes
6	Reaction of Costs	Yes	Yes	Yes	No	Yes	Yes	No	Yes
7	Spare capacity	Yes	No	Yes	Yes	Yes	No	No	Yes
8	The time frame of training	No	Yes	Yes	No	Yes	Yes	No	Yes

When all other demand determinants, such as income, consumer preferences, and the costs of substitute and complementary goods are held constant, the amount of a specific type of e-vehicle that buyers are willing and able to purchase at various prices is represented by a demand schedule, which is graphically represented as a demand curve. The demand curve always slopes downward in accordance with the law of demand, which means that when the price drops, more people will purchase the vehicle.

TREND FORECASTING

Is the process of using market research and consumer data to create predictions about customers' future buying habits and preferences? The impact presented in table 5.

Figure 5 shows the actual trend forecasting provides product designers with insight that may help them design an item that their target audience likes and purchases with respect to revenue generation for based on the trend equation as per the figure 6 and table 6.

EXPONENTIAL SMOOTHING

Sales Forecast

We will see one simple data to do the sales forecast for next year. We have 10 years of revenue data as per table 7.

Figure 3. E-Vehicles Sales

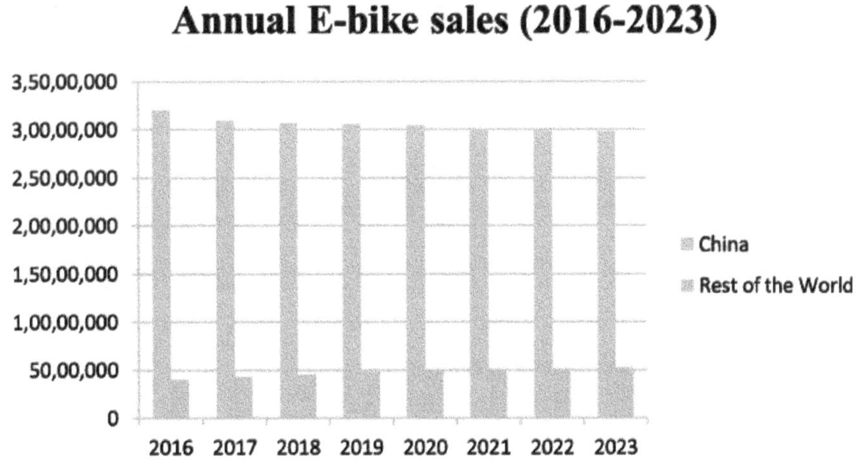

Using Exponential Smoothing, we need to forecast the revenue. The "Damping factor" will smooth out the graph. The values should be between 0 and 1. Technically it is 1 – α (alpha). We have mentioned 0.3 as the "Damping factor" as per table 8.

Using Exponential Smoothing, we need to forecast the revenue. The "Damping factor" will smooth out the graph. The values should be between 0 and 1. Technically it is 1 – α (alpha). We have mentioned 0.2 as the "Damping factor" as per table 9.

Forecast Trend at Different Damping Factors

Figure 7, 8 & 9 shows the forecast trend at different dumping factors - using monthly sales trend data with the help of table 10.

Figure 4. Price and Purchasing power of the product

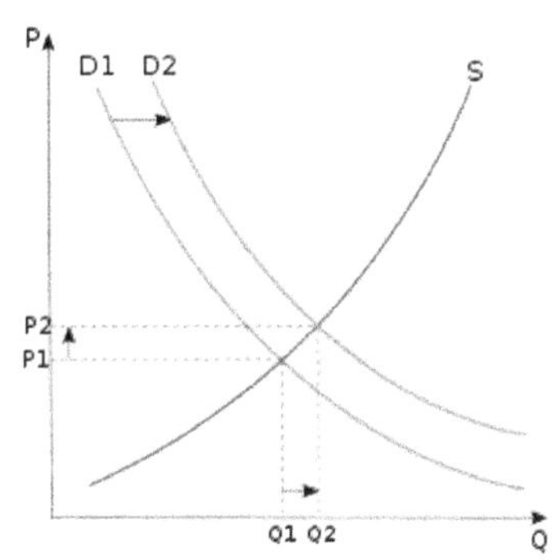

Table 5. Trend forecasting with years per e-vehicles

Year	e-Vehicles	Trend
2016	467000	#N/A
2017	759000	467000
2018	1372000	700600
2019	1546000	1237720
2020	2165000	1484344
2021	4899000	2028869

Figure 5. Exponential Smoothing Actual vs Forecast

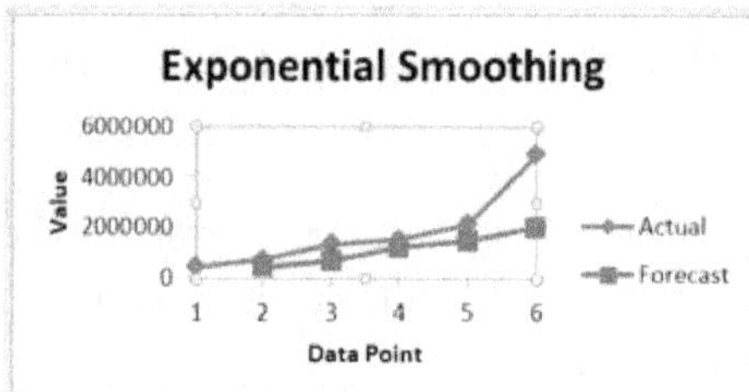

FUTURE RESEARCH DIRECTIONS

This type of growth may be frequent in the gadget industry, where new consumer items with quick growth rates like the iPhone are not uncommon. Such a phenomenon is, however, incredibly uncommon in the world of transportation. The majority of people can buy electric vehicle as their first mode of transportation; talking about a sizable population that is switching from their daily automobile commute

Table 6. Trend forecasting with month and revenue

Month	Revenue
January	275
February	350
March	450
April	4500
May	3745
June	8765
July	9935
August	6500
September	8798
October	???
November	???
December	???

Figure 6. Revenue with function

to a new mode of transportation. If the news about the electric vehicle is true, I believe it will be the biggest mode of transportation since the excitement around personal automobiles at the turn of the past century. But, few factors should be considered to prevent the damage of vehicle, or property, or person.

CONCLUSION

Electric vehicles are a cheap and efficient form of transportation that offers significant levels of personal mobility. Many people were able to buy electric vehicles thanks to the global economy's improvement. If they didn't have an electric bike, the typical user of electric bikes would utilise a bus or a bicycle. Compared to public transportation, it offers a better level of mobility in today's world. The fact that e-vehicle are already a popular trend around the world to serve as evidence that the aforementioned challenges can be addressed. The developed world may have various obstacles, but with the appropriate laws, city planning, technological advancements, and by investing money in the right places, miracles can happen practically immediately.

Table 7. Trend forecasting with years vs. revenue

Year	Revenue
2010	2500000
2011	3500000
2012	4500000
2013	3400000
2014	9800000
2015	8760000
2016	4560000
2017	45670000
2018	67000000
2019	85000000
2020	120000
2021	23000000
2022	45000000

Table 8. Forecasting with year's e-vehicles damping factor and generated revenue

Year	Revenue	Forecast
2010	2500000	#N/A
2011	3500000	2500000
2012	4500000	3200000
2013	3400000	4110000
2014	9800000	3613000
2015	8760000	7943900
2016	4560000	8515170
2017	45670000	5746551
2018	67000000	33692965
2019	85000000	57007890
2020	120000	76602367
2021	23000000	23064710
2022	45000000	23019413

Figure 7. Exponential Smoothing Actual vs Forecast Damping factor @ 0.3

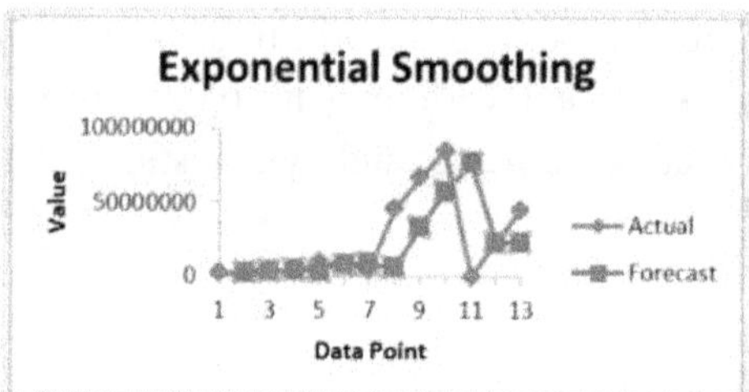

Table 9. Forecasting with year's e-vehicles damping factor 0.2 and generated revenue

Year	Revenue	Forecast
2010	2500000	#N/A
2011	3500000	2500000
2012	4500000	3300000
2013	3400000	4260000
2014	9800000	3572000
2015	8760000	8554400
2016	4560000	8718880
2017	45670000	5391776
2018	67000000	37614355
2019	85000000	61122871
2020	120000	80224574
2021	23000000	16140915
2022	45000000	21628183
2023	56540000	40325637

Figure 8. Exponential Smoothing Actual vs Forecast Damping factor @ 0.2

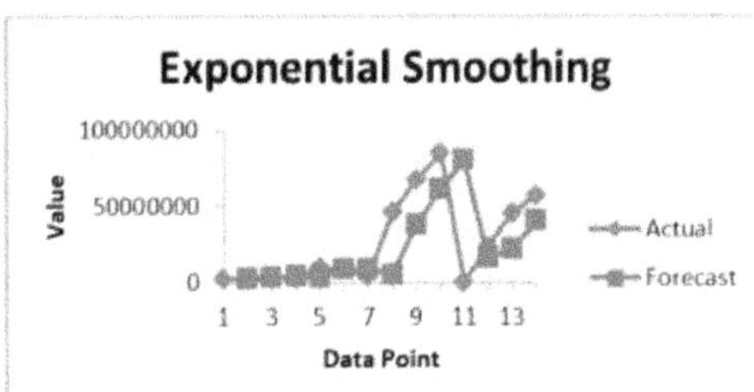

Table 10. Monthly Revenue with various Damping factors

Month	Revenue	Damping Factor @0.1	Damping Factor @0.5	Damping Factor @0.2	Damping Factor @0.9
January	275	#N/A	#N/A	#N/A	#N/A
February	350	275	275	275	275
March	450	342.5	312.5	335	282.5
April	4500	439.25	381.25	427	299.25
May	3745	4093.925	2440.625	3685.4	719.325
June	8765	3779.8925	3092.8125	3733.08	1021.8925
July	9935	8266.48925	5928.9063	7758.616	1796.20325
August	6500	9768.14893	7931.9531	9499.7232	2610.08293
September	8798	6826.81489	7215.9766	7099.9446	2999.07463
October	8900	8600.88149	8006.9883	8458.3889	3578.96717
November	9500	8870.08815	8453.4941	8811.6778	4111.07045
December	5600	9437.00881	8976.7471	9362.3356	4649.96341

Figure 9. Exponential Smoothing various Damping factor

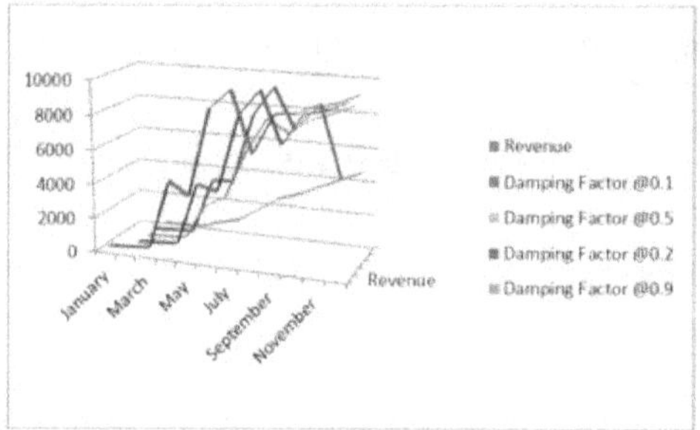

REFERENCES

Ahmed, M., Zheng, Y., Amine, A., Fathiannasab, H., & Chen, Z. (2021). The role of artificial intelligence in the mass adoption of electric vehicles. *Joule*, *5*(9), 2296–2322. doi:10.1016/j.joule.2021.07.012

Altuntas, S., Dereli, T., & Kusiak, A. (2015). Forecasting technology success based on patent data. *Technological Forecasting and Social Change*, *96*, 202–214. doi:10.1016/j.techfore.2015.03.011

Aristodemou, L., & Tietze, F. (2018). The state-of-the-art on Intellectual Property Analytics (IPA): A literature review on artificial intelligence, machine learning and deep learning methods for analysing intellectual property (IP) data. *World Patent Information*, *55*, 37–51. doi:10.1016/j.wpi.2018.07.002

Bhowmik, A., Castelli, I. E., Garcia-Lastra, J. M., Jørgensen, P. B., Winther, O., & Vegge, T. (2019). A perspective on inverse design of battery interphases using multi-scale modelling, experiments and generative deep learning. *Energy Storage Materials*, *21*, 446–456. doi:10.1016/j.ensm.2019.06.011

Boulanger, A. G., Chu, A. C., Maxx, S., & Waltz, D. L. (2011). Vehicle electrification: Status and issues. *Proceedings of the IEEE*, *99*(6), 1116–1138. doi:10.1109/JPROC.2011.2112750

Cowan, R., & Hultén, S. (1996). Escaping lock-in: The case of the electric vehicle. *Technological Forecasting and Social Change*, *53*(1), 61–79. doi:10.1016/0040-1625(96)00059-5

Dharmakeerthi, Mithulananthan, & Saha. (2011). Overview of the impacts of plug-in electric vehicles on the power grid. In *2011 IEEE PES Innovative Smart Grid Technologies*. IEEE.

Ernst, H. (2003). Patent information for strategic technology management. *World Patent Information*, *25*(3), 233–242. doi:10.1016/S0172-2190(03)00077-2

Falcão, E. A. M., Ana, C. R. T., & Sodré, J. R. (2017). Analysis of CO2 emissions and techno-economic feasibility of an electric commercial vehicle. *Applied Energy*, *193*, 297–307. doi:10.1016/j.apenergy.2017.02.050

Fujii, H., & Managi, S. (2018). Trends and priority shifts in artificial intelligence technology invention: A global patent analysis. *Economic Analysis and Policy*, *58*, 60–69. doi:10.1016/j.eap.2017.12.006

García-Olivares, A., Solé, J., & Osychenko, O. (2018). Transportation in a 100% renewable energy system. *Energy Conversion and Management*, *158*, 266–285. doi:10.1016/j.enconman.2017.12.053

Geels, F. W. (2012). A socio-technical analysis of low-carbon transitions: Introducing the multi-level perspective into transport studies. *Journal of Transport Geography*, *24*, 471–482. doi:10.1016/j.jtrangeo.2012.01.021

He, G., Zhang, H., Xu, Y., & Lu, X. (2017). China's clean power transition: Current status and future prospect. *Resources, Conservation and Recycling*, *121*, 3–10. doi:10.1016/j.resconrec.2016.11.003

Jahangir, H., Tayarani, H., Ahmadian, A., Golkar, M. A., Miret, J., Tayarani, M., & Gao, H. O. (2019). Charging demand of plug-in electric vehicles: Forecasting travel behavior based on a novel rough artificial neural network approach. *Journal of Cleaner Production*, *229*, 1029–1044. doi:10.1016/j.jclepro.2019.04.345

Karvonen, M., Kapoor, R., Uusitalo, A., & Ojanen, V. (2016). Technology competition in the internal combustion engine waste heat recovery: A patent landscape analysis. *Journal of Cleaner Production*, *112*, 3735–3743. doi:10.1016/j.jclepro.2015.06.031

Khan, Ahmad, & Alam. (2019). Fast EV charging station integration with grid ensuring optimal and quality power exchange. *Engineering Science and Technology, an International Journal, 22*(1), 143-152.

Lee, M. (2020). An analysis of the effects of artificial intelligence on electric vehicle technology innovation using patent data. *World Patent Information*, *63*, 102002. doi:10.1016/j.wpi.2020.102002

Lü, X., Wu, Y., Lian, J., Zhang, Y., Chen, C., Wang, P., & Meng, L. (2020). Energy management of hybrid electric vehicles: A review of energy optimization of fuel cell hybrid power system based on genetic algorithm. *Energy Conversion and Management*, *205*, 112474. doi:10.1016/j.enconman.2020.112474

Manzetti, S., & Mariasiu, F. (2015). Electric vehicle battery technologies: From present state to future systems. *Renewable & Sustainable Energy Reviews*, *51*, 1004–1012. doi:10.1016/j.rser.2015.07.010

Mihet-Popa & Saponara. (2018). Toward green vehicles digitalization for the next generation of connected and electrified transport systems. *Energies, 11*.

Mishra, M. K., & Karthikeyan, K. (2009). A fast-acting dc-link voltage controller for three-phase DSTATCOM to compensate ac and dc loads. *IEEE Transactions on Power Delivery*, *24*(4), 2291–2299. doi:10.1109/TPWRD.2009.2027501

Pfenninger, S., Hawkes, A., & Keirstead, J. (2014). Energy systems modeling for twenty-first century energy challenges. *Renewable & Sustainable Energy Reviews*, *33*, 74–86. doi:10.1016/j.rser.2014.02.003

Pilkington, A., Dyerson, R., & Tissier, O. (2002). The electric vehicle: Patent data as indicators of technological development. *World Patent Information*, *24*(1), 5–12. doi:10.1016/S0172-2190(01)00065-5

Rajagopalan, S. (2013). *Fast charging: An in-depth look at market penetration, charging characteristics, and advanced technologies. In 2013 world electric vehicle symposium and exhibition (EVS27)*. IEEE.

Rigas, E. S., Ramchurn, S. D., & Bassiliades, N. (2014). Managing electric vehicles in the smart grid using artificial intelligence: A survey. *IEEE Transactions on Intelligent Transportation Systems*, *16*(4), 1619–1635. doi:10.1109/TITS.2014.2376873

Rigas, E. S., Ramchurn, S. D., & Bassiliades, N. (2014). Managing electric vehicles in the smart grid using artificial intelligence: A survey. *IEEE Transactions on Intelligent Transportation Systems*, *16*(4), 1619–1635. doi:10.1109/TITS.2014.2376873

Sierzchula, W., Bakker, S., Maat, K., & van Wee, B. (2014). The influence of financial incentives and other socio-economic factors on electric vehicle adoption. *Energy Policy*, *68*, 183–194. doi:10.1016/j.enpol.2014.01.043

Skrúcaný. (2019). Impact of the electric mobility implementation on the greenhouse gases production in central European countries. *Sustainability, 11*.

Taiebat, M., & Xu, M. (2019). Synergies of four emerging technologies for accelerated adoption of electric vehicles: Shared mobility, wireless charging, vehicle-to-grid, and vehicle automation. *Journal of Cleaner Production*, *230*, 794–797. doi:10.1016/j.jclepro.2019.05.142

Xydas, E., Marmaras, C., & Cipcigan, L. M. (2016). A multi-agent based scheduling algorithm for adaptive electric vehicles charging. *Applied Energy*, *177*, 354–365. doi:10.1016/j.apenergy.2016.05.034

Yoldaş, Y., Önen, A., Muyeen, S. M., Vasilakos, A. V., & Alan, İ. (2017). Enhancing smart grid with microgrids: Challenges and opportunities. *Renewable & Sustainable Energy Reviews*, *72*, 205–214. doi:10.1016/j.rser.2017.01.064

Yong, J. Y., Ramachandaramurthy, V. K., Tan, K. M., & Mithulananthan, N. (2015). A review on the state-of-the-art technologies of electric vehicle, its impacts and prospects. *Renewable & Sustainable Energy Reviews*, *49*, 365–385. doi:10.1016/j.rser.2015.04.130

Chapter 10
Data Analytics in Electric Vehicles

Shyam Sihare
Dr. APJ Abdul Kalam Government College, Silvassa, India

ABSTRACT

Data analytics techniques are used to process the massive amounts of data produced by smart grid and electric vehicle technologies. Electric vehicles collect data from various sources, including sensors and trip records. Big data techniques can be used to analyze this immense quantity of data, which can then be used to implement rules for charging station placing, creating smart charging algorithms, addressing energy efficiency issues, evaluating the competence of power distribution systems to handle supplementary charging loads, and finally, forecasting the market effectiveness of the assets supplied by powered mobility. This chapter's data analytics landscape assessment addresses the incorporation of electric vehicles with green, smart urban environments. It serves as a blueprint for the objectives and alternatives for the incorporation of electric vehicles into smart future cities.

BACKGROUND

Data analysis is the science of analyzing raw data and making inferences from it (Aggarwal, C. C., 2011). Many data analysis techniques and procedures have been automated into mechanical workflows and algorithms in order to transform raw data for human understanding. There are five different types of analytics: descriptive, diagnostic, prescriptive, predictive, and cognitive (Amin, B. R. et al., 2021; Cao, Y. et al., 2018; Kambatla, K. et al., 2014).

Today, data analysts can gather data, store data, analyze data, and publish results using a range of software tools (LaValle, S. et al., 2011). Data collection, analysis, and presentation are all carried out with the help of data visualization and analysis tools such as Tableau and Power BI (Lee, J. et al., 2014; Runkler, T. A., 2020).

For the benefit of data analysts, more tools are being developed. Data analytics has been used by many businesses, including those with frequent quick turnaround periods like airlines and hotels. Creating a more economical method of conducting business and adopting it into the business model might aid in

DOI: 10.4018/978-1-6684-6631-5.ch010

cost reduction. Numerous quality control techniques, including the well-known Six Sigma systems, are supported by data analysis. A wide range of analytical techniques and methodologies can be employed by data analysts to extract information from huge volumes of unstructured, unprocessed data. Additionally, data analysts employ simple programming languages to update and alter databases (Russom, P., 2011, Wang, K. et al., 2017, Tsai, C. W. et al., 2015).

The acceptance of electric vehicles has grown steadily over time. The vehicle industry is using cutting-edge technologies to improve current prototypes and gain a deeper understanding of its consumers' preferences. At every stage of the product lifecycle, data science, artificial intelligence, and big data tools can help you enhance the performance of your final product and increase ROI (return on investment) (Yu, N. et al., 2015). A system-level knowledge of how these businesses operate can be unlocked by utilizing data. Stakeholders' decision-making is improved and risk is decreased via analysis and interpretation of EV market volatility and data. Today, data science combining manufacturing processes, marketing tactics, and charging point optimization is essential due to the electric vehicle industry's digital revolution and shifting consumer preferences (Elgendy, N. et al., 2014).

The availability of fossil fuels is already dwindling. Vehicles powered by internal combustion engines, on the other hand, only have an efficiency of 18–22%. However, there are a lot of difficulties with electric automobiles. Despite several difficulties and problems, moving to electric vehicles is better for the environment and more practical economically in the long run (Akhavan-Hejazi et al., 2014). Although the energy cost of producing an electric vehicle is also quite costly, given everything and how inexpensive it is to charge electric vehicles, EVs are a useful solution. The market for electric vehicles is expanding quickly (Cai, H. et al., 2014).

In order to maximize EV charging, big data analytics facilitate the mobility of electric vehicles (EVs). Big data analytics supports EV integration in a variety of ways, including improved charging, effective battery management, tracking EV status, and more. By conserving energy already present in the vehicle, electric vehicles' energy efficiency can be increased. Understanding issues like brake energy consumption, waste heat energy use, electric motor design, dependency on solar power supply, etc. would be necessary to achieve energy efficiency (Cheng, Z. et al., 2017). As a result, choosing environmentally beneficial options, like electric vehicles, is now required. The ideal levels of EV adoption depend on convenience, cost-effectiveness, and other considerations, but big data analysis can strengthen these levels. Amazon acquired an electric delivery van from Michigan-based Rivian to address climate change. By adopting an electric vehicle (EV), one can improve the health of society by reducing greenhouse gas emissions. These emissions could be further reduced if electric vehicles could be charged using renewable energy sources like solar or wind (Chen, T. D. et al., 2013).

At some point in their lives, everyone has come across the words "big data" or "data analytics," but it is less clear how these concepts are used in actual situations. Such applications include route search systems, from which EV routes can be selected according to user preferences, EV range estimations, and consumer characteristics. It also proposes original routes and selects the closest charging stations along the way (Diamantoulakis, P. D. et al., 2015; Lv, Z. et al., 2020).

By considering speed, temperature, and power use while driving, HMG (Hyundai Motor Group) calculates "big data-based energy consumption and range estimation." By analyzing data gathered from driving habits and vehicle condition information, automakers can better understand drivers and design vehicles that satisfy the needs of their customers (Hussain, M. M. et al., 2020).

Big data traffic analytics can promote the use of electric vehicles outside the house. The number of chargers required (and thus the installation cost), grid stress, and charging-related costs are all signifi-

Figure 1. (a) Short-term regional global EV market share of new passenger vehicle sales (b) Long-term regional global EV market share of new passenger vehicle sales

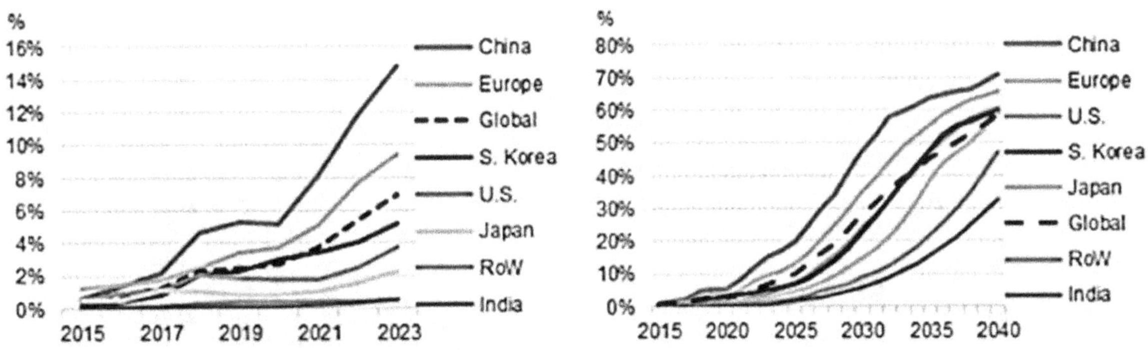

Source: BNEF. Note: Europe includes EU, U.K. and EFTA.

cantly impacted by the distribution of charging time. Different areas could charge electric automobiles differently during the day (Lee, C. H. et al., 2015; Wu, J. et al., 2016).

Instead of using gasoline, like its gasoline-powered predecessor, the electric vehicle uses electricity as its fuel. Battery-powered, hybrid, and plug-in hybrid electric vehicles are the three subtypes available. The market for electric vehicles has been segmented based on region, type, peak speed, and vehicle drive type. Businesses in the electric vehicle industry are prioritizing product development and product launch as their primary growth opportunities (Lee, J. et al., 2016, Northfield R., 2016).

Sensor information is in gigabytes or terabytes. The analysis of such a huge amount of information can be done only by big data. The heap of information is analyzed with the help of Big data and the desired information is obtained. The received information is sent for further action with the help of big data. Big data can be used for the following points: making policies for the location of the charging station; building smart algorithms for the charging station; solving the energy problem correctly; monitoring the electric battery during charging; monitoring a load of power distribution; distributing power at different charging stations as per demand; continuous monitoring of electric vehicles and related markets; monitoring of different parts of the vehicle; instructing the driver from time to time; Big data can be used to exchange information in real-time by creating a connection between the driver and the vehicle, etc. Apart from this, by analyzing the various devices that are installed in the electric vehicle and the information generated from them, it can be analyzed in real-time to tell the status of each component of the vehicle.

The demand for high-performance, fuel-efficient vehicles with low emissions is always increasing. As a result of growing gasoline costs, it is obvious that demand for EVs will continue to grow over time, as seen in Figure 1. The recent increase in the cost of gasoline and diesel has resulted in a significant demand for fuel-efficient vehicles. Electric vehicles convert more than 50 percent of grid electricity, whereas gas-powered vehicles convert only 17 to 21 percent (Kisacikoglu, M. C. et al., 2017; Li, B., Kisacikoglu et al., 2017).

The bulk of electric vehicles sold worldwide are passenger electric vehicles. The majority of DC charging stations are located in Asia, whilst the supercharging infrastructure in Europe and North America is expanding at the fastest rate. As China and the US retain these accords, the market is expanding even more (Figure 2).

Figure 2. Estimates of the annual sales of electric vehicles (Hussain, Md Muzakkir, et al., 2020)

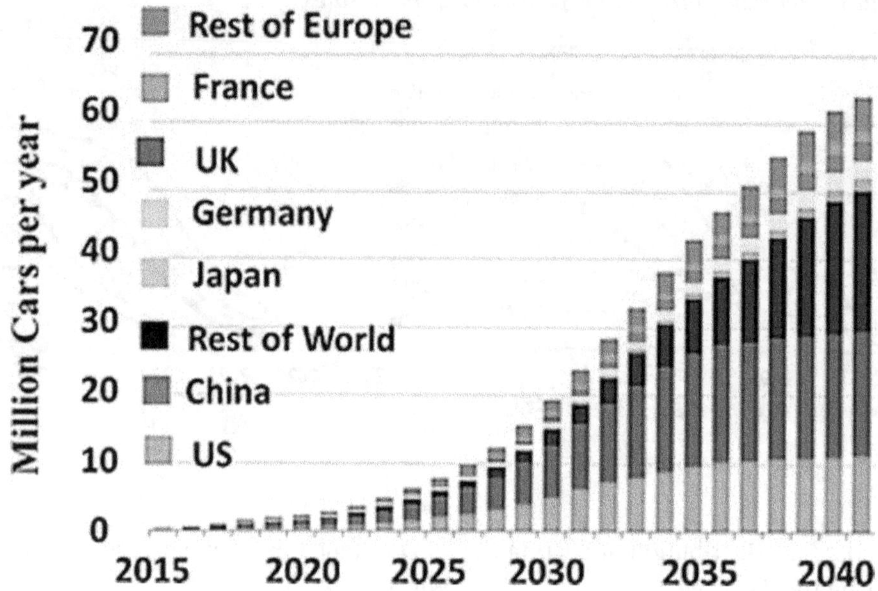

Figure 3. Passenger EVs are the main drivers of transport electrification outlays (courtesy of the BloombergNEF (BNEF) report)

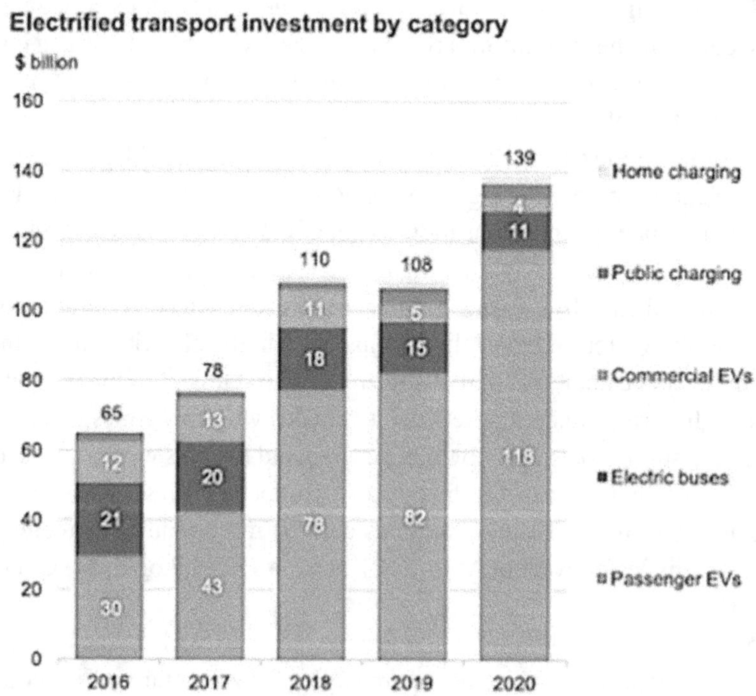

The number of electric vehicles sold worldwide will hit 10 million in 2020, making up 1% of the market. While Europe has the biggest yearly growth rate of 3.2 million, China has the largest total. In 2020, electric vehicle sales increased by 70%, for a record 4.6 percent. By 2020, investments in electric transportation are shown in Figure 3 by category (Rahimi-Eichi, H. et al., 2014; Vermesan, O. et al., 2021).

The category-wise investment in electric vehicles is outlined in Figure 3. This investment is only till 2020. After that, COVID-19 came, which badly affected all the manufacturing sectors. The automobile sector was also not untouched by this; this sector was the most affected. Figure 3 gives the investment in electric vehicles, which is till 2020. According to experts, this target has not been fully met in 2020 either. Due to COVID-19, the electric vehicle market has completely collapsed, and it may take time to recover. The electric vehicle market has gained momentum in 2022 as the demand for electric vehicles in China and the US is larger than expected.

The cost of gasoline is rising daily. The market for electric vehicles is expanding with time and will expand much more in the future, taking into account the price of gasoline. The primary factors driving the growth of electric vehicles are that they are more affordable than gasoline, do not harm the environment, and the nation's economy is in good shape.

The author has covered the electric vehicle in significant detail in this chapter. Additionally, other elements of the electric vehicle have been considered. The usage of artificial intelligence and deep learning is explored in the context of the topic. Additionally, one can upload information about an electric vehicle using cloud computing, as well as get precise position data and monitor the vehicle. Artificial intelligence allows for the synchronization of all the vehicle's parts and the timely delivery of appropriate instructions. By gathering all of the vehicle's data with the use of deep learning, users can perform an accurate evaluation. The data collected during the evaluation can be used in the future to benefit the vehicle and the driver. Furthermore, with the help of a flowchart, the author has attempted to illustrate how to update the algorithm in the charging station. With the assistance of intelligent charging algorithms, the author has concentrated on how to use the battery for extended periods of time.

Large-scale data processing can be completed quickly with the help of NoSQL. The author then proceeded into significant detail on the wireless battery charging point. such as battery maintenance, regularly paying attention to battery statistics, etc. Additionally, the chapter goes into great depth on how to locate the battery charging point, which is the most reasonable distance. Then details regarding smart meters and the smart grid are provided. In the conclusion, the author attempts to demonstrate how we can get data from electric vehicles and how we can utilize that data to make the vehicle and its driver safer.

The big data challenge is driven by the enormous quantities of information generated by machinery, vehicles, buildings, the power grid, and several other connected entities, as well as the growing data transmission rates. Smart grids and electric vehicles, among many other Internet-of-things industries, confront this dilemma due to their dual responsibilities as producers and consumers of large amounts of data. By cutting transportation-related emissions, electric vehicles could considerably contribute to the green economy of smart cities and play a key role in their expansion.

Every significant component of an electric vehicle has embedded sensors, which gather data from them and send it to a server linked to the cloud. Following transmission, the data is examined, and an appropriate judgment is formed. These results are transmitted to the designated electric vehicle, which interprets them as instructions and transmits them to the targeted component. This entire process takes only a few nanoseconds. Additionally, all of them require internet speeds that are several times higher than the existing ones. Achieving this is possible with 5G technology.

Sensors may track things like driving activities, battery health and maintenance, grid health and maintenance, and more. In order to track the vehicle's movements at all times, sensor data is relayed continually from the server to the vehicle and from the vehicle to the server. The sensor produces a large amount of data, which is then transformed into useful data by data mining and delivered to the server. The driver receives certain sensor data so that he may be informed of the vehicle's present state and take additional suitable action based on it. Some sensor data is approximated, whilst other data is merely utilized to relay it to the server. Information is gathered and anticipated future activity with the support of the electric vehicle's sensors.

Drivers may often link the sensor to their smartphone and monitor the vehicle's current condition. IoT can also encompass Internet of Energy and Internet of Vehicles. Meaning that all significant vehicle parts may be connected to the Internet in order to monitor their present state and get the appropriate instructions. That is, electric vehicles employ cloud computing, IoT, the Internet of Things, the Internet of Energy, high communication efficiency, big data, data mining, and data analytics. The threads of all of these are intertwined together in such a manner that the features of the vehicle can be quickly discovered and necessary action can be made on them to prevent further undesirable situations. Only with the aid of the server can data analytics be undertaken.

With ACC (Adaptive Cruise Control), your car's speed automatically adjusts to keep up with the vehicle in front of you. ACC can automatically slow down your vehicle according to the speed of the vehicle in front.

In traditional cruise control, the vehicle runs at the prescribed speed, while the speed in automatic cruise control automatically decreases according to the speed of the vehicle ahead. Even if the driver's attention is lost, the vehicle survives the accident. It provides additional safety. The speed of the vehicle in an automatic manner depends entirely on the speed of the other vehicle.

The IoT (Internet of Things) based vehicle parking system is operated by the Internet. It is a real-time-based system. It provides the driver with real-time parking information. This system uses sensors and microcontrollers. The sensor sends information about occupied or empty parking spaces to the cloud server, and the server sends this information to all electronic devices connected to the Internet. Different algorithms can be used to manage the automatic parking system. Round robin can perform load balancing, while the steady state evolutionary algorithm can optimize the use of parking space. Dijkstra and Ant Colony can algorithmically find the nearest vacant place for parking the vehicle.

Section 2 of this chapter discusses the benefits of EV vehicles. The challenges of EV vehicles are explained in Section 3. Section 4 then delves into the EV analytics tools and access data, such as tools and wireless access to data. In addition, electric vehicle components for data analytics are discussed in Section 4, such as smart grids, smart meters, electronics with intelligence, and phasor management units. Subsequently, policies for the analysis of data analytics are discussed in Section 5. Section 6 discusses the future of electric vehicle data analytics. Finally, in the conclusion section, all of the discussion is extracted.

BENEFITS OF EV VEHICLES

A few advantages of electric vehicles are as follows:

1. The use of more economical vehicles, such as hybrid and electric versions, boosts economics and helps diversify the fleet. Several different fuel sources provide consistent energy to the electrified sector of the transportation industry.
2. An electric automobile can cost significantly more to buy than a comparable conventional vehicle. Fuel cost reductions, tax credits, and state and utility subsidies can all help to cover the initial cost.
3. Compared to conventional vehicles, all-electric and plug-in hybrid vehicles consume less petrol. HEVs (hybrid electric vehicles) frequently beat comparable conventional vehicles in terms of fuel efficiency and gasoline cost.
4. Flexible charging brings the grid closer to where most people park, making it compatible with both plug-in hybrid electric vehicles and completely electric vehicles. Gas stations are more prevalent, while public charging outlets are less prevalent.
5. Pollution-free transportation is provided via plug-in hybrid and completely electric vehicles. The emissions of electric vehicles are influenced by the regionally unique power sources that are utilized to charge them.
6. The energy efficiency of an electric vehicle is higher than that of a petrol-powered vehicle and is more economical.
7. The pollution in the environment from electric vehicles is negligible because they run on electricity instead of petrol or diesel.
8. The performance of electric vehicles is much better than that of petrol or diesel vehicles. It also requires less maintenance than a conventional vehicle. If analyzed comparatively, it does not make a sound and, at the same time, its mileage is quite cheap.

CHALLENGES OF EV VEHICLES

A few challenges of electric vehicles are as follows:

1. Sales of electric vehicles are still in their infancy. The sector's shortage of skills will be a problem for foreign investors.
2. The industry's potential for expansion is significantly constrained by ambiguous consumer behavior and inadequate charging infrastructure.
3. Due to lithium scarcity, India is significantly dependent on imports for the manufacture of electric vehicles. For long-term success, foreign businesses are turning to consultants to research industry and technological trends and provide a variety of commercial options in India.
4. Due to overheating or overcharging, the batteries of electric vehicles have started exploding lately, which has led to information about the loss of life and property.
5. Despite the battery being fully charged, its mileage is much less than that of a conventional vehicle. As of today's date, the battery can travel only about 200 kilometers after a full charge. Some models claim to travel between 400 and 500 km on each charge.
6. It takes a long time for an electric vehicle's battery to be fully charged. It takes about eight hours to charge levels one and two. A fast charging station also takes 30 to 50 minutes.
7. Electric vehicles cost more than conventional vehicles. However, in the future, all the money is recovered due to lower maintenance costs and getting rid of expensive petrol and diesel.

8. Due to defects in the visibility of the display screen, sensors not working properly in different weather conditions, sudden failure of parts, and other reasons, customer seems to avoid buying an electric vehicle nowadays.

EV ANALYTICS TOOLS AND ACCESS DATA

The following conditions must be satisfied in order to analyze data from electric vehicles:

1) Tools for Data Analytics

Hadoop MapReduce and Pig Script

Hadoop is a framework that efficiently stores and analyzes large datasets ranging from gigabytes to petabytes. In this, instead of one large high-processing computer, many computers are connected in parallel, and in which massive datasets are stored and processed. Hadoop mainly consists of four modules, namely Hadoop Distributed File System, Yet Another Resource Negotiator, MapReduce, and Hadoop Common. The Hadoop Distributed File System works on low-end hardware. Yet Another Resource Negotiator does the job of monitoring, managing, and scheduling tasks. MapReduce is used to perform parallel computations while Hadoop Common provides the library to Java. MapReduce processes multi-terabyte datasets in parallel across thousands of nodes in a reliable and fault-tolerant manner.

Pig Script is a high-level scripting language used with Hadoop. It represents complex data easily. There is no need to understand the Java language to perform the transformation of complex data.

Hadoop is a new distributed computer cluster data processing technique that uses Hadoop. It is designed to assist with remote computing that requires a lot of data. For storing huge files, HDFS (Hadoop Distributed File System) is a native distributed file system solution. Hadoop could be employed to organize massive data streams generated by EV-related economic models. Hadoop manages the quantity and variety of spatial-temporal streams, whereas Pig provides a programming interface in the form of a data stream. Currently focusing on available state of charge currents and charger operating status records. The original data stream is now altered to generate new spatio-temporal sequences suitable for further processing, such as in artificial neural networks (Lee, J. et al., 2015). With the help of the MapReduce programming model, a distributed computer cluster is capable of processing enormous amounts of data concurrently. The MapReduce function takes input and outputs many key-value pairs. The values associated with a given key are added together by a reduction function. Pig is a cutting-edge platform or technology used to handle enormous datasets. It provides a very high level of abstraction for processing over MapReduce. Pig results are consistently stored in HDFS and available from any Hadoop cluster. In 2006, Yahoo researchers developed the Apache Pig open-source project. Pig Latin can write 200 lines of Java code in just 10 lines of code.

Cloud Computing

Cloud computing provides computing services over the Internet, including servers, storage, databases, networking, software, analytics, and intelligence, to economically reach these services.

There are three types of cloud computing, namely public clouds, private clouds, and multiclouds. Similarly, there are three types of services cloud computing provides, namely IaaS (Infrastructure-as-a-Service), PaaS (Platforms-as-a-Service), and SaaS (Software-as-a-Service). Cloud computing is used for backing up data, email, building and testing software, big data analytics, web applications, virtual desktops, etc.

Electric vehicles have lower mileage and more expensive operational costs. Electric vehicle manufacturers' software architecture has to be updated on the core service layer, user interface access layer, and service management layer. Businesses can develop and build their intelligence and intelligence platforms using EV cloud technologies. To assist in the expansion of the electric vehicle sector by enabling hitherto unheard-of levels of automation and mobility connectivity. The construction of simulated driving scenarios as well as the redistribution of work and fixes for issues encountered in the field are both made possible by cloud technology. Cloud technology offers an electric vehicle's long-distance solution. It enables companies to build and control networks with several chargers. The cost of maintenance is another appealing and useful component of EV cloud systems. Smartphones can be used to make payments, and the online payment platform can be connected to the charging station.

NoSQL

NoSQL is a database whose main use is to keep the data in the form of documents rather than in the form of relational tables. It can store data in different formats such as key-value, document, columnar, and graph formats. NoSQL is a flexible, scalable, non-relational, and distributed database. It can store massive data on different servers. If the information increases, it automatically adds a new server. It is used to analyze massive amounts of information efficiently and effectively, hence it is used for big data.

In the past, open source NoSQL technologies were used to build repositories because of their capacity to manage large volumes of unstructured data. NoSQL represents a completely new database architecture. Large volumes of data can be processed quickly and effectively because of NoSQL. Distributed NoSQL database technology is already used to manage some of the biggest data warehouses in the world (Bolly, V. K. et al., 2014).

2) Wireless Access to data

Wireless Battery Charging Point

Inductive charging of electric vehicles refers to the wireless transmission of power via electromagnetic induction. Wires that accept either direct current (DC) or alternating current (AC) power are frequently used to charge electric vehicles. Level 4 charging for electric vehicles can be finished quickly at public DC charging stations. It has a power output of about 120kW and can finish the battery's charge in 20 to 40 minutes. High-power inductive charging is used in electric vehicles to wirelessly recharge the battery. Mobile wireless networks are the main connectivity option for EV charging stations. Compared to wired networks, they can be set up more efficiently and for less money. Since 5G wireless supports up to a million devices per square kilometer, it is perfect for communities that want to install a substantial number of EV chargers (Wei, W. et al., 2016).

With the use of big data, it is possible to create electric vehicle charging stations, create intelligent charging algorithms, address problems with energy efficiency, and turn smart cities into green smart cities. A crucial component of the infrastructure for EV charging is becoming wireless connectivity. The most effective techniques to connect to the cloud for intelligent charging process control are currently Wi-Fi and cellular technologies.

ELECTRIC VEHICLES COMPONENTS FOR DATA ANALYTICS

1) Smart Grids

The energy balance of the grid will be more significantly maintained by plug-in electric vehicles (PEVs). In order to prevent interruptions, PEVs will work with smart grids to ensure that they are fully charged and ready to go when traffic is heavy. PEV can help the grid incorporate different energy sources, such as solar and wind. Using smart grid technologies, plug-in vehicle charging stations could be able to help with this problem. For EV adoption to increase, smart cities must have access to charging stations. An electrical distribution system known as a "smart grid" employs digital communications technology to monitor and react to changes in local demand. EVs can aid in lowering the load on the smart grid in addition to the infrastructure for smart grid charging. Smart charging can hold the key to increasing productivity in the energy sector and lowering emissions. The battery life of electric vehicles can now only be managed to a limited extent by engineers, who are working hard to discover solutions (Figure 4) (Munshi, A. A. et al., 2017, Ranganathan, P. et al., 2011).

Figure 4. Connection between an EV charging station and the smart grid

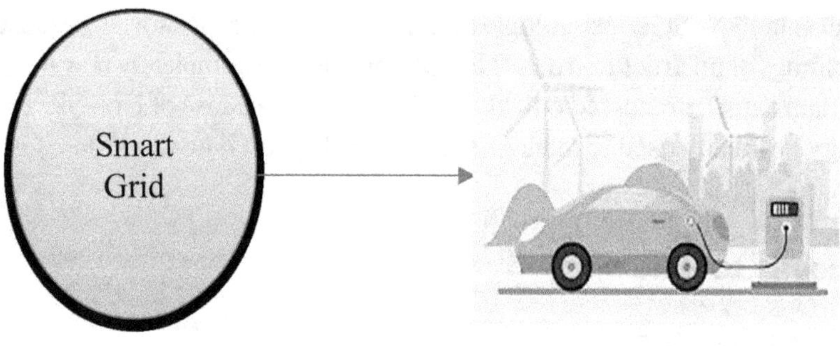

2) Smart Meter

Smart meters are digital meters that enable two-way communication between the utility provider and the client. Real-time data connectivity with cloud servers is made possible, both locally and remotely. The system contains features including two-way metering, remote communication, different electricity price billing, remote power outages, power quality monitoring, and user engagement. Electric vehicle charging stations are linked to grid firms by smart meters, which serve as ports and edge computing nodes.

Successfully developing a Smart Energy Meter (SEM) is essential for the power billing system for chargers for electric vehicles. Through an Android-based application, Bluetooth interoperability between

Figure 5. Connection of smart meters and charging stations to the grid, customers, billing systems, and utility providers

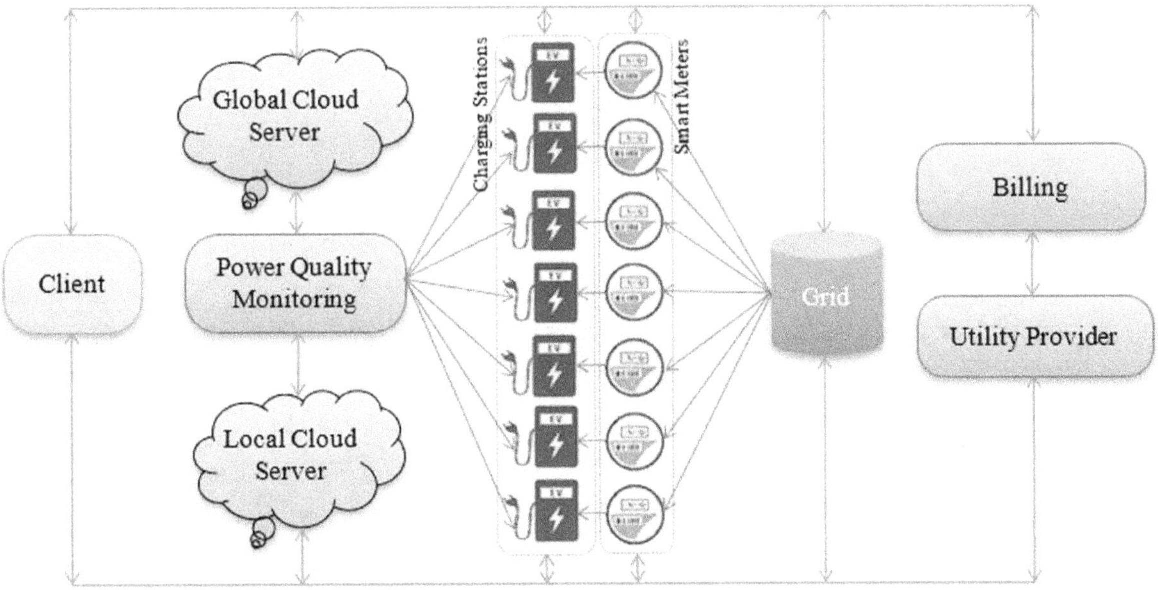

SEM users or operators is made feasible. Bluetooth reduces operator errors in voltage and current monitoring. Data from smart meters is comparable to personal data because it is kept on a smart meter (Figure 5).

3) Electronics with Intelligence

IEDs (Intelligent Electronic Devices) are digital protection relays since they employ microprocessors to carry out different protection, control, and associated duties. Voltage regulators, capacitor bank switches, closure controllers, tap changer controllers, and circuit breaker controllers are just a few examples. Developing commercial solutions at a reasonable cost is made feasible by M2M (Machine-to-Machine) communications and Intelligent Electronic Devices (IEDs). The IED prototype is made to be sufficiently functional to monitor important automobile parts. GPRS/GSM data transfer, signal processing, geolocation maps (GIS), and systems that combine GPS functions into a single device (Figure 6) (Kisacikoglu, M. C. et al., 2017).

4) Phasor Measurement Unit

A phasor measurement unit (PMU) is used to determine the magnitude and phase angle of electrical values in an electrical grid, such as voltage or current. The grid's frequency can be set using the PMU. The AC waveform's phase angle and voltage amplitude are measured by the PMU. The PMU can accept sources as long as they are all calibrated and functioning as a single unit, which frequently includes non-GPS references. PMU is a quantifiable indicator of the current's intensity and phase. The PMU can swiftly gather samples and rebuild values from waveforms, including data for angle and amplitude. A typical PMU can deliver measurements with a very high temporal resolution at a rate of up to 120 per second. A sinusoid is used to solve the mathematical fitting issue of fitting measurements. The phase representa-

Figure 6. Intelligent electronic components are the backbone of an intelligent cyber-physical power system (courtesy of Amin, B. R. et at., 2021)

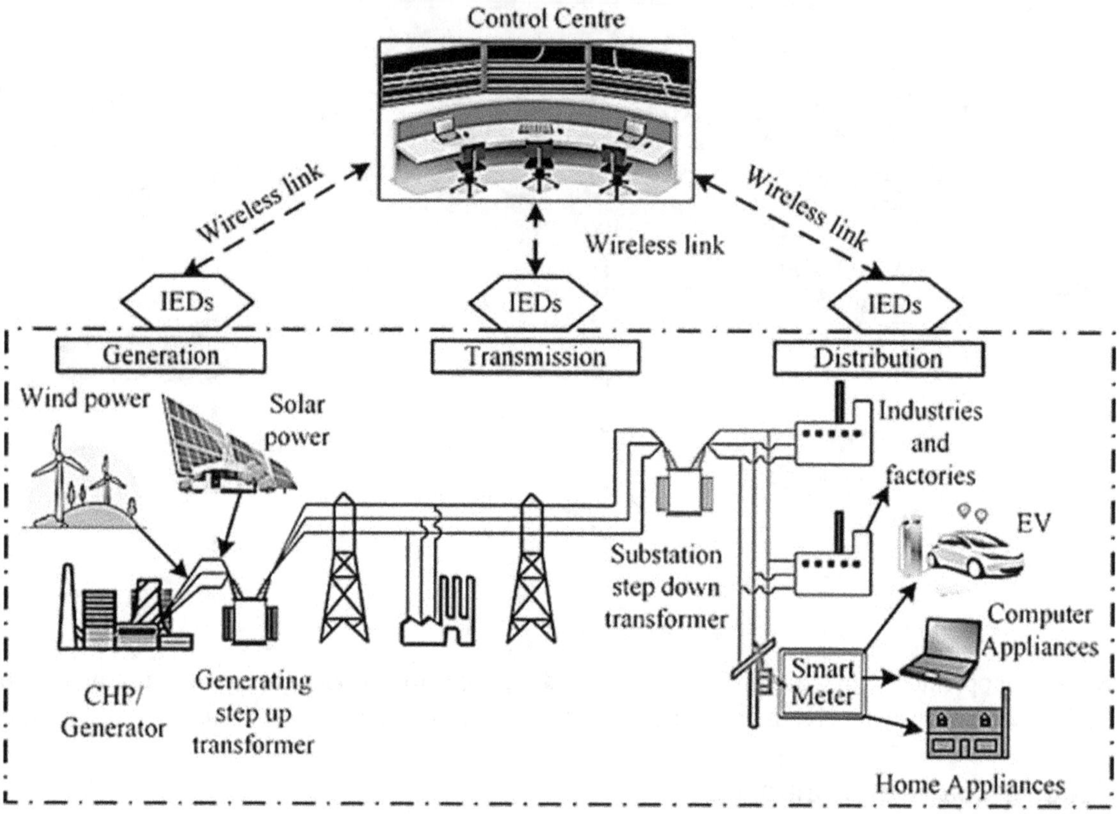

tion is inadequate when there are few sine waves available, as there are when there are voltage dips or grid behavior issues. The system's visibility and preventative management strategies might be greatly enhanced by this high degree of accuracy. The measurement error on the tiny PMU leads is decreased from 1° to 0.05° as a result of the phase angle. PMUs are referred to as "micro," which denotes preciser measurements. Phase measurements known as synchrophasors take place concurrently across any distance. PMUs measure voltage and current at significant substations (main grid intersections). Phase coordination is effective, enabling the simultaneous comparison of two variables in real time. Utilize phasor data concentrator technology to gather data locally or at a single site (Figure 7).

Figure 7. Phasor management unit sequence

POLICIES FOR ANALYSIS OF DATA ANALYTICS

5) Positioning of Charging Stations

Electric motors, which are found in electric vehicles, run on electricity. Due to the absence of a gasoline engine, driving an electric vehicle produces zero carbon dioxide. Charge times for electric vehicles are longer than for fuel vehicles. The placement of charging stations will become a significant issue as the EV industry expands. Through a number of studies, the best location for electric vehicle charging stations has been identified.

Meta-heuristics like genetic algorithms and neural networks are among them. The majority of these technologies are thought to save costs through shortening wait times, reducing travel time, and lowering destination or distance costs (Figure 8).

Currently, charging an EV is probably free, whether it be done privately or publicly through open charging stations offered by organizations like the government and enterprises. Commercial charging stations will need to be planned for and constructed as the EV market expands. It is vital to consider the interests of the constituency, which includes users, owners of charging stations, operators of power grids, local governments, etc., in order to build an effective administration and supervision of EV charging infrastructure. The ideal charging technique can be investigated and predicted using EVs from nested logistic models. A nested logit model could be used by service providers to gauge the charging demand at each possible charging station at the start of each phase. Bayesian games could be used to discover the best placement strategy. The QoS (Quality of Service) requirements for present charging stations could no longer be satisfied when EV adoption rates rise. As a consequence, the installation of the charging station will move into a new phase that is intended to be convenient for the user.

6) Intelligent Charging Algorithms

Employing a Markov decision process and a reinforcement learning solution model to examine how electric vehicle charging works and create an intelligent charging technique. The results show that the intelligent charging algorithm can successfully reduce expenses by about 30% on average when compared to traditional charging. Vehicle-to-grid technology, which enables electric automobiles to discharge their energy back into the grid to ease grid pressure brought on by increased demand, is a potential future solution (Lee, K. B. et al., 2020).

Effective control of auxiliary systems on a vehicle is likely to be crucial if an intelligent charging algorithm is to be employed on actual vehicles. The learning process of intelligent charging algorithms can take into account power costs as well as include strategies for forecasting uncertain electricity rates.

Evidently, the intelligent charging algorithm increases charging current while decreasing it for the subsequent action. There are at least two key reasons why this occurred. The first is that there are big disparities between a model and a reward signal. The second justification is that techniques for dynamic programming are supposed to include round-off mistakes in them.

There are four main modules in EVCNS. Communication with all of the EVCS located across the city is the CPM's primary duty. For each EVCS, the FEM (feature extract module) extracts the requested facility status, including the anticipated journey time, travel distance, and arrival time. The RCSM is responsible for choosing the best route and EVCS for each request for an EV charge. The ideal action-value function is estimated using a deep Q-network (DQN) that has been trained. Sequential decision

Figure 8. A flowchart for determining the shortest path to a charging station

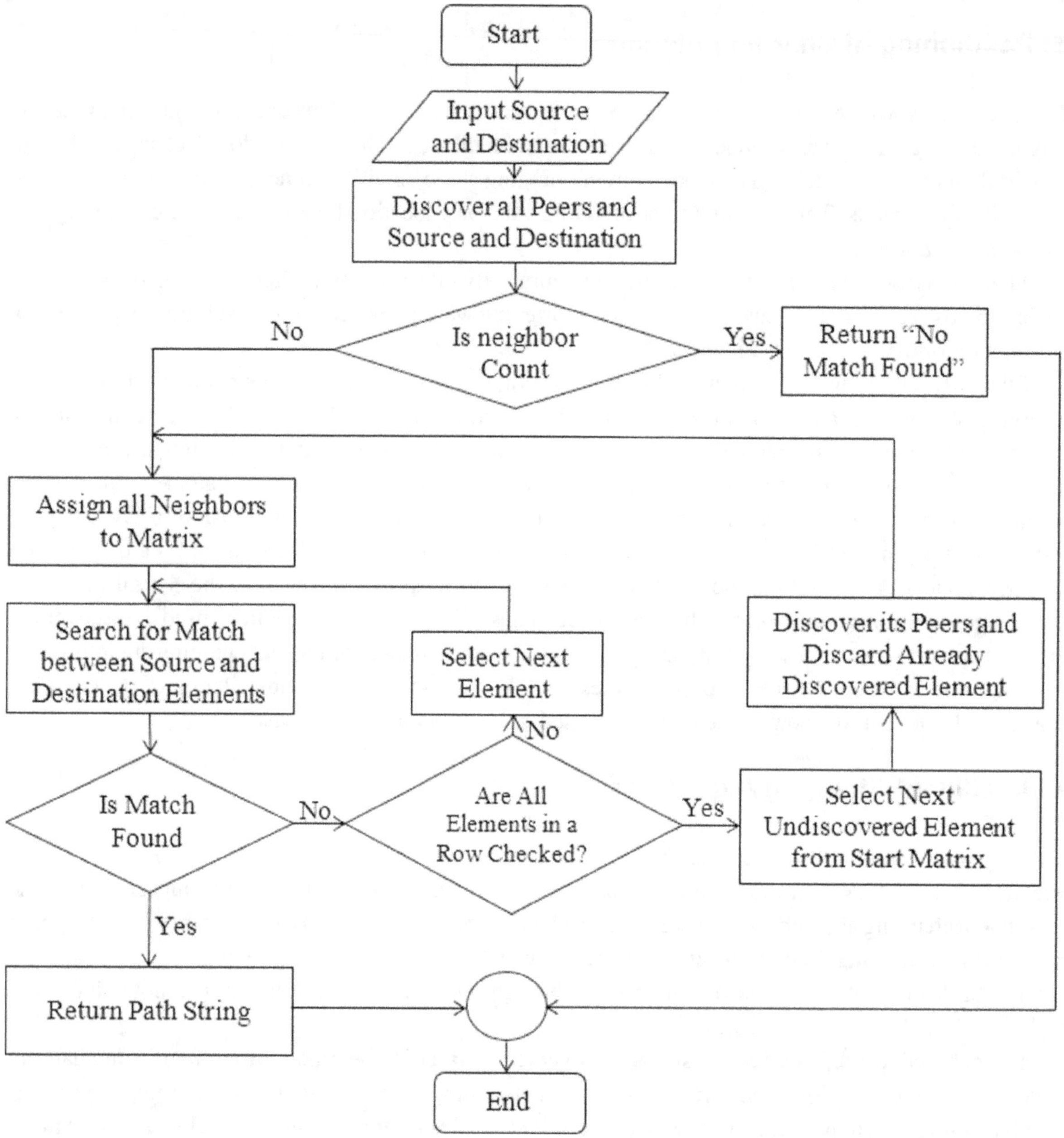

issues are frequently formalized using the MDP (Markov decision process) model (Figure 9) (Malek, Y. N. et al., 2021).

7) Solving Problems with Energy Efficiency

Through machine learning (ML) models, the electric vehicle manufacturer has been working to increase the energy efficiency and consumption of electric vehicles. AI systems can be very effective when ML

Figure 9. EV Charging Navigation System based on Deep Reinforcement Learning (adapted from (Lee, K. B., A et al., 2020))

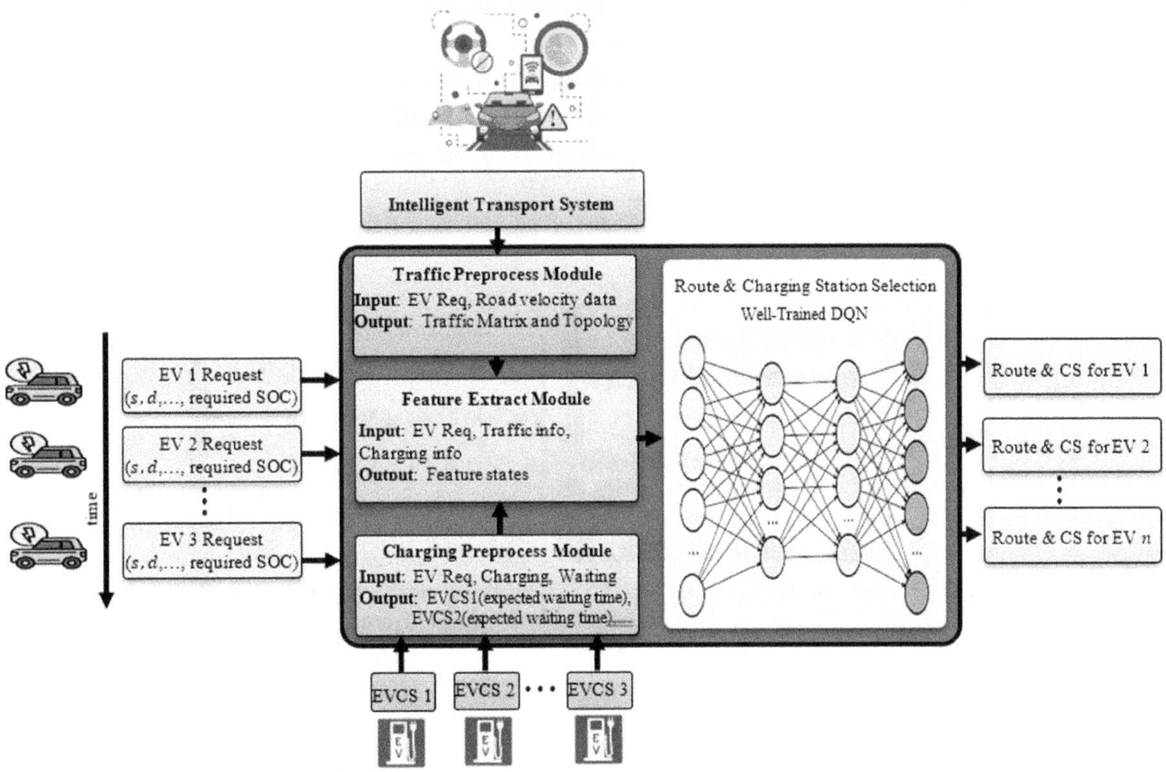

models are developed locally to improve vehicle energy efficiency and electric vehicle utilization. It offers data-centric solutions for effective contract execution, ongoing contract monitoring, and dependable, secure digital service exchange (Figures 10 and 11).

Companies like Tesla, Google, Bosch, and others are researching self-driving vehicles using AI. It can assist us in maintaining a connection with our vehicles as well as with another vehicle for the ease of routine maintenance, economical driving, or passenger safety. One vehicle can collaborate and exchange information with other vehicles on the road.

AI could enable vehicles to do much more than simply monitor the road; this could help drivers stay focused on the road by warning them when there are distractions around. In the case of accidents, AI may make "do it yourself" insurance claims possible. The size of the global battery market is rapidly expanding. Some predict that it will approach $250 billion annually by 2025. After taking a look at the EV market in India, Tesla decided to establish a supercharging station. The battery will fully charge using this method in a minimum of 75 minutes and a maximum of 3 hours. Rapid charging is not yet a reality, though, due to a number of issues. The complex problems relating to battery development can be handled more rapidly and affordably via data science. AI enables innovators to adjust more than one component at a time and review data more quickly, enabling them to achieve things that traditional statistical analysis cannot. Toyota said that it would speed up battery-related research utilizing artificial intelligence. It might be quite beneficial to know when to upgrade a battery or drive. A Stanford research team developed the machine learning-based method. It decreases testing time by 98% and reduces the number of tests from two years to 16 days. Researchers at the US Department of Energy are utilizing machine learning and artificial intelligence to find possible new battery options. They used a database of roughly 133,000 tiny organic compounds to assess possible electrolyte candidates in an effort to identify a solution to the charging problem for batteries.

EVCS- electric vehicle charging stations, EV-Electric Vehicle, SOC- state of charge, *s*-source, *d*-destination, CS-Charging Station, CPM-Charging Preprocess Module

Figure 10. Machine learning application in EV

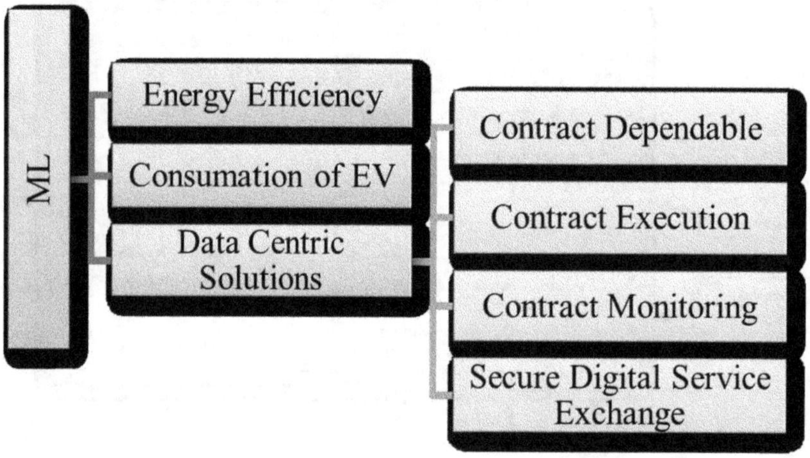

Figure 11. Artificial intelligence application in EV

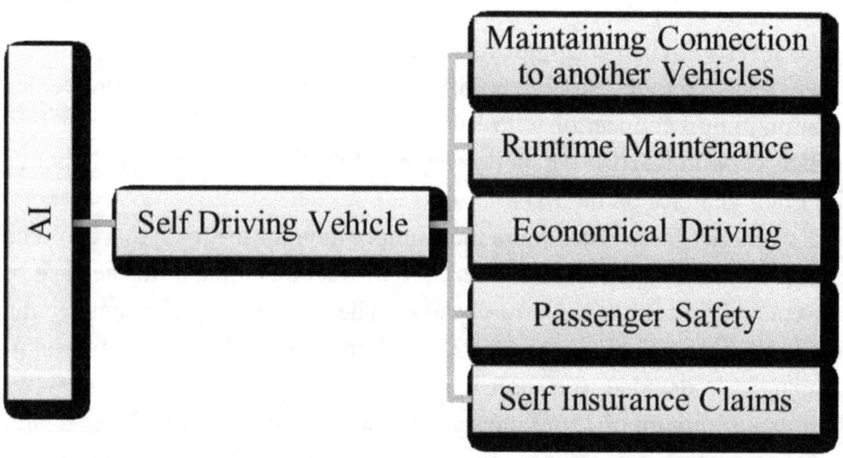

8) Analysis of Power Distribution System's Capacity To Cope With Additional Charging Load

The unpredictability of the system is increased by the growing usage of distribution grids for penetration of photovoltaic (PVRE) and electric vehicle (EV) charging. The input distribution model has to be improved for the physical purity of the load, PV generation, and EV charging. PV (photovoltaic) is the primary distributed VRE (variable renewable energy) source linked to the power distribution network. PV grid integration can result in a number of unfavorable outcomes, including high voltages from excess output, poor power quality, equipment damage, reliability issues, and protracted outage restoration periods. Correlation models are required to account for the correlations between the input variables (Figure 12).

FUTURE OF DATA ANALYTICS IN ELECTRIC VEHICLE

The information generated by electric vehicles could be a resource for engineers and developers. This could require modifying the electrical grid's capacity, picking locations for charging stations and expanding the range of feasible vehicles. Through these, during peak hours, the load on the grid could be minimal.

All-electric vehicles generate and retain enormous quantities of data, and this data can be utilized to significantly enhance driving. And there has never been a more exciting time to extract massive volumes of data from our roads than now, with the growth of electric, connected automobiles on the road.

Figure 12. An outline for modelling an EV-PV charging station that is connected to the grid

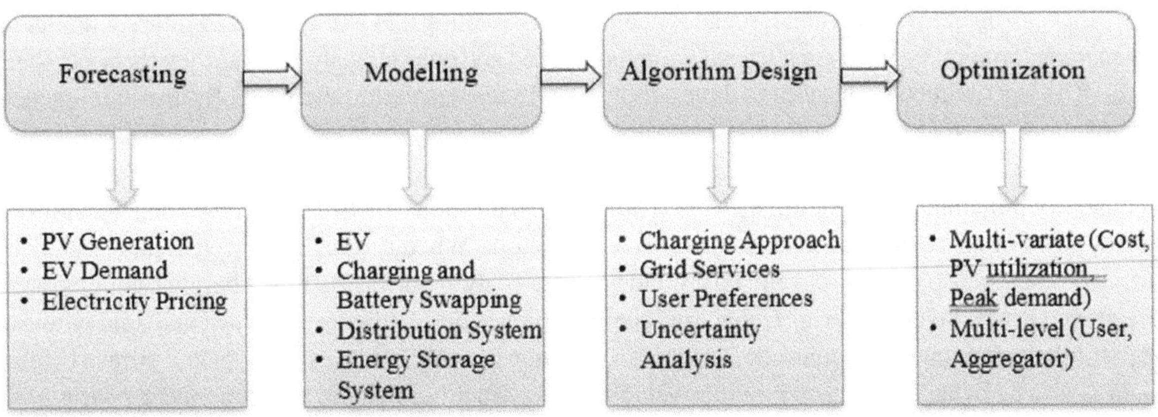

An improved understanding of electric vehicles can be achieved through data analytics. The following issues will be helped by data analysis in the future:

1. Determining the data analytics areas for electric vehicles
2. How can data analytics be applied to guide the driver?
3. How can the application of data analytics affect the vehicle's other components?

4. How can predictive analytics and data intelligence be used for the operation of a driverless electric vehicle?

Predictive maintenance makes it possible to have high battery efficiency and operational reliability. Utilizing data science, AI, and big data techniques will improve the performance of battery packs. The inherent potential and possibility of battery life cycle control have been taken advantage of due to the usage of machine learning. The secret to prolonging battery life is data. Predictive intelligence is used by machine learning, which integrates current electronics with IoT, data science, and digital twins, to estimate battery life, foresee future deterioration and failures and their causes, and stop delays and errors from occurring. ML could add an extra layer of intelligence after gathering and evaluating enormous quantities of data on battery life, performance, charge level, stress from abrupt acceleration and deceleration, temperature, number of charge cycles, etc. Engineers can update cloud-based software without consumers having to visit a service center by employing AI algorithms. Future all-electric vehicle scaling will be facilitated by algorithms for feed optimization, operator allocation, positioning, path planning, sensor fusion, autonomous stack algorithms, object recognition and perception, and computer vision (Lee, J. et al., 2017).

For electrically connected autonomous and shared (ECAS) vehicles, digitalization accelerates the technological convergence of vision, computer processing, communication, propulsion, and data fusion. Vehicles are becoming increasingly intelligent, connected, edge micro-servers on wheels, driven by sensors and actuators, hardware, software, and digital infrastructure. The basic five human senses of sight, hearing, taste, smell, and touch must be matched or surpassed by ECAS vehicles. They must possess the decision-making and problem-solving skills necessary to interact safely with other vehicles, other road users, and the physical environment. Future automotive advancements will rely on many AI types that can operate in individual, social, or collaborative settings. In a fully realized future, intelligent green transportation systems are intricately interconnected and continuously learn as they carry out their tasks. The fleet will be comprised of Internet of Vehicle Systems with integrated Swarm Intelligence and AI algorithms.

The use of next-generation ECAS vehicles to achieve sustainable green transportation requires dependability, reliability, legislation, and standardization. The future of ECAS vehicles depends on the use of AI to offer technologies across many vehicle sectors. When a vehicle is electrified, a complete infrastructure industry can be developed around wireless charging infrastructure, charging stations, and, to a greater extent, vehicle-sharing technology across communities. Utilizing two-way intelligent wireless and cellular communication, connected vehicles communicate with their surroundings and infrastructure. Data is shared by automobiles with other vehicles, infrastructure, people, and equipment both within and outside the vehicle. The requirement for safe, secure, dependable, and unhackable in-vehicle and edge/cloud-based components grows as the risk of malevolent hackers and compromised cyber security increases with increased connectivity.

Automotive driving automation systems that continuously carry out any or all of the dynamic driving tasks (DDT) are defined by SAE. The reliability of the system is crucial. For upcoming vehicles, the term "shared" also has a second connotation. Users may concentrate on their destination and how to combine a variety of mobility options to get there with the help of seamless multimodal experiences. At the component, embedded software, and system levels, vehicles require rigorous security and privacy measures.

ECAS vehicles are leading the way in utilizing AI to mimic, improve, and support human behaviors while simultaneously utilizing the quick response times of machine-based systems. Implementing highly automated and environmentally friendly green mobility applications requires the use of perception, sensor fusion, AI, edge computing, connectivity, virtualization of functions, and resource and information sharing. Manufacturers of ECAS vehicles are creating their own electric and electronics (E/E) designs. Applications and services that require automotive-grade embedded AI-based high-performance computing components with considerable compute capacity can access ECAS Vehicle Architecture Level 5 applications and services while improving energy economy and scalability. All architectural layers' design concepts and tactics have changed as a result of developments in e/e vehicle architecture and the use of AI techniques. All the components that enable vehicles to explore, comprehend, comprehend, think, connect, act, cooperate, and learn are organized and connected by the domain-based approach. The ECAS vehicle's architecture will develop into a unified vehicle/edge/cloud-centric design based on distributed intelligence at the infrastructure and vehicle levels. Architectural conceptions must incorporate the autonomous system qualities of composability and modularity.

The domain-based ECAS vehicle architecture uses separate, interoperable, and reusable building elements to provide scalability across each vehicle's domain. The creation and use of algorithms for autonomous driving are also made simpler. ECAS vehicles are increasingly using lithium-ion batteries as their power source. The development of new solid-state batteries that can offer safer, lighter, more energy-dense, and power-generating battery solutions is the focus of current innovations. The energy industry is undergoing fundamental change due to digitalization, and battery layouts have been optimized for sensors and sensing points. The state of charge (SOC) and other elements of the battery condition should be properly estimated by the BMS (battery management system). Due to their tremendous flexibility and capacity for self-learning, ML and DL techniques such as temporal convolutional neural networks and recurrent neural networks have shown their capacity to solve these issues.

In the energy domain, V2G (Vehicle-to-Grid) control and function interfaces are present. As a result, the interaction with the charging infrastructure is guaranteed to be secure, practical, affordable, and energy-efficient. An effective power management system makes it possible to link a large number of ECAS vehicles to the grid. For vehicles with internal combustion engines (ICE), hybrids, or electric motors, the powertrain is a component of the vehicle architecture. Various sensors, actuators, and auxiliary subsystems based on actuator technologies are used in vehicle dynamics to address the impact of forces and torques on the motion. The use of components that adhere to an automotive integrity level is required for the powertrain and vehicle dynamics. More and more ECAS vehicles are being powered by lithium-ion batteries. New solid-state batteries that can offer battery solutions that are safer, lighter, more energy-dense, and power-generating are the main focus of current innovations. With battery topologies tailored for sensors and sensing sites, digitalization is profoundly altering the energy sector. The BMS should provide an accurate estimation of the battery's state of charge (SOC) and other characteristics. Due to their tremendous flexibility and capacity for self-learning, ML and DL techniques such as temporal convolutional neural networks and recurrent neural networks have shown their capacity to solve these issues.

V2G control and function interfaces are present in the energy domain. As a result, the interaction with the charging infrastructure is guaranteed to be secure, practical, affordable, and energy-efficient. An effective power management system makes it possible to link a large number of ECAS vehicles to the grid. Automotive Safety Integrity Level (ASIL) D-compliant components must be used in the powertrain and vehicle dynamics domains. The wired and wireless interfaces that link the vehicle to its onboard computers, other vehicles, the internet, and infrastructure are all under the control of the connectivity domain. For ECAS vehicles, the autonomous driver and cognition domain includes planning, computing, cognitive, and decision-making capabilities. High-performance embedded edge computing and AI-based technologies have been used in its implementation. By establishing a connection to the edge/cloud computing infrastructure, the training and learning process can be carried out continually. The infotainment vehicle domain has grown into the vehicle experience domain, giving vehicles the potential to give passengers the desired onboard wellness, comfort, entertainment, and productivity. Passengers can produce and edit digital and virtual information, and it makes it easy to access digital content.

For ECAS vehicles, safety metrics and standards are being created. ODDs (operational design domains) for the processes of authentication and homogenization must be introduced in stages. The development of ODD-based autonomous vehicle safety standards faces additional difficulties with the introduction of AI-based tools and techniques. Even using dependable components won't eliminate flaws. It's crucial to keep track of the HW and SW components and how they work (Figure 13).

Figure 13. Information fusion is a method used by ECAS vehicles (adapted from (Vermesan, Ovidiu, et al., 2021))

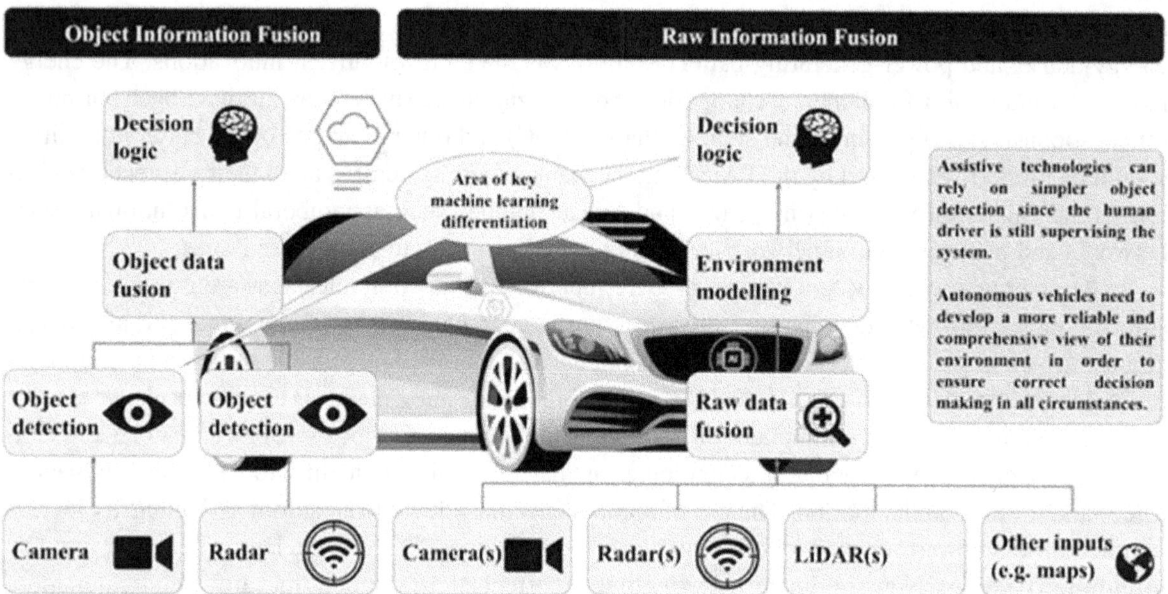

CONCLUSION

EVs are the way of the future for transportation and driving. They lower CO_2 emissions and offer considerable operating and maintenance cost reductions. The future need for millions of barrels of oil will decline as there are more electric vehicles on the road. By 2030, switching to electric vehicles would help India cut its CO_2 emissions by about one giga tonne. The air quality will improve as toxic emissions are reduced, which implies fewer health issues. Compared to internal combustion engines and their exhaust systems, electric motors are substantially quieter.

The Indian government is taking a number of significant initiatives to promote electric vehicles, and it is anticipated that they will continue to do so in the future, which will increase customer interest in purchasing electric vehicles. India's adoption of electric vehicles would lower India's capital expenditure, which would significantly boost the country's GDP. Along with this, the Indian government ought to offer incentives to overseas manufacturers of electric vehicles so they may establish operations in India.

REFERENCES

Aggarwal, C. C. (2011). An introduction to social network data analytics. In *Social network data analytics* (pp. 1–15). Springer. doi:10.1007/978-1-4419-8462-3_1

Akhavan-Hejazi, H., Mohsenian-Rad, H., & Nejat, A. (2014, September). Developing a test data set for electric vehicle applications in smart grid research. In *2014 IEEE 80th Vehicular Technology Conference (VTC2014-Fall)* (pp. 1-6). IEEE. 10.1109/VTCFall.2014.6965934

Amin, B. R., Hossain, M. J., Anwar, A., & Zaman, S. (2021). Cyber attacks and faults discrimination in intelligent electronic device-based energy management systems. *Electronics (Basel)*, *10*(6), 650. doi:10.3390/electronics10060650

Bolly, V. K., Springer, J., & Dietz, J. E. (2014, June). Using Open Source NoSQL technologies in Designing Systems for Delivering Electric Vehicle Data Analytics. In *2014 ASEE Annual Conference & Exposition* (pp. 24-1339). 10.18260/1-2--23272

Cai, H., Jia, X., Chiu, A. S., Hu, X., & Xu, M. (2014). Siting public electric vehicle charging stations in Beijing using big-data informed travel patterns of the taxi fleet. *Transportation Research Part D, Transport and Environment*, *33*, 39–46. doi:10.1016/j.trd.2014.09.003

Cao, Y., Song, H., Kaiwartya, O., Zhou, B., Zhuang, Y., Cao, Y., & Zhang, X. (2018). Mobile edge computing for big-data-enabled electric vehicle charging. *IEEE Communications Magazine*, *56*(3), 150–156. doi:10.1109/MCOM.2018.1700210

Chen, T. D., Kockelman, K. M., & Khan, M. (2013). Locating electric vehicle charging stations: Parking-based assignment method for Seattle, Washington. *Transportation Research Record: Journal of the Transportation Research Board*, *2385*(1), 28–36. doi:10.3141/2385-04

Cheng, Z., Chow, M. Y., Jung, D., & Jeon, J. (2017, June). A big data based deep learning approach for vehicle speed prediction. In *2017 IEEE 26th International Symposium on Industrial Electronics (ISIE)* (pp. 389-394). IEEE. 10.1109/ISIE.2017.8001278

Diamantoulakis, P. D., Kapinas, V. M., & Karagiannidis, G. K. (2015). Big data analytics for dynamic energy management in smart grids. *Big Data Research*, *2*(3), 94–101. doi:10.1016/j.bdr.2015.03.003

Elgendy, N., & Elragal, A. (2014, July). Big data analytics: a literature review paper. In Industrial conference on data mining (pp. 214-227). Springer. doi:10.1007/978-3-319-08976-8_16

Hussain, M. M., Beg, M. S., Alam, M. S., & Laskar, S. H. (2020). Big data analytics platforms for electric vehicle integration in transport oriented smart cities: Computing platforms for platforms for electric vehicle integration in smart cities. In *Cyber warfare and terrorism: Concepts, methodologies, tools, and applications* (pp. 833–854). IGI Global. doi:10.4018/978-1-7998-2466-4.ch051

Kambatla, K., Kollias, G., Kumar, V., & Grama, A. (2014). Trends in big data analytics. *Journal of Parallel and Distributed Computing, 74*(7), 2561–2573. doi:10.1016/j.jpdc.2014.01.003

Kisacikoglu, M. C., Erden, F., & Erdogan, N. (2017). Distributed control of PEV charging based on energy demand forecast. *IEEE Transactions on Industrial Informatics, 14*(1), 332–341. doi:10.1109/TII.2017.2705075

LaValle, S., Lesser, E., Shockley, R., Hopkins, M. S., & Kruschwitz, N. (2011). Big data, analytics and the path from insights to value. *MIT Sloan Management Review, 52*(2), 21–32.

Lee, C. H., & Wu, C. H. (2015, April). Collecting and mining big data for electric vehicle systems using battery modeling data. In *2015 12th International Conference on Information Technology-New Generations* (pp. 626-631). IEEE. 10.1109/ITNG.2015.104

Lee, J., Bagheri, B., & Kao, H. A. (2014, July). Recent advances and trends of cyber-physical systems and big data analytics in industrial informatics. In *International proceeding of int conference on industrial informatics (INDIN)* (pp. 1-6). Academic Press.

Lee, J., & Park, G. L. (2016). Electric vehicle charger management system for interoperable charging facilities. *Jurnal Teknologi, 78*(5-8).

Lee, J., Park, G. L., Cho, Y., Kim, S., & Jung, J. (2015, April). Spatio-temporal analysis of state-of-charge streams for electric vehicles. In *Proceedings of the 14th International Conference on Information Processing in Sensor Networks* (pp. 368-369). 10.1145/2737095.2742932

Lee, J., Park, G. L., Han, Y., & Yoo, S. (2017, May). Big data analysis for an electric vehicle charging infrastructure using open data and software. In *Proceedings of the eighth international conference on future energy systems* (pp. 252-253). 10.1145/3077839.3081670

Lee, K. B. A., Ahmed, M., Kang, D. K., & Kim, Y. C. (2020). Deep reinforcement learning based optimal route and charging station selection. *Energies, 13*(23), 6255. doi:10.3390/en13236255

Li, B., Kisacikoglu, M. C., Liu, C., Singh, N., & Erol-Kantarci, M. (2017). Big data analytics for electric vehicle integration in green smart cities. *IEEE Communications Magazine, 55*(11), 19–25. doi:10.1109/MCOM.2017.1700133

Lv, Z., Qiao, L., Cai, K., & Wang, Q. (2020). Big data analysis technology for electric vehicle networks in smart cities. *IEEE Transactions on Intelligent Transportation Systems, 22*(3), 1807–1816. doi:10.1109/TITS.2020.3008884

Malek, Y. N., Najib, M., Bakhouya, M., & Essaaidi, M. (2021). Multivariate deep learning approach for electric vehicle speed forecasting. *Big Data Mining and Analytics, 4*(1), 56–64. doi:10.26599/BDMA.2020.9020027

Munshi, A. A., & Yasser, A. R. M. (2017). Big data framework for analytics in smart grids. *Electric Power Systems Research, 151*, 369–380. doi:10.1016/j.epsr.2017.06.006

Northfield, R. (2016). Greening the smart city. *Engineering & Technology, 11*(5), 38–41. doi:10.1049/et.2016.0503

Rahimi-Eichi, H., & Chow, M. Y. (2014, October). Big-data framework for electric vehicle range estimation. In *IECON 2014-40th Annual Conference of the IEEE Industrial Electronics Society* (pp. 5628-5634). IEEE. 10.1109/IECON.2014.7049362

Ranganathan, P., & Nygard, K. (2011, October). *Smart grid data analytics for decision support. In 2011 IEEE Electrical Power and Energy Conference*. IEEE.

Runkler, T. A. (2020). *Data analytics*. Springer Fachmedien Wiesbaden. doi:10.1007/978-3-658-29779-4

Russom, P. (2011). Big data analytics. *TDWI Best Practices Report, 19*(4), 1-34.

Tsai, C. W., Lai, C. F., Chao, H. C., & Vasilakos, A. V. (2015). Big data analytics: A survey. *Journal of Big Data, 2*(1), 1–32. doi:10.118640537-015-0030-3 PMID:26191487

Vermesan, O., John, R., Pype, P., Daalderop, G., Kriegel, K., Mitic, G., ... Waldhör, S. (2021). Automotive Intelligence Embedded in Electric Connected Autonomous and Shared Vehicles Technology for Sustainable Green Mobility. *Frontiers in Future Transportation*, 18.

Wang, K., Yu, J., Yu, Y., Qian, Y., Zeng, D., Guo, S., Xiang, Y., & Wu, J. (2017). A survey on energy internet: Architecture, approach, and emerging technologies. *IEEE Systems Journal, 12*(3), 2403–2416. doi:10.1109/JSYST.2016.2639820

Wei, W., Minghao, A., Naishi, C., Xianjun, G., & Tianjiao, P. (2016). Multi-Level Feeder Queue Optimization Charging Model of Electric Vehicle and its Implementation of MR Algorithm. *International Journal of u-and e-Service, Science and Technology, 9*(3), 199–208.

Wu, J., Guo, S., Li, J., & Zeng, D. (2016). Big data meet green challenges: Big data toward green applications. *IEEE Systems Journal, 10*(3), 888–900. doi:10.1109/JSYST.2016.2550530

Yu, N., Shah, S., Johnson, R., Sherick, R., Hong, M., & Loparo, K. (2015, February). Big data analytics in power distribution systems. In 2015 IEEE power & energy society innovative smart grid technologies conference (ISGT) (pp. 1-5). IEEE. doi:10.1109/ISGT.2015.7131868

ADDITIONAL READING

Chan, C. C., Chau, K. T., & Chau, K. T. (2001). *Modern electric vehicle technology* (Vol. 47). Oxford University Press on Demand.

Dhameja, S. (2001). *Electric vehicle battery systems*. Elsevier.

Garcia-Valle, R., & Lopes, J. A. P. (Eds.). (2012). *Electric vehicle integration into modern power networks*. Springer Science & Business Media.

Hamut, H. S., Javani, N., & Dinçer, I. (2016). *Thermal management of electric vehicle battery systems*. John Wiley & Sons.

Larminie, J., & Lowry, J. (2012). *Electric vehicle technology explained*. John Wiley & Sons. doi:10.1002/9781118361146

KEY TERMS AND DEFINITIONS

Big Data: Using sophisticated analytical tools, big data analytics is the study of huge and varied data sets.

Cloud: Data on a private or public cloud can be subjected to cloud-based analysis algorithms to provide discoveries that are of interest.

Data Analytics: It involves examining a dataset to identify trends and form opinions about the data it contains.

Electric Vehicles: Instead of an electric motor, use a combustion engine.

Hadoop: A system that can analyses and store enormous datasets that range in size from gigabytes to petabytes and is free source.

MapReduce: The MapReduce programming model can be used on clusters to process large datasets using concurrent distributed algorithms.

Pig: A large-scale data analysis tool.

Smart Grid: Utilize complex analytics, including forecasting, optimization, and predictive and prescriptive analytics.

Chapter 11
Autonomous Electric Vehicles and Edge Computing

Amuthan Nallathambi
https://orcid.org/0000-0001-8830-3353
AMC Engineering College, Bengaluru, India

Sivakumar Nallathambi
https://orcid.org/0000-0002-6876-3253
Department of Mechanical Engineering, Nagercoil, India

Gopal Samy B.
https://orcid.org/0000-0001-8938-6023
VSB Engineering College, Karur, India

ABSTRACT

There is software that can be installed on a computer to control how an autonomous controller operates in a vehicle. The central processing unit (CPU) of the controller receives input from position sensors, processes that input, and then outputs operation control signals that indicate the updated course that the vehicle will take. A programming interface and operation controls are also included as parts of the controller. This interface helps to facilitate communication between the various controller components. Inputs from position sensors are normalized for the processor by the controller, which then sends outputs from itself as inputs into the operation control mechanisms. It is a stand-alone device that can be configured to function with a broad range of distant sensors and control mechanisms. As a direct consequence of the rapid spread of new technologies, edge computing has developed into a more widespread practice. Users may be able to have a more individualized and interactive experience in their day-to-day lives thanks to edge computing and edge devices.

INTRODUCTION

Over the last several years, there has been a notable rise in the number of electric vehicles (EVs) being

DOI: 10.4018/978-1-6684-6631-5.ch011

used in the transportation industry. As oil reserves go lower, the world's dependence on oil as a source of energy for passenger cars will have more negative effects on the global economy and politics. These effects are only going to get worse. As the world struggles with the effects of climate change brought on by greenhouse gas emissions from automobile tailpipes and concerns about oil depletion and supply security remain at an all-time high, there is a growing interest in non-traditional means of transportation. This interest comes at a time when the world is also suffering from the consequences of climate change. Electric cars, often known as EVs, are gaining popularity as a promising environmentally friendly technology that has the ability to reduce emissions of greenhouse gases and remove our reliance on oil imported from other countries (Indragandhi, Selvamathi, R., & Subramaniyaswamy, V., 2022). When it comes to the safe operation of a contemporary vehicle, the judgments that the driver makes about speed, steering, knowledge of obstacles and hazards, and the avoidance of such barriers and hazards, among other things, continue to be vital components of the vehicle's control system. It is conceivable for a driver's ability to do all of these jobs to be impaired by physical factors such as driver fatigue, impairment, or inattention, or by other variables such as poor eyesight. However, it is feasible for a driver to still be able to fulfil all of these duties. Because of this, the amount of time a driver has to respond in order to avoid being involved in an accident is cut down significantly. When travelling through hostile territory or doing business in places where there is a significant risk of exposure to chemical or radioactive dangers, such as may be the case during times of conflict, the driver is also placed in danger while doing so.

The capacity of the diver (or the captain) to recognize and react to risks that are coming or to which the ship is approaching would be hindered in an environment that is confusing or incapacitating. This would apply to both the diver and the threats. An illustration of this would be operating a car or steering a ship under weather conditions that include snow, fog, rain, and/or nighttime power outages. As a consequence of this, there is a need for both a system and a method for the identification of stationary and moving things that are in the path of a vehicle or that are going to enter that path via a vehicle. This identification must be able to take place while the vehicle is in motion. This is the case regardless of whether the goal is to perform operations in environmentally hazardous situations where human control is not desired or to compensate for the limitations that are inherent in human control of vehicles. This is because human control is not desirable in these conditions (Nallathambi, A. et.al, 2022).

ELECTRIC VEHICLE CHARGING INFRASTRUCTURE

The proliferation of a charging infrastructure that is both more widespread and more advanced for electric cars, as well as electric automobiles in general, has contributed to a rise in the desirability of these automobiles. Over the last few years, there has been a significant increase in the quantity of energy that has been produced as a result of the use of solar power. Over the course of the last several years, there has been a discernible increase in both public knowledge and concern over the damage to the natural world that is caused by the very polluting internal combustion engine. As a direct consequence of this, there has been a surge in the amount of research conducted with the goal of producing an electric-powered car that is not only feasible but also reasonably priced.

Electric vehicles have far less impact, both positively and adversely, on the surrounding natural environment. They cannot function without energy, yet the energy they use often comes from sources that contribute to environmental pollution. For instance, if the power comes from the community's utility system, around eighty percent of it is produced by the combustion of potentially harmful fossil fuels.

Only a very small amount of the electrical energy that is produced actually finds its way to the people who are in need of it. This is because so much of it is squandered during the process of manufacturing, transferring, and distributing electricity. If we want to see any positive effects on the environment as a result of the shift to electric vehicles, we will need a source of pollution-free electricity that is not only cheap but also uncomplicated, simple, easily accessible, and immediately available. If we are going to be successful, this is not only a desirable perk; rather, it is a need and an essential factor that must be satisfied before we can achieve that success.

The production of a charging station for electrically powered automobiles that is not only user-friendly but also highly effective in its use of available energy. Another significant objective of the plan is to clear the path for the development of an electric vehicle charging station that is friendly to the natural environment in its immediate vicinity. An additional factor that contributed to the evolution of this idea was the pressing need for a user-friendly charging station designed specifically for electric vehicles. In addition to this, the concept proposes the development of a charging station that is both completely automated and has the capability of being connected to an electric car. This innovation has several purposes, one of which is to provide an improved setup for electrically driven autos so that their batteries may be charged throughout the daylight hours.

A charging station that is built and structured in accordance with the principles includes a space for storing energy, a space for storing vehicles, and a solar panel array that is exposed to a source of radiation in order to convert the radiation into electrical energy. The charging station also includes a solar panel array that is exposed to a source of radiation in order to convert the radiation into electrical energy. Within the confines of the charging station, you may be found in each of these regions. The electrical power of the vehicle will be kept in reserve at the energy storage facility for as long as it takes to transport the vehicle back to the location where it was first kept. The motor vehicle is now wired into the source of electricity so that it may get a charge from the source. Because of this explanation, the goals, benefits, and qualities will be easier to understand.

Electric mobility experts are in agreement that, in the not-too-distant future, a distribution and charging station network will be required in order to serve the vast variety of customers and uses for electric automobiles. This is something that will be necessary in order to meet the demand for electric mobility. The i-ES charging station, which is powered by solar energy, has been designed with the intention of facilitating the recharging of electric cars in high-traffic areas in a manner that is effective, simple, and environmentally friendly. As a result of the modular nature of the present invention, it is feasible that the national grid may eventually include Electric Vehicles in addition to devices for the microgeneration of energy. This would be accomplished in a gradual but consistent manner.

The solution incorporates a solar photovoltaic system for the generation of renewable energy, a direct charging station for electric cars (which eliminates the need to exchange batteries), and an easy user interface (UI). In addition to that, there is a system for monitoring and keeping track of the amount of energy that is used. The solar power plant is built using a series of modules, each of which is designed to get the most possible sunlight while at the same time fitting in perfectly with the natural environment in which it is situated. The fact that all of its electrical components are housed inside a single hub is another facet that differentiates it from goods that are functionally similar to it. Both the solar panel racks and their associated connections to the grid, as well as the infrastructure for charging electric vehicles, need to be physically compatible with one another. This distinction is significant for a number of reasons, the most important of which are that it simplifies construction and reduces the expenses that are connected with power losses.

There is an urgent need for the creation of a network of charging stations for Electric Vehicles (EVs), namely those that are capable of "rapid charging" in a period of time ranging from twenty to thirty minutes. This is a necessity that must be met immediately. When the driver does not have the luxury of committing a substantial amount of time to the process of recharging his car, this method of charging is advantageous since it allows the vehicle to be charged in a shorter amount of time. This charging station is designed for usage in public and private areas, such as parking lots of public and private buildings, fossil fuel service stations, and other sites where it would be impossible to charge an Electric Vehicle at a slow rate. This is because it is expected that the bulk of the charging will take place throughout the night in these sorts of circumstances. As a consequence of this, it is feasible to quickly charge Electric Cars (EVs), in addition to having the added benefit of making use of renewable energy that is produced locally. This is achievable because of the combination of these two factors.

Figure 1. General Diagram of a Solar Powered Charging Station

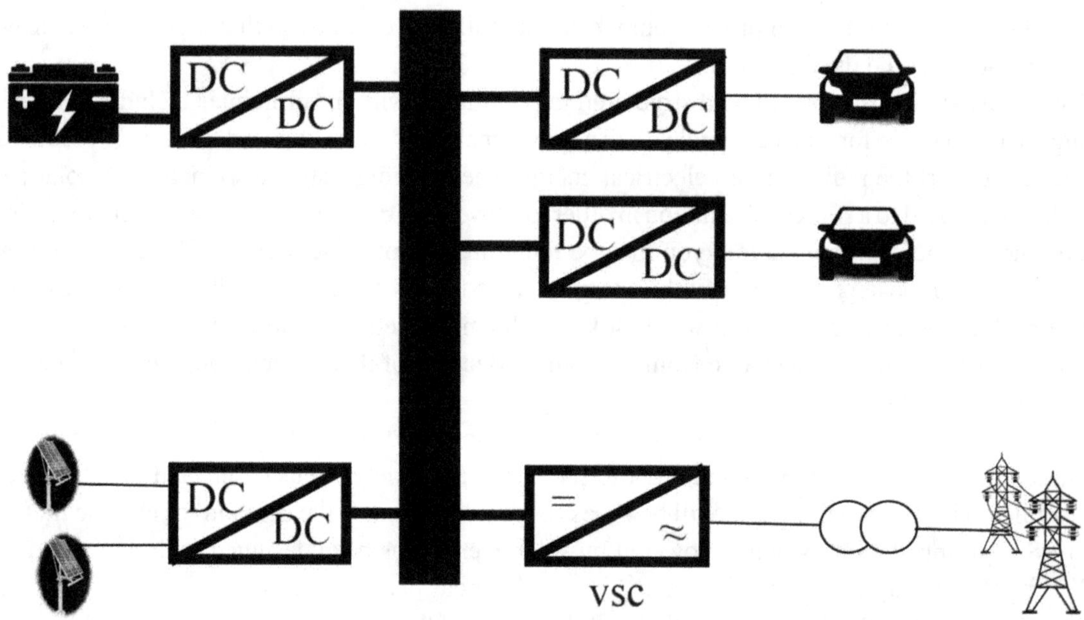

Figure 1. Shows the General Diagram of a Solar Powered Charging Station. Both the solar technology that is used for energy generation and the energy transformation that is used for rapid charging of the current DC batteries in the EVs can both be integrated into the station that is the subject of the present invention. This allows for the potential generation of the electricity that is required to power the Electric Vehicles (EVs) in the same location as the station itself. An integrated system enhances the energy efficiency of a network for charging electric automobiles when contrasted with an analogue network, in which there is no energy producing equipment for each charging station. This is shown in the following figure. A non-integrated analogue network stands in stark contrast to this digitally unified system. It is common knowledge that the distribution system takes a considerable hit whenever there is a need for the movement of energy from the place of origin to the location at which it will ultimately be used.

The quantity of expensive and cumbersome metal mechanical parts that are needed is cut down as a result of the framework that supports the manufacturing system while acting as a buffer and shade for the electric cars. It is possible to share and benefit from any and all wiring that is found in the connections that connect the production sites to the consumption sites because both types of equipment (for the production of energy and its consumption) are housed in the same location. This makes it possible to share any and all of the wiring that is found in the connections. The same technique should be used when it comes to securing the shade structure with concrete or other support base in order to hold solar panels. It is possible to execute this task without sending a representative to the location of the client, which is one way to make the most efficient use of the workforce that is already available.

Because of its adaptability, a solar power plant can be used to help with a variety of issues, such as decreasing reliance on energy sources that come from other countries (electric energy import), increasing sustainable urban mobility, compensating for the carbon dioxide emissions produced by electric vehicles, and increasing the availability of the national grid during times of peak demand. These are just a few of the many ways that a solar power plant can be used to help. On the other hand, it helps cut down on energy waste caused by transportation by putting generators in close proximity to consumers, which in turn helps decrease the amount of energy squandered. This is because placing generators in close proximity to customers helps minimize the amount of energy lost. In addition to enhancing the building's overall energy efficiency, it also makes it simpler to install photovoltaic (BIPV) systems on the roof of the structure. In order for the current invention to successfully meet the criteria of rapid battery charging, it may be implemented in either a public or private setting so long as sufficient amounts of power are available in both settings (either in the public network or in a private PT). Ideal alternatives for the placement of these sorts of facilities include public or commercial parking lots, existing fossil fuel service stations, grocery stores, shopping malls, hotels, hospitals, and schools.

This provides a solution to the problem of significantly reducing the amount of power that is lost and the amount of efficiency that is lost as a result. One of the key reasons for the creation of this technology was to accomplish this specific goal. It is generally known that the act of moving electrical current from one location to another results in the loss of some amount of energy. This method results in a significant reduction in expenses when compared to the traditional system, in which the charging station is connected to the distribution network. This is due to the fact that there is no waste of energy during the process of transmission or distribution using this method, which is why the traditional system results in a significant reduction in expenses. The quantity of energy that is utilized and produced by the electric grid would be in a condition of equilibrium if electricity were generated at the location of each consumer rather than at a central location (charging EVs). It takes about the same amount of energy that is used during the generation of decentralized electricity to charge an electric vehicle.

The electric equipment may be organized into either a radial or linear (of the backbone type) central core, with the connections to the grid, electric vehicles, and solar panels radiating outward from there. Alternatively, the electric equipment may be organized into a backbone-type radial core. This layout leads to savings in terms of the overall structural architecture of the station, as well as a large drop in the quantity of power that is lost. These benefits may be attributed to the station's ability to save electricity.

If there were additional stations like this, which were scattered in a radial fashion outward from a central point, then it would be easy to perceive the radial structure. The present invention generates technical results that are independent of the number of stations that are positioned around this core node, which is something that an individual who is well-versed in the relevant art would reasonably perceive to be a given.

When a sizeable number of stations of this kind are used, the linear structure may be seen in a manner that is much more transparent. After that, you should set up the stations such that they are longitudinally aligned and parallel to one another. Although it is possible to build the stations in a fashion that is longitudinal, it is preferable to organize them in a pair of parallel lines that run back-to-back with one another since this provides additional benefits.

The current technique is now in the process of being developed in order to improve solar power plants for the purpose of charging Electric Vehicles (EVs). It is planned that in addition to the typical function of charging electric vehicles, this infrastructure will also be used to compensate for the loss of renewable energy experienced by the public grid during the peak and full hours of the load diagram, as well as for the energy consumed by the EV from within this network. This is in addition to the energy that is consumed by the EV from within this network. As a result of this, it is feasible to decrease the amount of energy lost during transmission and distribution by 15–20% by using this method, and this reduction may begin after the battery of the electric car has been completely charged. The user is educated on the meaning of "green" energy by the solution's advocacy of environmentally responsible practices, such as the use of electric cars, which are examples of this kind of practice. This assists the user in developing an understanding of the notion of "green" energy.

The system for the generation of energy is provided by both photovoltaic solar technology and networked computer systems. This installation includes a DC/AC single-phase inverter that has a nominal power of 3.68 kW, all the DC wire that is required for the connectivity of this equipment, and an AC protection framework (class II insulation) that includes a differential switch and a bipolar circuit breaker. Additionally, this installation includes the entire set of DC wire that is required for the connectivity of this equipment. These individual parts come together to form the whole of this set. Even if the bulk of the solar system's output happens in the middle of the day, at the time of day when demand is at its highest point, this does not prevent any of the energy from being fed into the public grid. The inverter device, in addition to ensuring that the energy that is pumped is of high quality, guarantees that the network will remain in sync with both production and consumption, which results in no overall loss of energy. This is accomplished by ensuring that the network will remain in sync with both production and consumption (self-consumption regimen).

The present technology is the first of its kind, and it is what makes it possible for three distinct charging stations for vehicles to coexist in the same area at the same time. As a consequence of this, there is a charge control device that may be used. This device adjusts the circumstances under which charging occurs based on the number of vehicles that are now using the solar station. The rapid charging approach that is used in the implementation of the present invention involves charging via direct current (DC) with a maximum power of 43 kW, in either an alternating current (AC) or direct current (DC) regimen, and a charging duration of twenty to thirty minutes. This approach is called direct current (DC) charging (DC). The use of triphasic feeding is strongly suggested, and the present invention provides the necessary connection for using this strategy inside of a moving vehicle. It is the responsibility of the manufacturer of the electric car to provide the connection cable that is needed in order to connect the vehicle to the power grid. Because the connecting socket may be different depending on the manufacturer, this is something that has to be done.

In order to monitor the flow of energy between the different phases of a facility, makes use of bidirectional counters, particularly those having triphase direct connections. Because of this, it is possible to keep an exact record of the movement of energy. As a consequence of this, in addition to a counter that records the total amount of energy that is used by this innovation, there must also be counters (these specific people) that individually account for the amount of energy that is supplied to each charging station, or more specifically, to each electric vehicle. This is because the total amount of energy that is used by this innovation is being recorded by the counter. This is due to the fact that the overall quantity of energy that is used is completely unmatched. As a direct consequence of this, the energy flows as well as the overall amounts of energy that the current technology consumes and generates are entirely understood at every point (Nallathambi, A. et al., 2022a).

CYBER-PHYSICAL SYSTEM OF ELECTRIC VEHICLE CHARGING STATIONS

The functioning of a rising number of physical systems, such as those related to electric cars and charging stations, is increasingly being controlled via the use of online networks. This trend is expected to continue. As a direct consequence of this, cyber-attacks (such as those linked with computer viruses, harmful software, and so on) constitute a rising hazard to these control systems, which may afterwards lead to vehicle breakdowns or other difficulties that are comparable. The primary focus of the technologies that are currently available for intrusion detection is information technology (abbreviated as "IT"), which includes data storage, retrieval, transmission, and manipulation computers, as well as operation technology (abbreviated as "OT"), which includes direct monitoring devices and communication bus interfaces. These two types of technology are intertwined in order to provide comprehensive protection against unauthorized access. Despite the availability of these safeguards, the example of Stuxnet demonstrates that physical "domains" may still be penetrated by cyberattacks in spite of the fact that these safeguards are in place. These kinds of assaults have the ability to cripple a control system to the point where it either fully breaks down or experiences significant damage. During a cyber event, very little effort is now put towards automatically recognizing assaults at the domain layer, which is where the sensors, controllers, and actuators are situated. This is an area where there is currently very little work being done. It is not inconceivable that a string of assaults may take place in rapid succession to one another (for example, more than one actuator, sensor, or parameter within the control system devices can be maliciously altered by an unauthorized party at the same time). It is essential to bear in mind that the repercussions of some assaults, such as covert operations against the domain layer, may not be immediately apparent (e.g., when only one monitoring node, such as a sensor node, is used in a detection algorithm). The significance of knowing when a monitoring node goes down because of a true problem as opposed to being taken down as a consequence of an assault is highly significant. It may also be very necessary, depending on the circumstances, to know the particular nature of the problem that caused the node to go down. Because they are not intended to do so, existing means of securing an Electric Vehicle charging infrastructure, such as fault and diagnostic technologies, may not be able to adequately handle these issues. This is because these systems are not meant to automatically detect sneak attacks. This is especially true in the event that there are a number of attacks or failures taking place at the same time (Nallathambi, A et al., 2022b).

Figure 2. Shows the Architecture of a Cyber-Physical Electric Vehicle Charging Station. A rise in the energy levels that are used to charge Electric Vehicles (up to 350 kilowatts ("KW") in the case of exceptionally fast charging ("XFC")) may potentially enhance the cybersecurity threats that are posed to the increasing electrified electrical grid as well as the transportation system. Both the ephemeral nature of electric cars and the quickness with which data may be exchanged between vehicles owing to optional

Figure 2. Architecture of a Cyber-Physical Electric Vehicle Charging Station

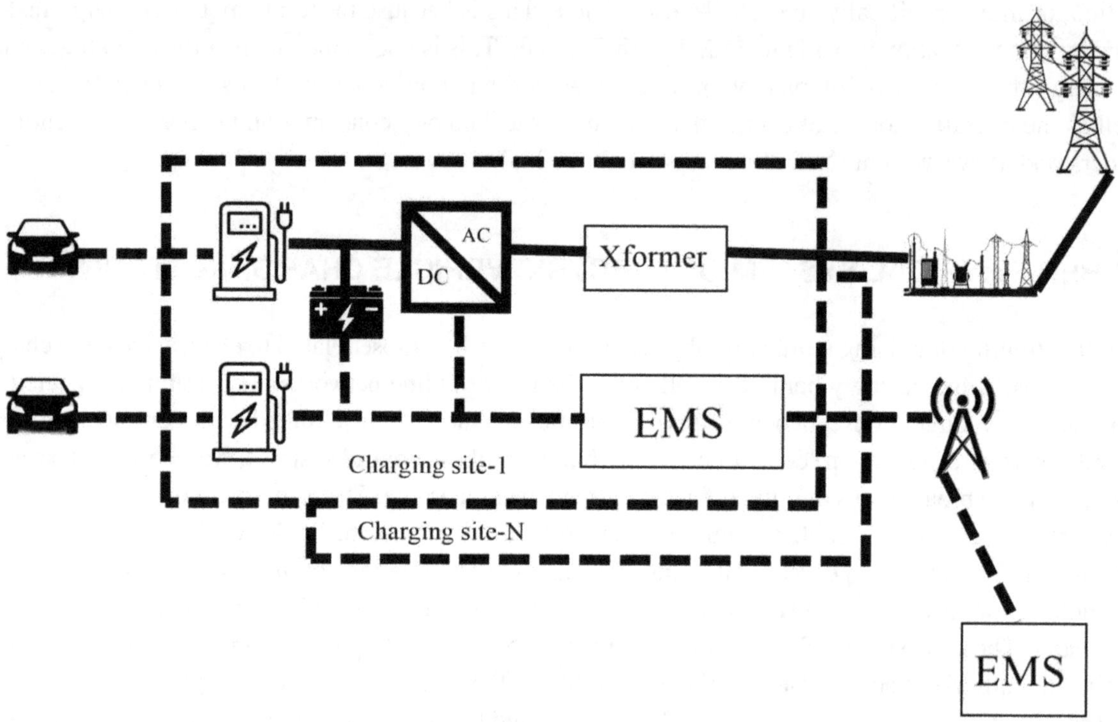

infrastructure both contribute to the increase of these hazards. Additionally, the growth of these risks is facilitated by the availability of optional infrastructure.

When it comes to the IT layer, for instance, malware that is installed on an Electric Vehicle (EV) or EV charging station has the capability of infecting other smart grid devices, which might eventually lead to widespread power interruptions in a particular place. This problem manifests itself at the OT layer when communications between an Electric Vehicle and an Electric Vehicle supply equipment (EVSE) are corrupted. An overcharged battery in an electric car has the potential to cause catastrophic harm to the vehicle. An assault on the physical layer of the electrical grid, which may take the form of the fast cycling of multiple charges between charging stations and high-powered electric cars, has the potential to produce catastrophic interruptions to the system. Information technology (IT), operational technology (OT), and physical layers of defense need to collaborate with one another so that cyberattacks on the infrastructure used to charge Electric Vehicles may be avoided. The creation of a "defense-in-depth" architecture that integrates protections at the information technology (IT), operational technology (OT), and physical levels is becoming an increasingly urgent need. This need is growing more and more urgent. It is essential to bear in mind that the particulars of the hardware implementation of a component may influence many of the security properties of that component. This is why it is so vital to keep this fact in mind. It is likely that the authenticity of a certain piece of hardware will be the decisive factor in determining whether or not you should invest your trust in a particular component. This is because authenticity is the hallmark of quality. Because Electric Vehicles and Electric Vehicle supply equipment (EVSEs) may be linked with a diverse group of manufacturers, it is easy to get counterfeit components. This makes it easier for criminals to exploit the market. Before putting your faith in a component, it is of the utmost importance to ascertain whether or not it is genuine (that is, not a fake). It is probable that depending on software to supply hardware identification will not be adequate since software has a history of being attacked in ways that lead to identity spoofing. This is due to the fact that software has a history of being attacked in ways that lead to identity spoofing. Therefore, it would be preferable to have an automated and precise method of protecting a charging infrastructure for Electric Vehicles from cyberattacks, even if the attacks penetrate the IT and OT layers and cause direct damage to the control systems. This would be the case even if the attacks penetrated the IT and OT layers. Even if the assaults were successful in penetrating the IT and OT levels, this would still remain the case.

The tactics that have been offered here may serve as a starting point for constructing a plan to foil a cyberattack on a charging station or on the whole power infrastructure. These strategies have been supplied for your convenience. The system may include features such as cyber attestation, protection, anomaly detection, intrusion prevention, and decision fusion. This is a distinct possibility. Platform certificates are able to provide two-way authentication between a vehicle and its charging infrastructure. These certificates may be given at both the IT and OT levels by cyber security and/or hardware sources of trust (as well as reference integrity specifications for verification of dynamic integrity). Manage power electronics and rapid dynamic power systems with the use of the "one shot" anomaly detection and localization approach. The physical layer is the location where this approach is carried out. An integrated decision that is made by a decision fusion element has the potential to be more accurate than a choice that is made by any individual module. This is because the decision fusion element is able to draw on information on detection and accommodation from the IT layer, the OT layer, and the physical layer.

When it comes to protecting the infrastructure that is used to charge electric vehicles, there are a few distinct approaches that a protective system may take if it were to be deployed. Charging stations for electric cars have the ability to transform the alternating current (AC) energy that is supplied by the grid into the direct current (DC) power that is required by the vehicles. At the charging station, there might be more than one monitoring node, and each of these nodes creates a series of values based on the current condition of the node that is being monitored. These values are determined by the state of the node that is being monitored in a time series that provides an accurate representation of the functioning of the infrastructure already in place to charge electric vehicles. An access requester for an electric car may send a request and a platform certificate to the fueling equipment communication controller (and, in some embodiments, a software identifier). The access requester and the secondary actor are allowed to continue their engagement at a higher level after the secondary actor policy decision point at the billing site has validated the identity of the access requester. Please be aware that the information about the current values of the monitoring nodes and/or the access request might potentially be recorded in a trustworthy distributed transaction ledger (e.g., an attestation blockchain).

The following elements are included in a variety of alternative approaches to the problem: Methods of obtaining data from a space-based source that are generally accepted for A data retrieval system is a kind of system that is able to obtain data sources from an infiltrating area for each of a number of monitoring nodes, where the data represents the regular functioning of an electric car or charging infrastructure.

Electric car charging infrastructure monitoring nodes, a time-series of attacked monitoring node values that reveals the state of the infrastructure that is being attacked, receivers, electric car charging infrastructure monitoring node values. Each of the countless monitoring nodes gathers its very own unique set of measurements from a data source that stands in for a fault space.

A method for automatically calculating and outputting, by a computer used for the construction of a multiclass classifier model, at least one decision limit for a multiclass classifier model based on the set of normal feature vectors, the set of attacked feature vectors, and the set of fault characteristics vectors: and values of fault monitoring nodes over time representing the fault operation of the Electric Vehicle charging infrastructure. This is accomplished by using a computer that is used for the construction of a multiclass classifier model.

In other implementations, components are provided for accessing at least one multiclass classifier model with at least one decision boundary, executing at least one multiclass classifier model, generating a set of vectors of current features, and receiving, from a plurality of monitoring nodes, a series of current monitoring node values over time representing the current operation of an Electric Vehicle charging infrastructure. Additionally, components are provided for generating a set of vectors of current features, which can be used to generate a set of These elements are put to use in order to carry out the following responsibilities: accessing at least one multiclass classifier model that is equipped with at least one decision boundary.

Monitoring the condition of the node to determine whether it is normal, under attack, or has failed, utilizing at least one multiclass classifier model, and providing a classification result based on the currently active feature vector set as well as at least one decision threshold. This monitoring is done in order to determine which of the three states the node is currently in. There are many different modalities that are addressed in this article. Some of the technological benefits of these modalities include enhanced security systems and processes, which can protect an electric car charging infrastructure from cyberattacks automatically and precisely (and, in some cases, failures). It is possible for the monitoring nodes to have either a homogenous or a heterogeneous composition; this will depend on the sort of decision that has to be made at this crossroads. Techniques such as Bayesian inference, fuzzy logic, heuristic-based (voting), and classical inference are examples of common approaches to decision-level sensor fusion. The most common use of the Bayes fusion method is outlined here, along with a description of its normal operating procedures. To get started, the information that was produced by combining the data layers should be utilized as the baseline information, and the correctness of that information should be checked whenever it is required to do so. The second step of the process of minimizing the impacts of an online cyberattack involves the determination of the historical distributions of the parameters. At the OT layer, which is concerned with the loading of website communications, the use of technologies such as attestation and blockchain may assist offer a first line of defense against assaults on software or firmware. Performing the loading of the website in a secure manner is one way to achieve this goal. The implementation of attestation servers that are based on the blockchain may, in some circumstances, make it possible to determine whether or not a security compromise has occurred. Platform certifications, which Hardware Roots of Trust is able to provide, enable electric cars and the charging stations that they use to participate in a two-way verification process. This makes it feasible for Electric Vehicles to be more environmentally friendly (as well as reference integrity specifications for dynamic integrity checking). Testing the component integrity in a continuous and centralized manner enables the prompt and reliable detection of malicious attempts to corrupt system firmware and software.

The information that is gathered through digital communications may then be analyzed for indications of infection. This information may also be analyzed for the stability of the power grid, the location of any abnormalities, and the techniques that are the most effective in order to accommodate them. Combining these three aspects into a single hack is a strategy that may be used to carry out complex intrusions into a network.

Both tactics for detection and accommodation, as well as strategies for selecting how to proceed based on an in-depth analysis of the environment in which one is operating, are included in this section.

Propagation of Malware, Monitoring and Stability Assurance, as well as Locating and Accommodating Vulnerabilities Anomalies are composed of three distinct modules that carry out their functions separately from one another. The results from the preceding three modules are combined in the Decision Fusion module, which then allows for more precise and dependable conclusions to be drawn. assurance against infections, stability, and monitoring, with the one and only goal of anomaly discovery and correction. The decision fusion module may provide actionable information in the form of alarms that are displayed in the user interface of a charging network and/or network operator and/or control instructions that are given directly to each load controller, EMS, and EVSE communication device. For instance, the alarms may be displayed in the user interface of a charging network. Alternatively, the network operator may see the alarms ("EVCC"). When drawing a final judgement, all of the information that was obtained from the various charging stations and the various devices that were connected to each charging station is taken into account. It is possible to identify assaults using this infrastructure, including clandestine operations and those that include a large number of stations working together in concert. If the charging station or EVSE layer is in charge of fusing the options, it is probable that this objective will not be attainable. Nevertheless, the ensuing delay in communication may make it more difficult to make decisions. In order to view and/or manage attack and failure information in accordance with any of the modalities that are described in the following paragraphs, a user may log into the system using any of the monitoring devices (such as a personal computer, also abbreviated as "PC"), including a tablet or smartphone. Other monitoring devices include personal digital assistants (abbreviated as "PDAs"). The authoring multiclass classifier model and the node classifier computer may be able to provide an interactive graphical display interface. The user may be able to configure and/or alter specific settings using this interface (such as the attack detection trigger levels), as well as submit or receive suggestions or automatically generated findings.

PEDESTRIAN DETECTION FOR AUTONOMOUS VEHICLE

It is feasible that driverless vehicles will have a considerable influence on both road safety and mobility, two basic issues that exist within the current transportation system. Engineers have had the ability over the course of the past decade to create and test a wide variety of prototypes for fully autonomous cars, while at the same time gradually implementing ADAS into production vehicles. This has allowed the industry to make significant strides in the development of fully autonomous vehicles.

Figure 3. Shows the Pedestrian Detection Systems in Autonomous Systems. One of the most fascinating developments that has taken place in the automobile industry in the most recent years is the creation of vehicles that can function independently of the involvement of a driver. Once the required infrastructure is in place, it is likely that self-driving cars may completely revolutionize the way that people and businesses travel about. Driving conditions will soon improve as a direct result of this, which will ultimately make them safer. There will be a decrease in the number of people who are killed or injured as a consequence of traffic accidents, and the effect that vehicles have on the environment will become less severe. Both of these outcomes are expected to take place. Everyone who is engaged in the movement of products and people would benefit from increased convenience, reduced costs, and a more enjoyable experience if drivers were not necessary as often. If these networks were implemented, there would be less congestion, which would, in turn, result in lower rates for automobile insurance. The author considers both the good and negative elements of a newly created technological development

Figure 3. Pedestrian Detection Systems in Autonomous Systems

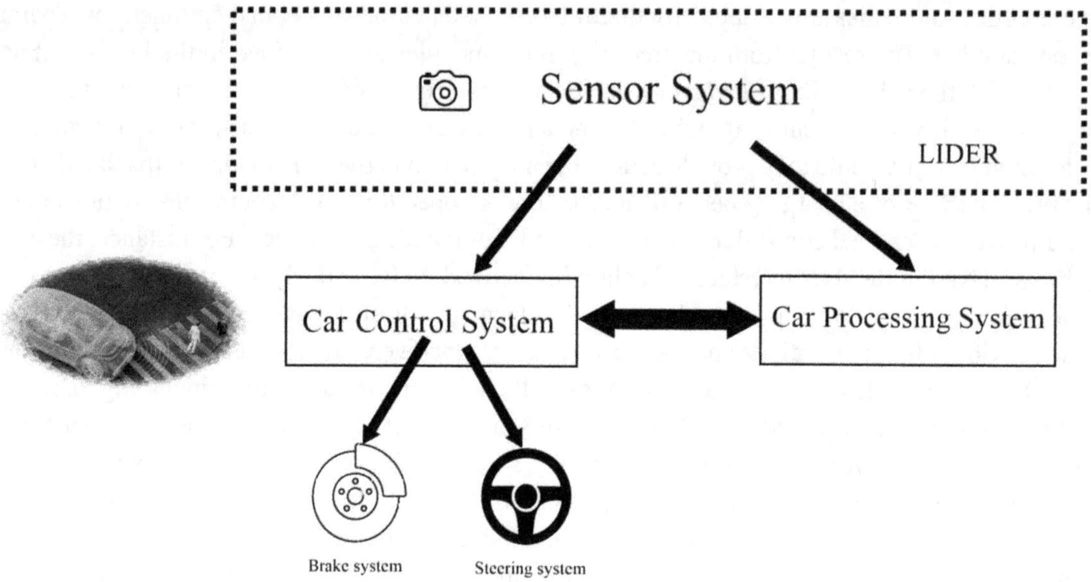

before making a case for its widespread use in society. Huge automobile manufacturers, in conjunction with large information and communications technology firms such as Google, are spending millions of dollars in research and development with the goal of bringing fully autonomous cars to market by the end of the year 2020. In spite of the fact that it was anticipated that driverless vehicles will become mainstream by the year 2040, there are a number of variables, including those connected to technology and the economics, that suggest this is still a very long way off in the future. As a consequence of this, situational awareness systems have a larger responsibility to pay attention to the environment that is directly around them. It is essential to have the ability to recognize, monitor, and, if at all possible, predict the movement of the vehicles in order to ensure both the safety of people and the public's faith in self-driving cars. This applies to both the protection of people and the public's faith in self-driving cars (AVs). When it comes to recognizing individuals, autonomous cars face a huge obstacle. This is a significant difficulty (Eswarawaka, R. et.al 2021).

EDGE COMPUTING

Its central processing unit (CPU) and the fact that its interface may be customized by the user are what set it apart from other products. According to the findings of a recent survey, an increasing number of individuals are interested in "pocketable" devices that are able to interpret intricate commands and respond to inputs via voice, touch, vision, and movement. This interest has been on the rise in recent years. Newer techniques of data processing, such as computing at the network's edge, have been developed in recent years in order to make the internet operate at a faster speed. Because serverless cloud computing, also known as edge computing, does not require the use of dedicated servers, deployment may be completed more rapidly, and the system may be able to support a greater number of users. It is required for programmers to be able to write, deploy, and execute their code at the edge of the network in order for them to be able to put together such systems. Real-time computing performed at the edge is often employed as the underlying structure of real-time applications that deal with a great deal of data. Customers positioned in close proximity to the place at which data or computationally heavy resources are required would benefit from this service. Contrary to popular belief, the Edge Network provides a connection that is both encrypted and secure between your device and the edge node that is geographically closest to you. This connection is made possible by the Edge Network. Users won't have to worry about the high data charges that are often associated with accessing wireless networks since this connection is wired rather than wireless. In order to make use of the Edge Network, it is not essential to join up for or become a member of anything in order to do so. Customers have access to a connection to the larger world wide web that is both very quick and secure via The Edge Network. At the same time, this connection is completely isolated inside the private network of the building in which The Edge Network is housed. Because encryption is turned on by default, sending any kind of data is entirely secure and does not require the use of a password at any point throughout the process. Users are able to check in using their badges in a fast and easy way owing to the Edge Network, which gives them safe access even when they are not physically present in the building. This enables users to check in at any time, regardless of whether or not they are in the building. On The Edge Network's very own wiki page, which may be visited via this link, you will discover further information on the service. Edge computing also has uses in the Internet of Things (IoT), in which it would collect data from sensors, transform that data, and then transfer it to the cloud for additional processing before delivering the results back to the edge device where it originated. Because of this method, the data can be analyzed and converted into insights that can be put into practice a great deal closer to the moment at which they were collected, which is just one of the numerous advantages that this method provides.

Numerous sensor systems, some of which have been connected to one another, have been put into use in order to realize the goal of being able to recognize humans who are seated inside of cars. This procedure is carried out with the aid of a variety of sensors, including cameras, among other devices. For instance, it is very difficult to keep up a high level of performance for an extended length of time under real-world driving conditions because of the many obstacles that might arise.

In recent years, a considerable amount of both time and money has been put into research and development efforts relevant to optical pedestrian identification. These activities have been carried out in a number of different locations. When dealing with direct methods, such as pedestrian segmentation and tracking, it is typically necessary to use strong CV algorithms. In order to properly recognize pedestrians that are situated in close proximity, a diverse range of visual cues, including stereoscopic data, movement, and appearance, have been used (for example, local features such as the Histogram of Oriented Gradients, which are used to detect pedestrians). Because of this advancement, the use of a histogram as a solution to the problem of algorithmically identifying pedestrians has made substantial progress. When LIDAR sensors were first made available on the market, those responsible were unable to find anything with them. A LIDAR sensor will often consist of a revolving array of individual laser diodes as its primary component. Despite the fact that there are a few drawbacks associated with employing LIDAR sensors for autonomous navigation, these sensors' accuracy has enabled them to achieve a high level of success in a wide range of other contexts, which has contributed to their rising popularity. The modelling of subterranean subway tunnels is one example of this; another example would be the detection of road markings; another example would be the practice of precision agriculture.

The Benefits That Can Be Obtained Through the Implementation of Edge Computing (Yeruva. V,2021, September 27).

1. You can enhance the network's performance and speed for end users by processing data at a location that is geographically closer to its source, where the data may have a better chance of benefiting from reduced network latency.
2. Keeping information secure while it is being sent through a computer network Since more data is handled locally, there are less possibilities for possible security breaches while it is being transported across a network. This is because of the distributed nature of the network. In addition to this, it considerably reduces the danger of any possible security breaches that may occur as a result of the fact that processing, storage, and applications are distributed over a huge number of distinct devices.
3. Computing near the network's edge offers a number of benefits, including a reduction in the amount of time spent waiting for data to come from across the network and cost savings.
4. The dependability of the system as a whole is significantly improved as a direct consequence of the use of locally stored data and processed information. This eliminates the possibility of frequent interruptions, such as those that are brought about by the loss of an internet connection to the cloud.
5. Edge computing enables businesses to expand the range and depth of the solutions they provide in a manner that is both fast and effective. This is accomplished by consolidating the capabilities of processing, storage, and analytics into a single device.

EDGE COMPUTING IN AUTONOMOUS ELECTRIC VEHICLES

Connecting autonomous cars around the perimeter offers a variety of benefits, including enhanced reliability, reduced amounts of downtime, fewer accidents, and less congestion in the area. These automobiles are packed with a myriad of sensors, and the information that these sensors produce must be analyzed as rapidly as humanly feasible.

A major upheaval is going to take place all across the planet. Edge computing will play an essential role in the tremendous expansion of the automobile sector, which is one of the businesses that is increasing at the fastest rate in the economy. The transportation system is set to undergo a transformation from a technology-driven system into a data-driven system as a direct result of developments in information technology. These developments include the introduction of new sensors as well as advances in data processing and control. In the past, it was a system that was driven by technology. Artificial intelligence will have far-reaching consequences outside of the transportation industry because of the large volumes of real-time data created by self-driving vehicles, driver monitoring systems, and security cameras. This data will be generated by systems that monitor drivers. The healthcare sector as well as the entertainment business will both be affected as a result of these repercussions. Computing at the edge of the network and other network enhancements have been shown to be helpful for a diverse range of commercial use cases and circumstances (Shurpali, S. 2020, January 24).

The phrase "edge computing" refers to a group of technologies that include compute, data storage, administration, analysis, and networking. These technologies are referred to together as "edge computing." Applications and devices now have the capability to respond rapidly to data. Real-time processing takes place at the edge of the network. When autonomous cars are connected to the edge of the network, there is the potential for an increase in safety and efficiency, in addition to a decrease in the number of accidents and a reduction in the amount of traffic congestion. These vehicles are loaded with a wide

Figure 4. General Architecture of Edge Computing with Autonomous Vehicles

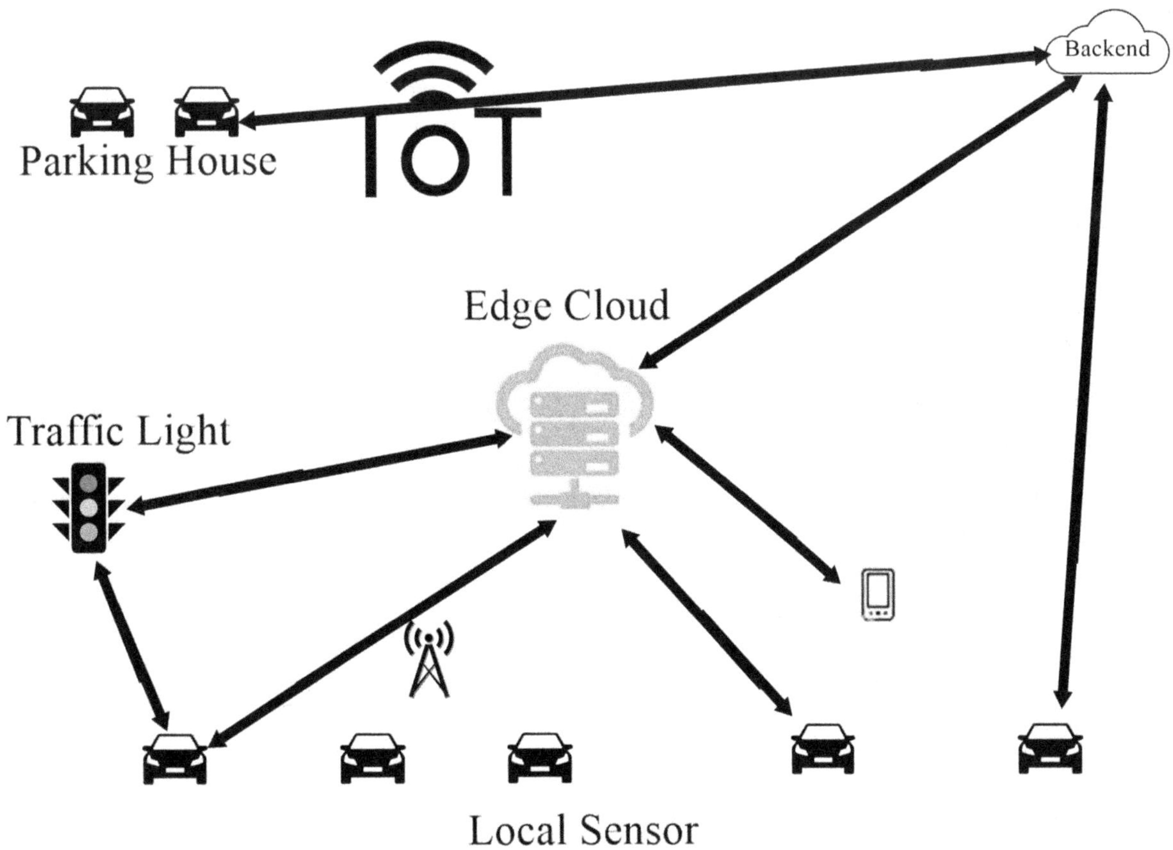

range of sensors, each of which produces a significant quantity of data that must be analyzed in a very short period of time.

Figure 4. Shows the General Architecture of Edge Computing with Autonomous Vehicles. To evaluate the current conditions of the road and choose the action that will have the most positive impact, self-driving vehicles employ algorithms that are designed for machine learning. These algorithms investigate enormous amounts of data in the expectation of discovering recurrent patterns. This may include considerations like as the placement of pedestrians in the near vicinity, the intensity of the illumination in the surrounding region, the visibility of the road in front of you, as well as the presence of other vehicles or impediments. The confluence of all these variables leads in the development of vast volumes of data, which has to be controlled locally, often known as “at the edge.”

CONCLUSION

It’s possible that the use of edge computing may lead to major gains for the automotive industry, encompassing all of the various subsystems that make up that sector. As the number of connected and autonomous vehicles continues to increase, it will be even more important to collect and examine enormous amounts of data to arrive at conclusions that are helpful in the fight to save lives. In the years to come, computers located on the periphery of networked systems will play an important role in making all of this possible.

There is little question that intelligent autos, particularly driverless vehicles, will shake up a variety of different economic sectors and have a huge impact on how we go about our day-to-day lives. A substantial amount of work in the automobile industry has been done to implement edge computing.

REFERENCES

Eswarawaka, R. (2021). IoT and AI based Pedestrian Detection for Autonomous Vehicles, Indian Patent Publication, 202141056543.

Indragandhi, S. R., & Subramaniyaswamy, V. (2022). Electric Motor Drives and their Applications with Simulation Practices. Indian Patent Publication,202241023774.

Nallathambi, A., ... (2022a). Design and Development of Advanced Solar Based Electric Vehicles Charging Power Station, Indian Patent Publication,202241023774.

Nallathambi, A., ... (2022b). Framework for the protection of the cyber-physical system of electric vehicle charging stations and the electricity grid, Indian Patent Publication, 202241045388.

Nallathambi, A., ... (2022). Autonomous Electric Vehicles Auto Monitor and Control, Indian Patent Publication, 202241015720.

Shurpali, S. (2020, January 24). *Role of Edge Computing in Connected and Autonomous Vehicles*. https://www.einfochips.com/blog/role-of-edge-computing-in-connected-and-autonomous-vehicles/

Yeruva, V. (2021, September 27). Council Post: Edge Computing's Applications in Autonomous Driving and Business at Large. *Forbes*. https://www.forbes.com/sites/forbestechcouncil/2021/09/27/edge-computings-applications-in-autonomous-driving-and-business-at-large/?sh=68b9426a96aa

Section 4

Latest Trends in Electric Vehicles

Chapter 12
Vehicle-to-Grid and Electric Vehicle-Integrated Demand Response Management

Bavly Hanna
University of Technology Sydney, Australia

ABSTRACT

The safe and efficient functioning of microgrids (MGs) has always depended on the best scheduling of MGs with renewable uncertainty. Electric vehicles (EVs) can back-feed power to the electrical grid to provide a variety of ancillary services, including demand response (DR), peak-load management, voltage support, and frequency regulation. This technology also enables EV holders to reduce their energy costs and make money through price arbitrage. Vehicle-to-grid (V2G) technology is developing quickly, and research is being done to improve its usability and use. The benefits of V2G include ancillary services, active power support, reactive power compensation, and promoting the use of renewable energy sources. In contrast, the challenges of V2G include battery degradation, effects on distribution equipment, elevated investment cost, and social obstacles.There are three main algorithms to optimize EV DR management: the artificial neural network, machine learning, and nature-inspired algorithm. The successful V2G-based DR strategies for energy management are covered in this chapter.

1. INTRODUCTION

The fundamental idea behind V2G power is that EVs power the grid when they are parked. The EV can be a "plug-in hybrid," a "fuel cell vehicle," or a "battery EV." Battery EVs may discharge when electricity is not needed and charge during periods of low demand. Fuel cell EVs use gaseous or liquid fuel to produce electricity. Both modes are functional in plug-in hybrid EVs.

Three essential components must be included in every vehicle: a grid connection for the flow of electrical energy, a control or logical link required for communication with the grid operator, and controls and metering inside the vehicle. These components change a little depending on the business model (Kempton & Tomic, 2005).

DOI: 10.4018/978-1-6684-6631-5.ch012

The EVs' unpredictability impacts the system's planning in terms of route, numbers, and sizes. As a result, the system performance is impacted by EVs' best scheduling, size, and positioning. EV charge stations, state of energy (SoE), and EV trajectory and location can all have an impact on the MG's primary performance in terms of power consumption and exchange patterns in relation to loads and the main grid. EVs' SoE may be divided into "charging" and "discharging".

In order to increase stability, the MG is frequently linked to the "main grid", and the amount of power transferred between the MG and the main grid significantly impacts the system's cost profile. When the MG's energy consumption is excessive, purchasing electricity from the "main grid" is necessary, which raises the "cost function". EVs can assist the MG during this time while using less energy obtained from the "main grid". Additionally, surplus energy may be sold to the main grid to boost MG's profitability.

V2G refers to "the performance of energy flow from EV to the main grid," while grid-to-vehicle (G2V) refers to the other direction. Also known as "vehicle-to-load (V2L)," this method of supplying energy from EVs to loads during peak hours. The dependability and profitability of the MG can be increased by proper analysis of these V2G, G2V, and V2L patterns. Figure 1 displays the difference between V2G and G2V.

Figure 1. V2G vs G2V

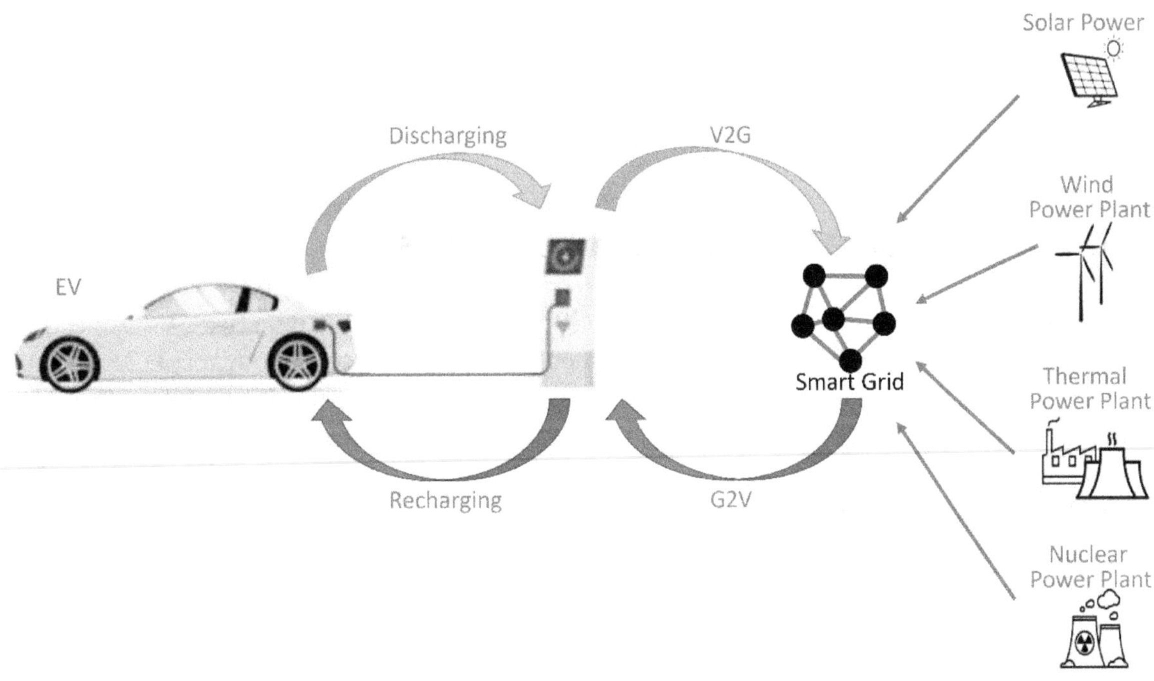

On the other hand, the DR programme enables consumers to voluntarily lower their energy use, which enables the system to sustain more crucial loads and lightens the pressure on the MG. DR, or customer cooperation in increasing energy consumption, is promoted as a useful strategy for energy conservation. DR is used in two ways, as a "price-based" programme and an "incentive-based" programme, both of which work with the MG to lower energy prices or provide incentives to programme participants (Asadinejad & Tomsovic, 2017).

Utilizing quick ramp resources like EVs, energy storage devices, and DR programmes is one method to boost the flexibility of smart grids (SGs). In order to assist the system sustain more important loads and lessen the stress on the MG, DR programmes are utilized to encourage consumers to lower their power use on a voluntary basis. DR refers to customer cooperation in increasing energy consumption and is promoted as a useful strategy for energy conservation.

DR is used in two ways, as a "price-based" programme and an "incentive-based" programme, both of which work with the MG to lower energy prices or provide incentives to programme participants (Asadinejad & Tomsovic, 2017). By lowering power consumption, particularly during on-peak hours, DR can improve the performance of the MG while preserving service quality and lowering system costs overall.

The fundamental idea behind V2G electricity is that EVs power the grid when they are parked. Three essential components must be included in every EV: a grid connection for the flow of electrical energy, a control or logical link required for communication with the grid operator, and onboard controls and metering.

EVs have been referred to be a power grid (PG) flexibility tool (Zhang & Kezunovic, 2016). The authors of (Roukerd et al., 2019) provided a framework for assessing the power system's flexibility utilizing DR and renewable energy sources (RESs). The authors of (Nikoobakht et al., 2019) described a method for evaluating the degree of flexibility of the power system employing energy storage, DR, and generating units with quick ramps.

The rest of the chapter is designed as follows: section 2 discusses the benefits and impediments of vehicle-to-grid. Section 3 analyses the approach to demand response programmes. This is followed by algorithms to optimize electric vehicles' demand response management in section 4 and conclusions in section 5.

2. VEHICLE-TO-GRID BENEFITS AND IMPEDIMENTS

In order to optimize the supply of local renewable energy and lower infrastructure costs, V2G directs the charging and discharging of EV batteries based on user needs and the grid's electricity supply. Meanwhile, EV owners can benefit from greener, more cost-effective electricity consumption while also earning money for supporting the grid. Figure 2 shows the benefits and challenges of V2G.

2.1 Vehicle-to-Grid Benefits

2.1.1 Ancillary Services

Plug-in EVs (PEVs) are able to offer auxiliary services, which results in a more stable operation of the power system, a decrease in the operation of protective relays, and perhaps a reduction in the effect of some eventualities. The extended service of demand will raise the yearly load payment even though these services might be credited with raising social welfare. PEVs can be used to control frequency either as a power source or as a load (Tan et al., 2016).

Figure 2. Benefits and challenges of V2G

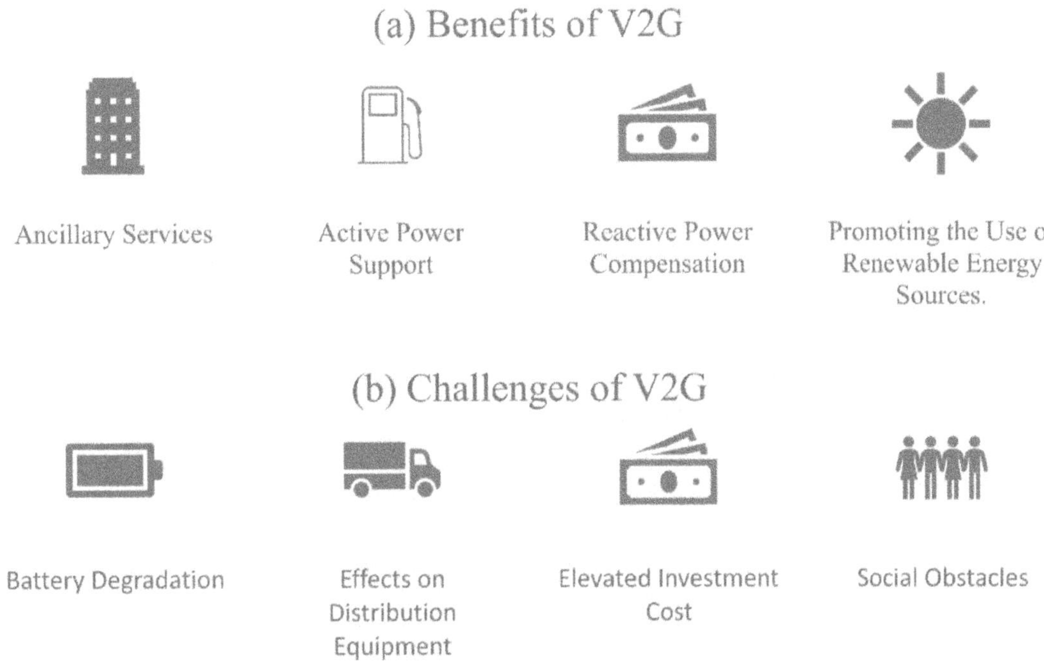

2.1.2 Active Power Support (APS)

Another V2G service makes use of the surplus EV energy to give the PG APS. Bidirectional V2G, not unidirectional V2G, is required for APS since EVs must "discharge the batteries' energy". This service aims to "peak load-shave" and "load-level" in order to flatten the grid load profile.

Throughout the day, "peak power" is often required for a brief amount of time. As a result, it would be further cost-effective to meet the need for "peak load" from "distribution sources," such as "grid-connected EVs" (Tu et al., 2011). In order to reduce the peak demand (PD), EVs can be used to contribute electricity to the PG at that time. This can lessen the stress placed on the power system's components during PD, and EV holders will be compensated with a higher energy rate. EV users may charge their batteries for less money during off-peak hours.

2.1.3 Reactive Power Compensation (RPC)

Voltage control in the electrical grid is accomplished through the use of RPC (Bolognani & Zampieri, 2013). Power factor adjustment, another benefit of reactive power support, lowers generational current flows and power line losses (Vovos et al., 2007). Additionally, by reducing the load on the equipment, this service can improve the operational efficiency of the electrical system.

According to the authors of (Yong et al., 2015), a unique RPC control and EV rapid charging control are presented. The RPC control adjusts the PG voltage during the EV rapid charging process. Therefore, the suggested RPC management of the EV "charging station" may address the grid voltage drop issue caused by EV charging operations. In the meanwhile, the V2G charger is built with a decoupled active and reactive power control that can provide the PG with the right amount of reactive power based on the utility's "reactive power command signal". A 12.5 kVA charger is used to evaluate the RPC control experimentally (Kesler et al., 2014).

2.1.4 Promoting the Use of Renewable Energy Sources

PEVs can be used to "buffer and store energy" produced by sporadic wind and solar power sources (Pillai & Bak-Jensen, 2011). In many markets, peak solar radiation, for instance, happens several hours prior to the peak energy draw (Clement-Nyns et al., 2011). Wind power is more complicated and highly intermittent due to unanticipated fluctuations in wind speed, which causes imbalances (Al-Awami & Sortomme, 2012). Investigating whether V2G might manage the erratic in-feed of wind electricity according to (Kempton et al., 2008).

The authors of (Guille & Gross, 2009) assessed the beneficial impact of PEV on "wind power operations" and suggested a model predictive control system. For the unanticipated RESs' sporadic availability, the authors' "autonomous distributed V2G" control approach developed a "distributed spinning reserve" (Ota et al., 2012). Different methods for incorporating EVs into a "wind-thermal power system" are suggested by (Goransson et al., 2010). The authors of (Wang et al., 2011) have presented a DR and wind power integration combination.

In order to maintain equilibrium, "centralized power plants" have to reduce production, or distributed generating units have to be scaled back if the amount of RE power on the grid is too high. By discharging and charging, vehicles can aid in balancing generation and consumption. PEVs have the ability to store extra RE. This "stored energy" may later be used to power the grid or for transportation requirements (Clement-Nyns et al., 2011). V2G boosts the grid's flexibility, enabling it to use intermittent RESs more effectively.

The authors of (Saber & Venayagamoorthy, 2011) conducted a simulation analysis of "an independent system operator of a ten-unit system with 50,000 PEVs" that are loads for RESs and sources for the grid. According to this study, an SG model that uses RE may cut emissions and save the power and transportation sectors at least $3.58 per day for each car.

2.2 Vehicle-to-Grid Impediments

2.2.1 Battery Degradation

Battery deterioration is a major hurdle to EV adoption. The battery system accounts for one-third of the total cost of EVs. An electric scooter prototype based on the design of a battery is presented according to (Affanni et al., 2005). A Li-on battery was used to create this one. Electric scooters are quite effective. Each battery has a set number of cycles for charging and draining. After then, the battery's efficiency starts to decline over time. The irreversible chemical process decreases the battery's capacity.

The pace of charging and draining affects battery aging. Additionally, it is affected by Depth of Discharge (DOD), temperature, and voltage. The V2G technology requires a battery with a maximum number of charging and discharging cycles. Modeling of the lithium-ion utilized in EVs is analysed by (Millner, 2010).

2.2.2 Effects on Distribution Equipment

Distribution equipment is probably going to be significantly impacted by PEV charging (Desbiens, 2012). Levels 2 and 3 battery chargers have the potential to significantly overload local distribution infrastructure, depending on the PEV penetration situations (Bae & Kwasinski, 2012). They increase PD, harmonic distortion, voltage variations, and distribution transformer losses (Raghavan & Khaligh, 2012). This necessitates additional expenditures for larger overhead, subsurface power lines, and higher transformer capacity (Lassila et al., 2012). Due to a potential loss in transformer life, the cost may substantially influence the dependability, security, efficiency, and economics of newly developed SGs (Yunus et al., 2012).

By employing a regulated charging method, degradation throughout the course of a conventional distribution transformer may be significantly decreased (Gómez & Morcos, 2003). A thermal model was used to study various PEV penetrations dependent on the lifespan of the transformer insulation (Gong et al., 2012). The findings demonstrated that a high PEV penetration - especially when charging schedules are not well coordinated - can have significant effects on the electricity system.

At a 50% PEV penetration, uncontrolled charging reduces transformer life by 200–300% compared to the base scenario; regulated charging extends life by 100–200% compared to uncontrolled charging (Moghe et al., 2011). Research that focuses on "the aging of a 160kVA MV/LV transformer installed in France" models and simulates "the aging rate of low-voltage transformers" with a significant PEV penetration (Turker et al., 2012).

2.2.3 Elevated Investment Cost

One of the main obstacles to the introduction of V2G is the high investment cost. V2G requires a more modern power system. Hardware and software infrastructure needs to be enhanced for the V2G deployment. Each electric vehicle (EV) taking part in the V2G technology needs a "bidirectional battery charger". High tension wire and a complicated controller are features of the bidirectional battery charger. Safety is required in the event of high-tension cable. Energy conversion occurs during the charging and discharging cycles, which causes energy losses in the power system. Investment in the American EV Market is discussed in the report (USEV Market, 2016).

2.2.4 Social Obstacles

The public's adoption of V2G systems presents another significant obstacle. EV users will keep some energy in reserve for unforeseen circumstances and long trips. Range anxiety is a concern brought on by the grid system's energy sharing. The absence of a charging station makes the matter worse.

3. EV-BASED DEMAND RESPONSE MANAGEMENT

3.1 Demand Response Management

DR main objectives for the "electric grid" are to lower peak loads and improve the stability and dependability of the system. On the other hand, DR will energize and involve consumers, encouraging them to

Figure 3. Approaches of DR programmes

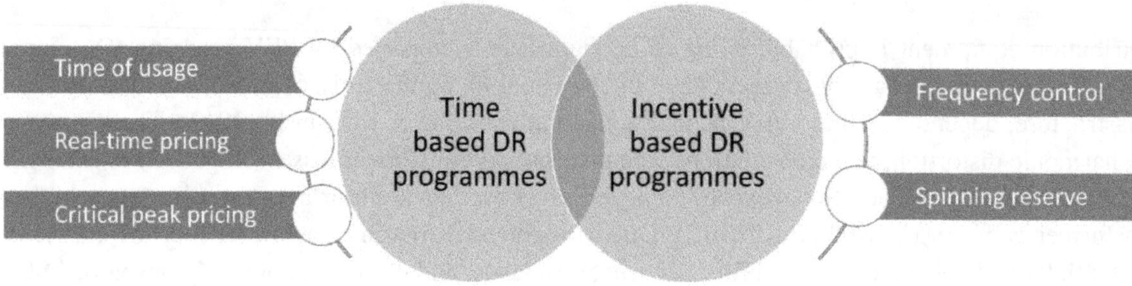

take an active role in the "energy market." The scheduling of use can be optimized either "centrally," allowing "grid operators" to manage peak loads, or locally, allowing residential customers to actively plan their usage. Accordingly, depending on their level of involvement in demand-side management (DSM), "future smart users" may be described as "energy consumers" or "energy citizens" (Goulden et al., 2014).

The SG might significantly alter the social dynamics of the energy system. It is possible to actualize and construct two competing visions. The first vision, referred to as "centralized" DSM, is predicated on a "centralized and hierarchical paradigm based on energy systems" from the previous century. Providing customers with precise use data, as well as "dynamic pricing tariffs" and "remote control" of electrical load devices to expand "generator control of consumption," are the goals of centralized DSM.

The second vision entails obfuscating the line between "generators" and "end users" and constructing distributed generation MGs, which are MGs that are achieved through distributed generation and "proper self-management". As a result, the paradigm for energy consumer framing is the same as that for centralized DSM. On the other hand, the foundation of energy citizens is based on distributed generating MGs.

In many sources, various perspectives on flexibility and resilience have been examined. According to the authors of (Lannoye et al., 2012), flexibility refers to the network's capacity to provide unmet energy needs. The network's capacity to respond to abrupt and significant load changes has been termed flexibility by (Papaefthymiou et al., 2014). EVs have been referred to as a PG flexibility tool by (Zhang & Kezunovic, 2016). The authors of (Roukerd et al., 2019) provided a framework for assessing the power system's flexibility utilizing DR and RESs. The authors of (Nikoobakht et al., 2019) described a method for evaluating the power system's degree of flexibility, employing energy storage, DR, and generating units with quick ramps.

3.2 Approaches of Demand Response Programmes

The fundamental goal of DR is to persuade consumers of electricity to control their usage at times of high demand or when the security of the system is in jeopardy (Kathan et al., 2012). The two types of DR programmes are time-based and incentive-based. Figure 3 shows the different approaches to DR programmes.

Time of usage (TOU) (Cao et al., 2012), real-time pricing (Sarker et al., 2015), and critical peak pricing (Fan, 2012) are three significant time-based DR schemes. Such schemes frequently include pricing for power that changes over time. The price will be greater during peak hours, while during off-peak hours, it will be lower.

By redistributing the load between peak and off-peak periods, the time-varying pricing seeks to level the load (Zhao et al., 2013). As a result, the PG's generating costs significantly decline, and EV owners incur lower charging prices.

When taking part in time-based DR programmes, the EVs will only be used in G2V mode. The TOU is acknowledged as the most effective time-based tariff for lowering the costs and emissions of EV charging (Sioshansi, 2012).

Frequency control and spinning reserve are two incentive-based DR systems for EVs (Habib et al., 2015). Although these applications have not yet been extensively used, they offer enormous potential (Damiano et al., 2014). Both initiatives fall under the category of auxiliary services, which assist the steady flow of electricity and stability of the PG so that supply (S) and (D) are always met.

The regulation service balances power output and load in real-time to ensure that the frequency is kept within a certain range of the normal frequency (e.g., 60 Hz). When there is an imbalance between the load and the power supply, the frequency departs from its nominal value.

When taking part in incentive-based DR programmes, EVs can be used in the G2V and V2G modes. When the power generation is more than the load, the regulation down service is executed in the G2V mode, and when the power production is inadequate to meet the demand, the regulation up service is implemented in the V2G mode (Kempton et al., 2008).

3.3 Demand Response Management Related Literature

The widespread use of EVs and innovative energy management techniques are necessary to effectively balance demand and supply as a result of the transition from fuel-based energy sources to RESs and the installation of electrified transportation networks (Bibak & Tekiner-Moğulkoç, 2021). One of the primary factors driving the increase in power demand in the next years will be the electrification of the transportation fleet (Aguilar-Dominguez et al., 2020). Balancing the generated electricity and the power demand will be essential since both the PG and energy consumption are variable and stochastic. The penetration of EVs and V2G can be seen as green technologies to support the PG because it is difficult and expensive to store electricity on a big scale for an extended period of time.

The V2G technology, also known as "mobile energy," "smart energy," or "virtual power plant (VPP)," is a reliable method for modifying the demand for electricity (Gough et al., 2017). The V2G technology has great potential for making use of several underutilized electric cars to serve the power network. In order to balance the supply and demand for electricity, the V2G creates a bidirectional conduit for energy, data, and money between electric cars, aggregators, and the network (Nunes & Brito, 2017). The most difficult problem for each electricity grid, which directly impacts the effectiveness and dependability of the system, is the PD time. The overall amount of electricity consumed during periods of high demand is fairly close to the total amount of power generated.

Therefore, any power plant malfunction that reduces the amount of electricity delivered might pose major challenges to meeting the necessary demand. Cutting the PD and shifting the resulting amount into off-peak times is one of the most effective ways to lower the potential threats to the power supply. To increase the system's dependability and sustainability, the V2G system can act as a link between the electrical grid and EVs. Electric batteries are employed in the V2G system to store power, and any surplus or unused electricity is subsequently supplied back to the network during peak hours (Sovacool et al., 2018). The negative environmental consequences of the transportation and electricity networks may also be lessened by the V2G system (Meelen et al., 2020).

Electric utilities all across the globe have implemented numerous DR programmes over the past few decades to lower their PDs during times of stress (U.S. Department of Energy, 2006). The creation of DR systems may be more innovative and adaptable with a wide range of alternatives thanks to modern sensor and control technology (Zerriffi et al., 2005). There is a summary of DR tactics used in commercial structures (Electric Power Research Institute, 2009). The technologies currently in use for residential DR were described and reviewed in a scoping study (National Energy Technology Laboratory, 2007).

DR applications in the industrial and commercial sectors have been thoroughly investigated by (Joo et al., 2007), but a thorough examination of the residential DR strategy that considers customer comfort is still needed. Indicators to gauge the effects of DR on customer convenience are also lacking.

The worldwide trend has led to the use of RESs, such as wind and solar power plants, because of the conventional resources' incapacity to adapt to the network and the irreversible impacts of their usage on environmental concerns. Due to the demand for active power networks, there has been a surge in interest in using RESs in power systems. RESs have the potential to improve the flexibility and resilience of the power system with good management and in the absence of adverse weather conditions. Additionally, the effectiveness of RE resources in lowering costs is growing (Lehtola & Zahedi, 2019).

The integration of RESs with conventional energy sources has been debated by (Yang et al., 2017). An innovative approach has been researched by (Shahbazitabar & Abdi, 2018) in relation to the difficulties of accurate forecasting of RE and synchronized charging of plug-in electric cars. These techniques take into account a wide range of scenarios for the generation of RE resources as well as the DSM of the load from plug-in electric cars. RE resources function as demand providers in these articles, but the authors of (Alirezazadeh et al., 2020) served as a thermal unit reserve. The thermal unit system operator is connected to the aggregator of the SG that houses RESs.

Numerous studies on the advantages of DSM and SG modeling may be found in the literature. A voluntary residential "load shedding model" is investigated by the authors of (Gyamfi & Krumdieck, 2011) to maintain the system under secure conditions during times of high demand. Other concerns with "scheduling appliances" based on "pricing models" are also covered. The advantages of DSM for customers and utilities are demonstrated using two distinct techniques by (Kinhekar et al., 2014), highlighting the significance of identifying the "flexible loads." In contrast to earlier studies, this article assumes that a portion of the "initial demand" is shiftable rather than taking into account how appliances operate. In contrast to other research that exclusively emphasized the role of consumers, our formulation also includes agents with generating or storage assets.

Other authors investigate the distribution of demand using optimization issues based on utility functions and other particular restrictions. According to the authors of (Conejo et al., 2010), the suggested model aims to optimize a consumer's utility under the constraints of a minimum daily consumption level, limitations on load levels, and pricing unpredictability. The authors of (Samadi et al., 2010) took into account the issue of optimizing societal welfare, which is calculated as the total of consumer utilities less the expenses of the energy provider. The authors of (Jiang & Low, 2011) suggested a similar strategy, but uncertainty in the production of RE is addressed, and S and D are managed over two separate time frames.

According to the authors of (Mercurio et al., 2012), the integrated consideration of "distributed" and "renewable generation units," "storage devices," "active demand," and EVs are investigated with an expansion to multiple energy sources. Furthermore, the authors of (Jiang & Low, 2011) analyzed societal welfare or utility functions, whereas the suggested model considers maximizing the agents' gain. Finally, some SG-relevant components are represented by (Mercurio et al., 2012).

4. ALGORITHMS TO OPTIMIZE EV DEMAND RESPONSE MANAGEMENT

The adoption of the carbon-neutral policy has led to an increase in the sales of electric vehicles (EVs). Therefore, it is critical to regulate EV charging and discharging behavior carefully. There are three main algorithms to optimize EV DR management: artificial neural network, machine learning, and nature-inspired algorithm. Figure 4 displays the techniques to optimize EV DR management.

Figure 4. Algorithms to optimize EV DR management

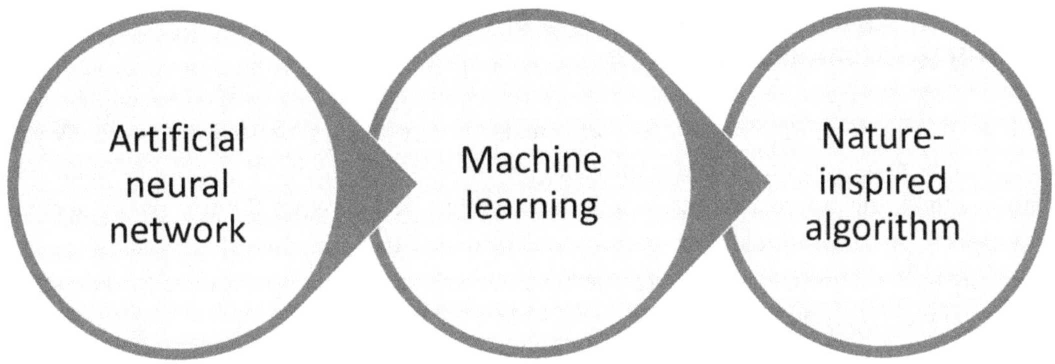

4.1 Artificial Neural Network-Based Modeling of Electric Vehicle Energy Demand Response

Using Reinforcement Learning, the authors of (Alfaverh et al., 2021) suggested an efficient DR technique for managing V2G and vehicle-to-home (V2H) energy (RL). Making the best choices for charging or delaying the charging of the EV battery pack and/or sending the stored power back to the grid without sacrificing driving demands is done using Q-learning, an RL technique based on a reward mechanism. Simulations are used to show how the suggested DR technique may efficiently control the EV battery's charging and discharging schedule.

An artificial neural network-based strategy to execute electric DR algorithms was developed by (Zhou et al., 2020). They employ a feed-forward artificial neural network technique to categorize hourly system loads based on customers' possible involvement in DR programmes and determine the efficient times of participation. In addition, a mathematical model based on incentives, penalties, and demand-price elasticity is created to determine the interruptible and curtailable loads for DR programmes.

By including the DR of EV, the authors of (Li & Li, 2019) suggested a "bi-level" programming approach for "scheduling" isolated MGs with renewable production. In order to solve the model, a "hybrid solution algorithm" called "JAYA-IPM" is created. The simulation findings show that EV operators may actively engage in MG scheduling with the help of DR from EVs in order to accomplish "peak load shaving," which provides a "win-win" approach to balance the interests of MG and EV operators. Table 1 shows artificial neural network-based modeling of EV energy DR.

Table 1. Artificial Neural Network-Based Modelling of Electric Vehicle Energy Demand Response

Reference	Year	Technique(s)	Objective(s)
(Alfaverh et al., 2021)	2021	Q-learning, an RL strategy	Smoothly distributed residential load, reduced power costs, and increased revenue.
(Zhou et al., 2020)	2020	A feed-forward ANN	Choose the right hourly loads before implementing DR programmes.
(Li & Li, 2019)	2019	Hybrid solution algorithm JAYA-IPM	Utilize MG scheduling to reduce peak load

4.2 Machine Learning-Based Modelling of Electric Vehicle Energy Demand Response

A multi-layer machine learning (ML) system was created by (Lin et al., 2020) to produce all EV journeys over the entire region. In contrast to the conventional Monte Carlo technique and the enhanced Latin hypercube method, this approach considers both the connection between the trip parameters and the temporal aspects of the historical data. It may be determined that the suggested strategy can forecast more correctly based on the actual trip data. Each parameter's sensitivity is examined to understand how the EVs' travel parameter prediction accuracy affects DEVA bidders.

For recognizing homes with EV, the authors of (Nandkeolyar & Ray, 2022) suggested a Hidden Markov Model (HMM)-based technique that is akin to ML. This has been done in a non-intrusive manner with the intention of identifying shiftable or time-flexible loads for DR involvement. At various sample rates, information from the smart meters of over 1000 houses has been considered. A few of these homes also had rooftop PV installed to further their independence.

In order to decide the time to charge the EV while connecting with others, the authors of (Lopez et al., 2018) offered an "intelligent charging strategy" utilizing ML techniques. In order to reduce the total cost of vehicle energy, this is accomplished by making "real-time" charging decisions based on a variety of "auxiliary data", such as "driving", "environment", price, and "demand" time series. Dynamic programming is used in the initial phase of the methodology to determine the best solution for previous connection sessions. Then, ML models are trained to understand how to make the best judgments in "real-time" without the information of "future energy" costs and automobile usage using these optimum decisions and other "historical data". Table 2 shows ML-based modeling of EV energy DR.

4.3 Nature-Inspired Algorithm-Based Modelling of Electric Vehicle Energy Demand Response

In order to allow a fleet of carsharing EVs to participate in the ancillary service market, the authors of (Falabretti & Gulotta, 2022) devised an optimization model with natural processes as inspiration. The centralized scheduler minimizes an EV fleet's variance from the power commitment stated in the Day-Ahead Market and gives the grid access to tertiary reserves and control over the power balance. The rule is implemented by maximizing each vehicle's initial charging time in accordance with a time frame decided upon by the carsharing operator.

For managing "energy flow" in the smart house that includes solar production, combined with ESS and EV, the authors of (Wang & Abdalla, 2022) provided optimum "energy management" based on the "Jaya algorithm". The suggested energy management aims to keep power costs down while still provid-

Table 2. Machine Learning-Based Modelling of Electric Vehicle Energy Demand Response

Reference	Year	Technique(s)	Objective(s)
(Lin et al., 2020)	2020	A Hidden Markov Model (HMM) based NILM algorithm	This technique aids in the operational desegregation of EV charging using energy consumption data from smart meters as well as the identification of their charging/discharging status and SOC estimate.
(Nandkeolyar & Ray, 2022)	2022	A multi-layer ML algorithm	Improving the ideal bidding strategy for EV service providers.
(Lopez et al., 2018)	2018	k-Nearest Neighbors (KNN), Shallow Neural Network (SNN), and Deep Neural Network (DNN)	Reduce the cost of charging and affect how much power is used.

ing for residential load demand and EV travel distance energy requirements. In addition to regulating the purchase and sale of energy from and to the grid from excess "PV generation and ESS", the "Jaya algorithm" also controls the modes of "home-to-vehicle (H2V)" and "V2H".

Table 3. Nature-Inspired Algorithm-Based Modelling of Electric Vehicle Energy Demand Response

Reference	Year	Technique(s)	Objective(s)
(Falabretti & Gulotta, 2022)	2022	Hybridization of the ABC algorithm (h-ABC)	Improve the scheduling of an electric carsharing fleet's charging demands
(Wang & Abdalla, 2022)	2022	Jaya algorithm	Lower the cost of power and increase grid stability
(Sowmya & Sankaranarayanan, 2022)	2022	IMPA	By taking into account the battery capacity, C-rating, and other physical limits, the overall cost of the power price for charging and discharging may be reduced.

By applying an "opposition-based learning" strategy to solve the proposed charge/discharge scheduling model using a recent "marine predator algorithm", the authors of (Sowmya & Sankaranarayanan, 2022) introduced a new "Improved Marine Predator Algorithm (IMPA)", and the performance is compared with other cutting-edge algorithms. The scheduling strategy is reiterated, taking uncertainty in the final SoC mismatch and power price into account. Table 3 shows nature-inspired algorithm-based modeling of EV energy DR.

5. CONCLUSION

The charging of many EVs will seriously impact the electricity system. EVs may also use DSM; if necessary, price signals may cause them to change their location or consumption pattern. The total power generated must be sufficient to fulfill the load demand, with spinning reserves for unanticipated emergencies, in order to ensure a steady supply of energy. It is impossible to prevent load curtailments if a power system is unable to meet all or a portion of its loads. However, load curtailments can be minimized or even prevented by deploying DR procedures. The efficient and accurate identification of suitable hourly loads for DR programmes is essential for the efficacy of DR programmes to reduce/avoid load curtailments.

The literature on DSM benefits and SG modeling has a lot of publications. Some of them deal with scheduling issues for appliances using price models; for example, the authors of (Gyamfi & Krumdieck, 2011) investigated a "voluntary household load shedding model" to maintain the system secure in the face of demand surges. The advantages of DSM are demonstrated by (Kinhekar et al., 2014) using two alternative approaches, emphasizing the significance of finding flexible loads. A portion of the initial demand is shiftable rather than taking into account how appliances operate. Some research exclusively emphasized the role of consumers.

Artificial neural networks (ANN) have been used in recent research on modeling transport energy demand for long-term energy demand forecasting at the national level while taking into account a variety of socioeconomic and transport-related data (Brand et al., 2019). Other studies either focus on real-time, short-term projections of EV energy usage (Shankar & Marco, 2013), EV power consumption on the various road network sections (De Cauwer et al., 2017), or are based on a single parameter that takes into account the relationship between "vehicle energy consumption" and "vehicle mass and driving cycle" (Delogu et al., 2016).

REFERENCES

Affanni, A., Bellini, A., Franceschini, G., Guglielmi, P., & Tassoni, C. (2005). Battery choice and management for new-generation electric vehicles. *IEEE Transactions on Industrial Electronics*, *52*(5), 1343–1349. doi:10.1109/TIE.2005.855664

Aguilar-Dominguez, D., Dunbar, A., & Brown, S. (2020). The electricity demand of an EV providing power via vehicle-to-home and its potential impact on the grid with different electricity price tariffs. *Energy Rep.*, *6*, 132–141. doi:10.1016/j.egyr.2020.03.007

Al-Awami, A. T., & Sortomme, E. (2012). Coordinating Vehicle-to-Grid Services With Energy Trading. *IEEE Transactions on Smart Grid*, *3*(1), 453–462. doi:10.1109/TSG.2011.2167992

Alfaverh, F., Denaï, M., & Sun, Y. (2021). Electrical vehicle grid integration for demand response in distribution networks using reinforcement learning. *IET Electr. Syst. Transp.*, *11*(4), 348–361. doi:10.1049/els2.12030

Alirezazadeh, A., Rashidinejad, M., Abdollahi, A., Afzali, P., & Bakhshai, A. (2020). A new flexible model for generation scheduling in a smart grid. *Energy*, *191*, 191. doi:10.1016/j.energy.2019.116438

Asadinejad, A., & Tomsovic, K. (2017). Optimal use of incentive and price based demand response to reduce costs and price volatility. *Electric Power Systems Research*, *144*, 215–223. doi:10.1016/j.epsr.2016.12.012

Bae, S., & Kwasinski, A. (2012). Spatial and Temporal Model of Electric Vehicle Charging Demand. *IEEE Transactions on Smart Grid*, *3*(1), 394–403. doi:10.1109/TSG.2011.2159278

Bibak, B., & Tekiner-Moğulkoç, H. (2021). A comprehensive analysis of vehicle to grid (V2G) systems and scholarly literature on the application of such systems. *Renew. Energy Focus*, *36*, 1–20. doi:10.1016/j.ref.2020.10.001

Bolognani, S., & Zampieri, S. (2013). A distributed control strategy for reactive power compensation in smart microgrids. *IEEE Transactions on Automatic Control*, *58*(11), 2818–2833. doi:10.1109/TAC.2013.2270317

Brand, C., Anable, J., & Morton, C. (2019). Lifestyle, efficiency and limits: Modelling transport energy and emissions using a socio-technical approach. *Energy Effic.*, *12*(1), 187–207. doi:10.100712053-018-9678-9

Cao, Y., Tang, S., Li, C., Zhang, P., Tan, Y., Zhang, Z., & Li, J. (2012). An optimized EV charging model considering TOU price and SOC curve. *IEEE Transactions on Smart Grid*, *3*(1), 388–393. doi:10.1109/TSG.2011.2159630

Clement-Nyns, K., Haesen, E., & Driesen, J. (2011). The Impact of Vehicle-to-grid on the Distribution Grid. *Electric Power Systems Research*, *81*(1), 185–192. doi:10.1016/j.epsr.2010.08.007

Conejo, A., Morales, J., & Baringo, L. (2010). Real-time demand response model. *IEEE Transactions on Smart Grid*, *1*(3), 236–242. doi:10.1109/TSG.2010.2078843

Damiano, A., Gatto, G., Marongiu, I., Porru, M., & Serpi, A. (2014). Vehicle-to-Grid technology: State-of-the-art and future scenarios. *J Energy Power Eng*, *8*(1), 152.

De Cauwer, C., Verbeke, W., Coosemans, T., Faid, S., & Van Mierlo, J. (2017). A Data-Driven Method for Energy Consumption Prediction and Energy-Efficient Routing of Electric Vehicles in Real-World Conditions. *Energies*, *10*(5), 608. doi:10.3390/en10050608

Delogu, M., Del Pero, F., & Pierini, M. (2016). Lightweight Design Solutions in the Automotive Field: Environmental Modelling Based on Fuel Reduction Value Applied to Diesel Turbocharged Vehicles. *Sustainability*, *8*(11), 1167. doi:10.3390u8111167

Desbiens, C. (2012). Electric Vehicle Model for Estimating Distribution Transformer Load for Normal and Cold-Load Pickup Conditions. *Rec. IEEE PES Innovative Smart Grid Tech. Conf. (ISGT)*. 10.1109/ISGT.2012.6175606

Electric Power Research Institute. (2009). *Report to NIST on the Smart Grid Interoperability Standards Roadmap*. Author.

Falabretti, D., & Gulotta, F. (2022). A Nature-Inspired Algorithm to Enable the E-Mobility Participation in the Ancillary Service Market. *Energies*, *15*(9), 3023. doi:10.3390/en15093023

Fan, Z. (2012). A distributed demand response algorithm and its application to PHEV charging in smart grids. *IEEE Transactions on Smart Grid*, *3*(3), 1280–1290. doi:10.1109/TSG.2012.2185075

Gómez, J. C., & Morcos, M. M. (2003). Impact of EV Battery Chargers on the Power Quality of Distribution Systems. *IEEE Transactions on Power Delivery*, *18*(3), 975–981. doi:10.1109/TPWRD.2003.813873

Gong, Q., Mohler, S. M., Marano, V., & Rizzoni, G. (2012). Study of PEV Charging on Residential Distribution Transformer Life. *IEEE Transactions on Smart Grid*, *3*(1), 404–412. doi:10.1109/TSG.2011.2163650

Goransson, L., Karlsson, S., & Johnsson, F. (2010). Integration of Plug-in Hybrid Electric Vehicles in a Regional Wind-thermal Power System. *Energy Policy*, *38*(10), 5482–5492. doi:10.1016/j.enpol.2010.04.001

Gough, R., Dickerson, C., Rowley, P., & Walsh, C. (2017). Vehicle-to-grid feasibility: A techno-economic analysis of EV-based energy storage. *Applied Energy*, *192*, 12–23. doi:10.1016/j.apenergy.2017.01.102

Goulden, M., Bedwell, B., Rennick-Egglestone, S., Rodden, T., & Spence, A. (2014). Smartgrids, smart users? The role of the user in demand side management. *Energy Research & Social Science*, *2*, 21–29. doi:10.1016/j.erss.2014.04.008

Guille, C., & Gross, G. (2009). A Conceptual Framework for the Vehicle-to-grid (V2G) Implementation. *Energy Policy*, *37*(11), 4379–4390. doi:10.1016/j.enpol.2009.05.053

Gyamfi, S., & Krumdieck, S. (2011). Price, environment and security: Exploring multimodal motivation in voluntary residential peak demand response. *Energy Policy*, *39*(5), 2993–3004. doi:10.1016/j.enpol.2011.03.012

Habib, S., Kamran, M., & Rashid, U. (2015). Impact analysis of vehicle-to-grid technology and charging strategies of electric vehicles on distribution networks review. *Journal of Power Sources*, *277*, 205–214. doi:10.1016/j.jpowsour.2014.12.020

Jiang, L., & Low, S. (2011). Multi-period optimal energy procurement and demand response in smart grid with uncertain supply. *50th IEEE conference on decision and control and European control conference (CDC-ECC)*, 4348-53. 10.1109/CDC.2011.6161320

Joo, J., Ahn, S., Yoon, Y., & Choi, J. (2007). Option valuation applied to implementing demand response via critical peak pricing. *Proc. IEEE Power Energy Soc. Gen. Meet.*, 24-28. 10.1109/PES.2007.385559

Kathan, D., Aldina, R., Lee, M.P., Medearis, L., Sporborg, P., & Tita, M. (2012). *Assessment of demand response and advanced metering*. Federal Energy Regulatory Commission Report.

Kempton, W., & Tomic, J. (2005). Vehicle-to-grid power implementation: From stabilizing the grid to supporting large-scale renewable energy. *Journal of Power Sources*, *144*(1), 280–294. doi:10.1016/j.jpowsour.2004.12.022

Kempton, W., Udo, V., Huber, K., Komara, K., Letendre, S., Baker, S., Brunner, D., & Pearre, N. (2008). A test of vehicle-to-grid (V2G) for energy storage and frequency regulation in the PJM system. *Results Ind Univ Res Partnersh*, 32.

Kesler, M., Kisacikoglu, M. C., & Tolbert, L. M. (2014). Vehicle-to-grid reactive power operation using plug-in electric vehicle bidirectional offboard charger. *IEEE Transactions on Industrial Electronics*, *61*(12), 6778–6784. doi:10.1109/TIE.2014.2314065

Kinhekar, N., Padhy, N. P., & Gupta, H. O. (2014). Multi objective demand side management solutions for utilities with peak demand deficit. *International Journal of Electrical Power & Energy Systems*, *55*, 612–619. doi:10.1016/j.ijepes.2013.10.011

Lannoye, E., Flynn, D., & O'Malley, M. (2012). Evaluation of power system flexibility. *IEEE Transactions on Power Systems, 27*(2), 922–931. doi:10.1109/TPWRS.2011.2177280

Lassila, J., Haakana, J., Tikka, V., & Partanen, J. (2012). Methodology to Analyze the Economic Effects of Electric Cars as Energy Storages. *IEEE Transactions on Smart Grid, 3*(1), 506–516. doi:10.1109/TSG.2011.2168548

Lehtola, T., & Zahedi, A. (2019). Solar energy and wind power supply supported by storage technology: A review. *Sustainable Energy Technologies and Assessments, 35*, 25–31. doi:10.1016/j.seta.2019.05.013

Li, Y., & Li, K. (2019). Incorporating Demand Response of Electric Vehicles in Scheduling of Isolated Microgrids With Renewables Using a Bi-Level Programming Approach. *IEEE Access: Practical Innovations, Open Solutions, 7*, 116256–116266. doi:10.1109/ACCESS.2019.2936487

Lin, J., Dong, P., Liu, M., Huang, X., & Deng, W. (2020). Research on Demand Response of Electric Vehicle Agents Based on Multi-Layer Machine Learning Algorithm. *IEEE Access: Practical Innovations, Open Solutions, 8*, 224224–224234. doi:10.1109/ACCESS.2020.3042235

Lopez, K. L., Gagné, C., & Gardner, M. A. (2018). Demand-Side Management Using Deep Learning for Smart Charging of Electric Vehicles. *IEEE Transactions on Smart Grid, 10*(3), 2683–2691. doi:10.1109/TSG.2018.2808247

Meelen, T., Doody, B., & Schwanen, T. (2020). Vehicle-to-grid in the UK fleet market: An analysis of upscaling potential in a changing environment. *Journal of Cleaner Production*, 125203.

Mercurio, A., Di Giorgio, A., & Quaresima, A. (2012). Distributed control approach for community energy management systems in presence of storage. *20th Mediterranean conference on control automation (MED)*, 1303-8. 10.1109/MED.2012.6265819

Millner, A. (2010). Modeling lithium ion battery degradation in electric vehicles. *IEEE Conf. Innov. Technol. an Effic. Reliab. Electr. Supply*, 349-356. 10.1109/CITRES.2010.5619782

Moghe, R., Kreikebaum, F., Hernandez, J. E., Kandula, R. P., & Divan, D. (2011). Mitigating Distribution Transformer Lifetime Degradation Caused by Grid-Enabled Vehicle (GEV) Charging. *Proc. IEEE Energy Conversion Cong. (ECCE)*, 835-842. 10.1109/ECCE.2011.6063857

Nandkeolyar, S., & Ray, P. K. (2022). Identifying households with electrical vehicle for demand response participation. *Electric Power Systems Research, 208*, 107909. doi:10.1016/j.epsr.2022.107909

National Energy Technology Laboratory. (2007). *A system view of modern grid*. Author.

Nikoobakht, A., Aghaei, J., Shafie-Khah, M., & Catalao, J. P. S. (2019). Assessing increased flexibility of energy storage and demand response to accommodate a high penetration of renewable energy sources. *IEEE Trans Sustain Energy, 10*(2), 659–669. doi:10.1109/TSTE.2018.2843161

Nunes, P., & Brito, M. C. (2017). Displacing natural gas with electric vehicles for grid stabilization. *Energy, 141*, 87–96. doi:10.1016/j.energy.2017.09.064

Ota, Y., Taniguchi, H., Nakajima, T., Liyanage, K. M., Baba, J., & Yokoyama, A. (2012). Autonomous Distributed V2G (Vehicle-to-Grid) Satisfying Scheduled Charging. *IEEE Transactions on Smart Grid, 3*(1), 559–564. doi:10.1109/TSG.2011.2167993

Papaefthymiou, G., Grave, K., & Dragoon, K. (2014). *Flexibility options in electricity systems. Ecofys*. European Copper Institute.

Pillai, J. R., & Bak-Jensen, B. (2011). Integration of Vehicle-to-grid in the Western Danish power system. *IEEE Trans. Sustainable Energy, 2*(1), 12–19.

Raghavan, S. S., & Khaligh, A. (2012). Impact of Plug-in Hybrid Electric Vehicle Charging on a Distribution Network in a Smart Grid Environment. *Rec. IEEE PES Innovative Smart Grid Tech. Conf. (ISGT)*. 10.1109/ISGT.2012.6175632

Roukerd, S. P., Abdollahi, A., & Rashidinejad, M. (2019). Probabilistic-possibilistic flexibility-based unit commitment with uncertain negawatt demand response resources considering Z-number method. *International Journal of Electrical Power & Energy Systems, 113*, 71–89. doi:10.1016/j.ijepes.2019.05.011

Saber, A. Y., & Venayagamoorthy, G. K. (2011). Plug-in Vehicles and Renewable Energy Sources for Cost and Emission Reductions. *IEEE Transactions on Industrial Electronics, 58*(4), 1229–1238. doi:10.1109/TIE.2010.2047828

Samadi, P., Mohsenian-Rad, A.-H., Schober, R., Wong, V., & Jatskevich, J. (2010). Optimal real-time pricing algorithm based on utility maximization for smart grid. *First IEEE international conference on smart grid communications (SmartGridComm)*, 415-20. 10.1109/SMARTGRID.2010.5622077

Sarker, M. R., Ortega-Vazquez, M. A., & Kirschen, D. S. (2015). Optimal coordination and scheduling of demand response via monetary incentives. *IEEE Transactions on Smart Grid, 6*(3), 1341–1352. doi:10.1109/TSG.2014.2375067

Shahbazitabar, M., & Abdi, H. (2018). A novel priority-based stochastic unit commitment considering renewable energy sources and parking lot cooperation. *Energy, 161*, 308–324. doi:10.1016/j.energy.2018.07.025

Shankar, R., & Marco, J. (2013). Method for estimating the energy consumption of electric vehicles and plug-in hybrid electric vehicles under real-world driving conditions. *IET Intelligent Transport Systems, 7*(1), 138–150. doi:10.1049/iet-its.2012.0114

Sioshansi, R. (2012). OR Forum-Modeling the impacts of electricity tariffs on plug-in hybrid electric vehicle charging, costs, and emissions. *Operations Research, 60*(3), 506–516. doi:10.1287/opre.1120.1038

Sovacool, B. K., Noel, L., Axsen, J., & Kempton, W. (2018). The neglected social dimensions to a vehicle-to-grid (V2G) transition: A critical and systematic review. *Environmental Research Letters, 13*(1), 13001. doi:10.1088/1748-9326/aa9c6d

Sowmya, R., & Sankaranarayanan, V. (2022). Optimal vehicle-to-grid and grid-to-vehicle scheduling strategy with uncertainty management using improved marine predator algorithm. *Computers & Electrical Engineering, 100*, 107949. doi:10.1016/j.compeleceng.2022.107949

Tan, K. M., Ramachandaramurthy, V. K., & Yong, J. Y. (2016). Integration of electric vehicles in smart grid: A review on vehicle to grid technologies and optimization techniques. *Renewable & Sustainable Energy Reviews, 53*, 720–732. doi:10.1016/j.rser.2015.09.012

Tu, Y., Li, C., Cheng, L., & Le, L. (2011). Research on vehicle-to-grid technology. *Proceedings of the CDCIEM, international conference on computer distributed control and intelligent environmental monitoring*, 1013-1016.

Turker, H., Bacha, S., Chatroux, D., & Hably, A. (2012). Aging Rate of Low Voltage Transformer for a High Penetration of Plug-in Hybrid Electric Vehicles (PHEVs). *Rec. IEEE PES Innovative Smart Grid Tech. Conf. (ISGT)*. 10.1109/ISGT.2012.6175651

U.S. Department of Energy. (2006). *Benefits of demand response in electricity markets and recommendations for achieving them*. A report to the United States Congress pursuant to Section 1252 of the Energy Policy Act of 2005, U.S. Dept. Energy.

USEV Market. (2016). *Compatibility and Investment*. Author.

Vovos, P. N., Kiprakis, A. E., Wallace, A. R., & Harrison, G. P. (2007). Centralized and distributed voltage control: Impact on distributed generation penetration. *IEEE Transactions on Power Systems, 22*(1), 476–483. doi:10.1109/TPWRS.2006.888982

Wang, J., Liu, C., Ton, D., Zhou, Y., Kim, J., & Vyas, A. (2011). Impact of PHEVs on Power Systems with Demand Response and Wind Power. *Energy Policy, 39*(7), 4016–4021. doi:10.1016/j.enpol.2011.01.042

Wang, M., & Abdalla, M. A. A. (2022). Optimal Energy Scheduling Based on Jaya Algorithm for Integration of Vehicle-to-Home and Energy Storage System with Photovoltaic Generation in Smart Home. *Sensors (Basel), 22*(4), 1306. doi:10.339022041306 PMID:35214211

Yang, Z., Li, K., Niu, Q., & Xue, Y. (2017). A comprehensive study of economic unit commitment of power systems integrating various renewable generations and plug-in electric vehicles. *Energy Conversion and Management, 132*, 460–481. doi:10.1016/j.enconman.2016.11.050

Yong, J. Y., Ramachandaramurthy, V. K., Tan, K. M., & Mithulananthan, N. (2015). Bi-directional electric vehicle fast charging station with novel reactive power compensation for voltage regulation. *International Journal of Electrical Power & Energy Systems, 64*, 300–310. doi:10.1016/j.ijepes.2014.07.025

Yunus, K. J., Reza, M., Zelaya-De La Parra, H., & Srivastava, K. (2012). Impacts of Stochastic Residential Plug-In Electric Vehicle Charging on Distribution Grid. *Rec. IEEE PES Innovative Smart Grid Tech. Conf. (ISGT)*. 10.1109/ISGT.2012.6175691

Zerriffi, H., Dowlatabadi, H., & Farrell, A. (2005). Incorporating stress in electric power systems reliability models. *Energy Policy, 35*(1), 61–75. doi:10.1016/j.enpol.2005.10.007

Zhang, B., & Kezunovic, M. (2016). Impact on power system flexibility by electric vehicle participation in ramp market. *IEEE Transactions on Smart Grid, 7*(3), 1285–1294. doi:10.1109/TSG.2015.2437911

Zhao, J., Kucuksari, S., Mazhari, E., & Son, Y. J. (2013). Integrated analysis of high-penetration PV and PHEV with energy storage and demand response. *Applied Energy, 112*, 35–51. doi:10.1016/j.apenergy.2013.05.070

Zhou, Z., Zhang, L., Zhu, F., Tu, W., Yuan, Z., & Gou, X. (2020). *Dynamic Compensation Pricing Scheme for Demand Resources Based on Deep Reinforcement Learning. In IEEE/IAS Industrial and Commercial Power System Asia*. I&CPS Asia.

Chapter 13
Hybrid Optimization Based on Genetic Routing and Scheduling Scheme With Secure Transmission in Wireless Sensor Networks

Kavipriya P.
Sathyabama Institute of Science and Technology, India

Ebenezar Jebarani M. R.
Sathyabama Institute of Science and Technology, India

Lakshmi S.
Sathyabama Institute of Science and Technology, India

G. Jegan
Sathyabama Institute of Science and Technology, India

ABSTRACT

Nodes in networks have limited power for their important tasks because node energy exhaustion reduces network connection. Power-related node failures create system failures, reducing network connection. Furthermore, node mobility and congestion generate frequent connection failures and packet losses, affecting the QoS performance of the protocol. The authors used a hybrid optimization system (HOS) in this research to improve wireless network routing and transmission. To maintain network lifetime and consistent communication, wireless sensor networks need an energy-efficient protocol. This may be accomplished by using a high-quality routing approach.

DOI: 10.4018/978-1-6684-6631-5.ch013

INTRODUCTION

Wireless Sensor Networks

Wireless sensor networks (WSN) are becoming more significant in the growing Internet of Things (IoT). The data collected by remote sensors about hardware or the environment in which it is being used may be very useful. When the sensor is a long distance away or the organization is retrofitted into an existing activity, the information is obtained remotely since linkages are costly to establish. Sensor networks are groups of transducers that communicate with one other to monitor and record environmental variables in various locations (Bagheri, T, 2012).

WSNs were originally designed for military use, but their applications have since expanded to include health, traffic, and a slew of other consumer and mechanical areas. WSNs The number of sensor hubs in a WSN might range from a few hundred to several thousand. Microcontrollers and electronic circuits may be found in the sensor hub gear along with a radio receiver. A power supply is often provided by a battery. As with sensor hub size, shoebox-sized sensors are possible, as are residue-sized sensors.

Network Routing Scheme

Routing is the process of selecting a path for traffic inside a network or between or across networks. Routing is performed by Layer 3 (or network) devices to transport a packet from one network to another by picking the most effective path. Routing is used in many different networks, including circuit-switched networks like the PSTN and computer networks like the Internet. Routing is a higher-level decision-making process in packet switching networks that directs network packets from their origin to their destination through intermediate network nodes utilizing specialized packet forwarding mechanisms. Routing tables may be created manually by a system administrator, discovered by network traffic analysis, or automatically generated by routing protocol assistance. Bridging, on the other hand, is a more general word for a method of connecting two or more networks. Assumptions underlying IP routing include the idea that network addresses are hierarchical and that similar addresses indicate closeness. Routing tables may indicate the path to a set of devices using structured addresses. Structural addressing (routing) is superior than unstructured addressing in big networks (bridging). Router-based addressing is now the most used method of addressing on the internet. Local area networks continue to rely heavily on bridging, (Balamurugan, 2014). Routing schemes differ in how they deliver messages:

A message is delivered to a single node through unicast.
•broadcast sends a message to all network nodes.
Multicast sends a message to a set of nodes that have shown an interest in receiving it.
•Anycast sends a message to any of a collection of nodes, usually the one closest to the source.
•Geocast is a method of delivering a message to a set of nodes depending on their geographic position.

On the Internet, unicast is the most common method of message transmission.

Genetic Optimization Algorithm

As an adjustable heuristic request computation, Genetic Algorithms (GA) fit the bill for the majority of formative estimates. Trademark decisions and intrinsic qualities are considered in inherited calculations.

In order to arrange the chase into a space that allows for better execution in game plan space, these are clever abuses of discretionary request equipped with recorded data, (Shabana Mehfuz, 2014). They are mostly utilized to resolve difficulties and search issues by delivering exceptional solutions. For the sake of addressing a problem, they essentially imitate "survival of the fittest" among individuals of successive generations. In each generation, there are an infinite number of people, each of whom symbolizes a potential solution to a given problem. Any one of these four types of data may be used to represent a person. The Chromosome is a good analogy for this string. (Shaveta, 2014).

LITERATURE REVIEW

Energy Aware Probabilistic Broadcasting for Mobile Ad hocNetwork

For Mobile Ad hoc Networks, Energy-Aware Probabilistic Broadcasting Mobile unusually called association uses broadcasting as the main foundation for a variety of limitations, including delivering data and exposing students to course material. Fundamental flooding is a broadcasting approach that is rapid yet has the potential to derail an association's efforts.

Energy Efficient Probabilistic Broadcasting for Mobile Ad-HocNetwork

The flooding mechanism of Mobile Ad-Hoc Networks (MANETs) is used to broadcast a course interest (RREQ) pack starting with one center and progressing to the next center for course disclosure. This is the easiest approach for broadcasting RREQ packages; however it often results in broadcast storms, initiating accidents, and bundles being blocked in the association, (Gyanappa A, 2012).

Distance based Cluster Head Selection Algorithm for Wireless SensorNetwork

We present a distance-based cluster head selection approach for extending the lifespan of sensor networks. This protocol achieves a high lifespan by balancing the energy demand among all nodes. This clustering strategy contributes to the extended life of wireless sensor networks, particularly in hostile environments where battery replacement of individual sensor nodes is not feasible after deployment in the target region, (Yeim-Kuan Chang, 2007). Thus, the suggested approach for delegating the job of cluster head (CH) to wireless sensor nodes within the same cluster is critical for extending the network's lifespan. This algorithm selects cluster heads using a distance-based strategy. Additionally, clustering algorithms provide effective load balancing and data aggregation inside the network, (Mandeep Kaur Gulati, 2015).

Detection and Prevention Mechanism for Black hole Attack in Wireless Sensor Network

The impact of Blackhole attack on association limitations is evaluated in this article, followed by the recommendation of a unique approach for recognizing and avoiding Blackhole assault in WSN. Occasionally, sensitive data is sent to the intended centre point through an unstable medium. Thus, WSNs are vulnerable to Denial-of-Service (DoS) assaults, which result in data loss and massive energy consumption. Similarly, establishing relationships is critical when establishing a sensor association. Additionally, a black hole assault is a denial-of-service attack, (Shabana Mehfuz, 2016).

The Security Issue of WSNs Based on CloudComputing

The Security Risks Associated with Cloud-Based Wireless Sensor Networks. Due to the low-power, small-size characteristics of WSNs, the combination of WSNs with distributed computing effectively resolves challenges such as hoarding, continual preparation aporia, and universal availability. In any case, there are still some distinct challenges, most notably security concerns, (Chan H, 2004).

In this study, we provide a bar/sub structure for the blend worldview and discuss its merits and drawbacks, including security concerns. Finally, the solutions to the security problem will be revealed. As cloud registration becomes more widespread in the future, experts evaluate the possibility of combining WSN with distributed computing to address the challenges implied by WSN's unique feature.

Cluster Head Selection Protocol using Fuzzy Logic for Wireless Sensor Networks

One approach for prolonging the life of a network is clustering. Clustering methods connect physically adjacent nodes into virtual groups called "clusters." One of the nodes is selected as the cluster head, while the other nearby nodes are identified as cluster members, (Sumit Kumar, 2016).

CHEF: Cluster Head Election mechanism using Fuzzy logic in Wireless Sensor Networks

As the sensor node's battery life is limited, energy saving is the top goal while designing Wireless Sensor Networks. Numerous investigations have been done to address this deficiency. Clustering is a common tactic. During clustering, cluster heads gather data from nodes, consolidate it, and broadcast it to the base station, (Bhawnesh Kumar, 2012).

Consequently, the sensor nodes may avoid the communication overheads that would result if each sensor node transmitted discovered data independently to the base station. LEACH is a popular clustering method. It selects the cluster leader using a probabilistic method. Since LEACH does not consider the position of sensor nodes or the amount of energy left in each node, this method may result in a shorter network lifetime. Nonetheless, clustering with location and energy information may incur substantial costs, (J. M. Kim, 2008).

ACE: An Emergent Algorithm for Highly Uniform Cluster Formation

We describe ACE, a cluster formation technique that produces very uniform clusters with a packing efficiency comparable to that of hexagonal close packing. By using the self-organizing qualities of three rounds of feedback between nodes, the technique induces the emergent development of clusters that efficiently cover the network with substantially less overlap than previous algorithms' clusters. The procedure is scale-independent; regardless of the total number of nodes in the network, it completes in a time proportional to the deployment density of the nodes. ACE is not location-aware and needs just a minimal constant amount of transmission overhead, (Sumit Kumar, 2019).

Intelligent probabilistic broadcasting in mobile ad hoc network: a PSO approach

In two-dimensional PSO, each particle stands for a potential solution to the numerical optimization problem in that dimension. Assuming the preceding theory is true, we may formulate the following description of the PSO procedure. Maximizing an object function is achieved via the act of bird flocking. Each agent reports its current most important value (p best) and its position in X and Y. In addition, every agent knows that there is one excellent bargain that stands head and shoulders above the others. Using its current speed and the gap between pbest and gbest, each agent modifies its location. The necessary increases or decreases in speed may be used to quantify the adjustments. Each agent's speed may be adjusted using the following equation.

Existing Methodology

Existing Methodologies A system-free and self-planned remote association to screen the environment or conditions, such as temperature, sound, and dampness, and to flexibly pass their data collected through the association to a central region or sink (base station) so that the data can be analysed for additional preparation. WSNs. WSN is often transmitted in risky or frightening situations.

Figure 1. Wormhole tunnel constructed between nodes X and Y

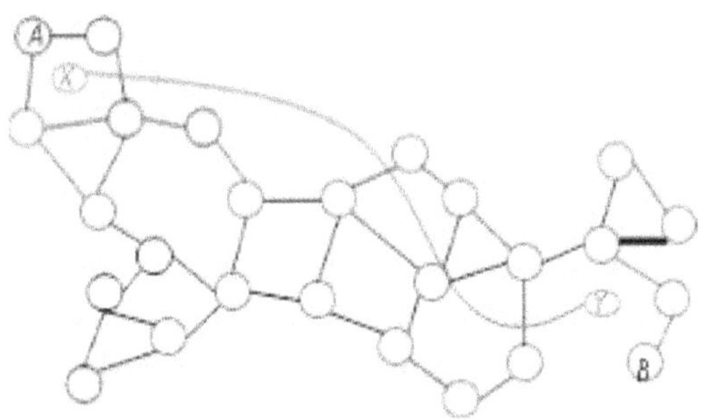

WSN has a huge amount of constraints that create new issues. The sensor centres have a dishonest communication channel and ridiculous resource constraints that make it very difficult to deliver security instruments, (SachinGajjar, 2014). The vast majority of previous WSN presentations assumed that all centres would be dependable and pleasant. Some sensor network applications, however, do not have this problem. WSN attacks such as wormhole are also conceivable, (A. Pratapa Reddy, 2016).

Wormhole Resistant Hybrid Technique (WRHT)

In the suggested method, WRHT, canine and Delphi are monitored simultaneously. System Delphi's gatekeeper canine and RTT setup is based on the concept that a course's bundle drops and RTT are directly connected to the evaluation of its HC (ricochet check) and distance. Realistic WSN circumstances have the potential to result in a high evaluation of RTT and pack drop due to congestion and several other factors, such as a short route and low HC. There is the possibility that a path in the sensor network that

Figure 2. Route Discovery Process of WRHT

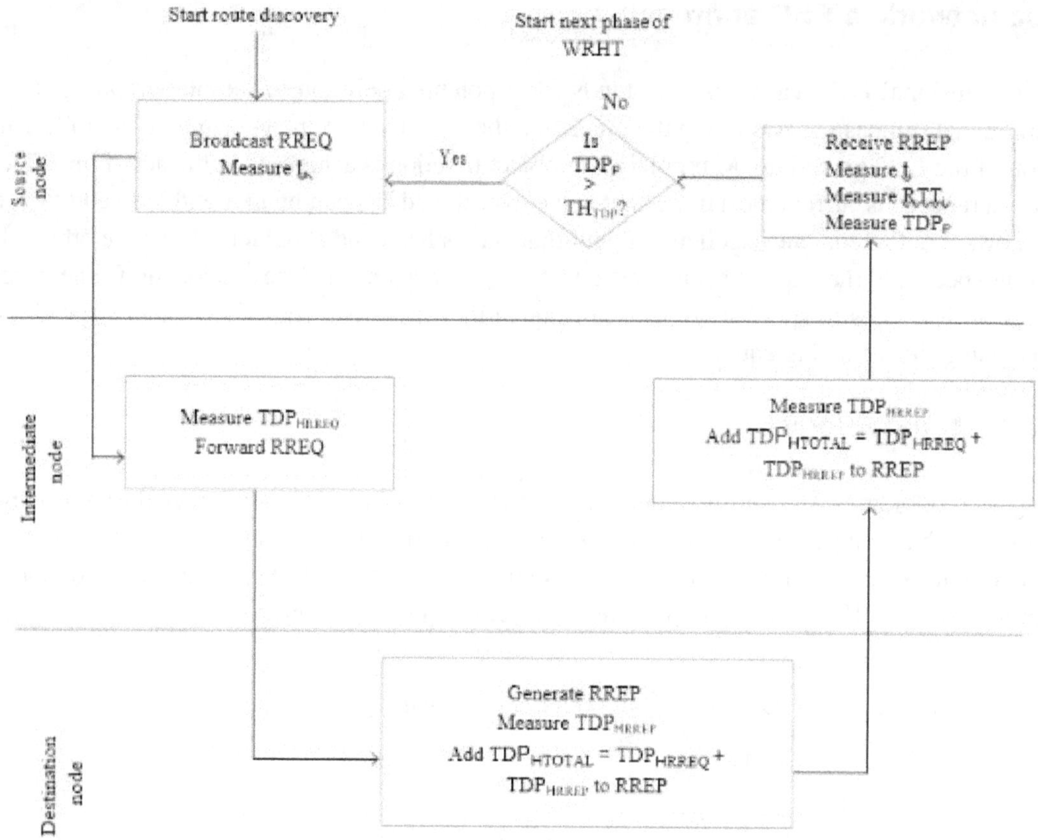

has been contaminated by wormholes might yield a poor rate of return on investment (RTI). That tiny bundle of deferrals may be a direct effect of all the centers getting ready to delay, (Yuanpeng Xie, 2013). Additionally, AODV may be less likely to be prompted to take the wormhole-affected path if an attacker drops fewer packages. Due to the fact that free monitoring canine and Delphi are employed under suitable WSN settings, wormhole responsiveness execution is sometimes compromised. Information about pack drop and delay is used by WRHT in the sensor association for all-out course. It is the goal of WRHT to create a wormhole ID system that can handle all types of wormholes and can be implemented in a WSN device and under the associated conditions without incurring astronomical computing costs.When the sender delivers the RREQ (route request) packet, the receiver receives the RREP (route reply) packet at time ti from its nearby node. We may use the following formula to get the delay/hop value for the path to the receiver through node I: DPH i=(RTTi)/2hi=(tits)/2hi, where RTTi=(tits)/2hi is the sensor network's round trip time. In this example, hi represents the hop count field in nodei's RREP. TDPH is the likelihood of time delay caused at each hop along the defined path from source to destination. To compute TDPH, subtract the time the packet took to travel from the time it reached at its destination.

However, several other suggested solutions explore reducing the amount of energy used to detect such attacks while also testing improved execution. In this research, we describe a lightweight multi-bounce direction convention for 802.15.4 WSN that allows us to decrease energy consumption while

Figure 3. Worm hole Detection Process of WRHT

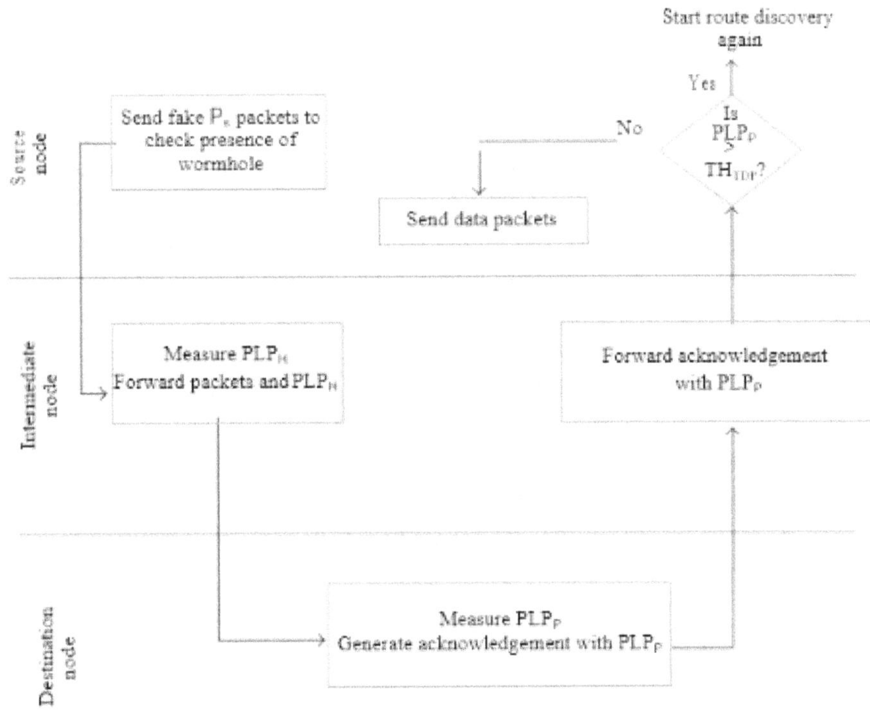

also identifying wormhole attacks, (Mohammad Wazid, 2013). The simulation findings show that our MAC Centralized Routing Protocol (MCRP) outperforms other similar standards.

Proposed Methodology

Hybrid Optimization System (HOS) was employed in this study for directing and transmitting information for long-distance relationships. Dynamic Opportunistic Routing, for example, is one way in our hybrid setup.

It is possible to route over network restrictions using a combination of a multipath scheduling scheme and robust transmission.

Using our suggested strategy, we want to improve the network's end-to-end connectivity while also reducing the number of failures at the connection or nodes themselves. Using an energy-efficient neighbour node selection process, a number of possible routes from the source to multicast destinations are formed.

At the node, load is balanced and a stable path between the source and destination that meets the delay requirement is determined.

The proposed protocol excels in all of these areas: packet delivery ratio, throughput, routing overhead, and average end-to-end latency.

Methodologies:

Initialization

Protocol Initialization:

Dynamic source directing (DSR), an on-request responsive steering standard, is meant to limit the amount of transmission capacity used by control bundles by getting rid of the table update messages that are needed in table-driven proactive technique. Instead of relying on the direction tables at each half-hub, it makes use of source steering.

DSR does not need periodic high parcel broadcasts since it does not have a beacon.

In order to transfer goods from a source to an Anycast hub, hub strength measures are used to identify stable hubs. The self-solidity of a hub and the soundness of its neighbours are the two metrics used to assess the dependability of hub growth as a network characteristic. Choice of neighbour hub based on lingering energy In order to construct stable multicast channels in WSN using processed force level of a hub, an energy-productive neighbour hub choice mechanism is used. Neighbor hubs are hubs that are within the transmission range of one another. There are two parts to this segment: defining a hub energy model to process hub degree and pruning the hubs that use more energy, both of which are discussed in this section. The model of a hub's energy use A WSN hub may function in four different ways.

Node Stability:

Packet delivery services will be improved if the nodes in the forwarding group are more reliable. We can get an indication of a node's stationary attribute by looking at how it moves about in relation to its present location. Using node stability measures, a route from a source to an Anycast node may be identified that is stable. The quality of connection may be measured using two metrics: self-stability and neighbour node stability. Residual energy-based neighbour node selection

Residual energy-based neighbour node selection

The selection of neighbour nodes is dependent on residual energy.

WSN multicast routes are established utilizing a node's calculated power level as a node's neighbour node selection method. Neighbor nodes are any two nodes that are inside each other's transmission range. Defining a node energy model to calculate node degree and then trimming the nodes that spend the most energy are the two components of neighbour node selection in this section.

Energy model of anode

There are four modes of operation for WSN nodes. Modes of transmission, reception, idle, standby, and sleep are all included. A node in sleep mode uses a little amount of energy, compared to the massive amounts used by a node in transmission. Nodes in idle mode use 1:1.2:1.8 kilowatt-hours of energy compared to nodes in transmitting and receiving modes, according to recent research. The anode's remaining energy is calculated using the following energy model, which is based on assumptions and a study of the literature about energy transactions at a node, (Raghavendra, 2011).

Computation of node degree

In multicast routing, the degree of a node is the number of neighbours it can handle. In a multicast session, di (t) is the degree of node I in the network. If Nn (T) is the total number of nodes, then di (t) shows that node I can send packets to d I (t)–1 nodes at the same time, because one link is already used to receive packets. The transmission range and bandwidth of a network decide how many nodes it can have. Differential dominant set theory is used to prune nodes.

Node which wishes to send a data to the destination in a network, it identifies the neighbors.

By regularly flooding HELLO packets (including its own table) and collecting information about all neighbours, a node that desires to transfer data to its destination in the network first determines its neighbours. This information includes the calculated power level and node density of each neighbour. In order to prune its neighbours, the source node or node in question must meet the threshold residual energy value. Stable multicast routes with no pruned nodes with lower node density are what we're aiming for. Make a note of the current node and put it in the queue. The adjacency list may be used to locate all of the nodes that are next to a given node. While the queue is not yet empty, remove a node from it. Retrieve the nodes that are next to the dequeued node, that is to say, the source node. Mark the nearby node as visited and add it to the queue if it hasn't been 11. Remove the last visited node if the neighbouring node is the last node and it is not a destination node in the adjacency list. In such case, return true; otherwise, go on with the Depth First Search. It is false if step 7 is not met.

Balanced Multipath Routing &Scheduling

Stable multicast routes are made using a method called "neighbour node selection." This method chooses the next node based on the amount of energy left and how dense it is. The next section goes into more detail about the RQ and RP packet formats, the request and reply phase, the process for setting up a route, and the process for managing a route. A source node uses RQ packets to figure out how to get to a set of destinations, (Mohammad Sazzadul Hoque, 2012).

Request Phase:

During the route request phase, the following steps take place in the order listed. Node densities and residual energy are included in an RQ packet that is prepared in phase(1). Neighbors that meet the selection criterion and the node density requirement may receive RQ packets from Inphase. If the RQ packet has already been received, the intermediate receiving node discards it in step (3). (Using sequence number in RQ packet). Phase (4) examines the Routing Information table for the availability of the route, if possible, and generates an RP packet and begins propagating the reply to the source.

Duplicated RQ packets in step (5) should be deleted and transmission stopped. Sending the RQ packet to its neighbours and updating its fields (route record, computed power level, node density, time stamp, and next hop address) as in step 2 if it's not duplicated and there's no route in the Routing Information table. Phase 8 sends the RQ packet back to the source node if the recipient can't be reached within a defined number of hops.

Reply Phase: The reply phase is initiated by the multicast destinations. When RQ packets reach their destination nodes, the reply phase commences with the steps listed below. For example, in phase 1, the following modifications are made: destination and source node addresses are swapped, route records are reversed, visited history records are updated as power level, and node density is updated. In phase 1, RP packets are created from RQ packets. The target node's route data is then updated with fresh information such as the IP address, path, voltage, node density, and current time. If the power level and node density are fulfilled, RP packets are delivered to the next hop node in phase 3. If the available power level and node density match the threshold value, the routing table is changed using RP packet contents.

The routing table will be updated if the current time is later than the time previously recorded in the table. If the amount of available energy is less than the threshold value and the number of nodes is more than the threshold value, the RP packet propagation will be stopped and RQ packets will be sent to the destination and any intermediate nodes that have been visited. It is necessary to repeat steps 3 and 4 till the phase (5). If the source cannot be located, the receiver will get an RQ packet and, in step (7), will choose a backup route based on the one with the highest residual energy and the lowest node density.

Figure 4. Multiple multicast routes

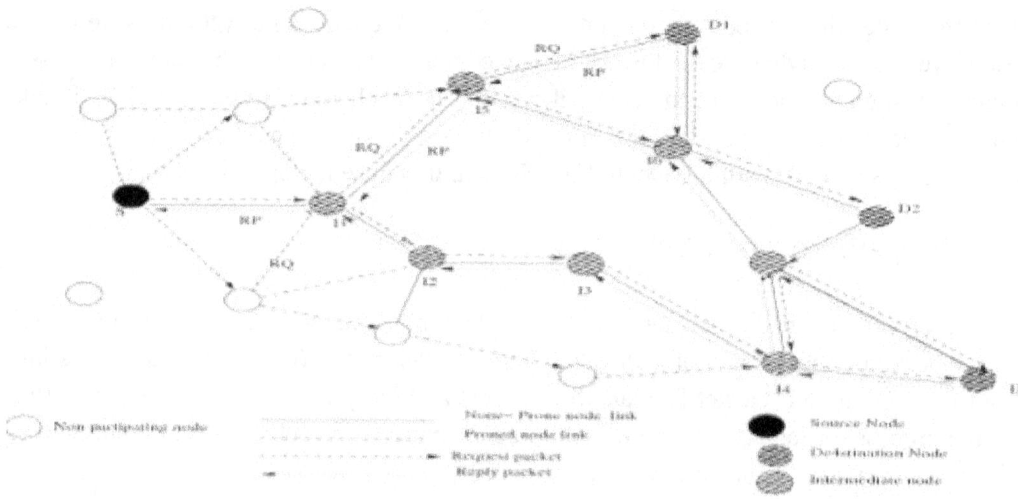

Route maintenance is necessary in the case of a connection or node failure. Because our technique creates a reliable path from source to destination in order to improve total network connection, the likelihood of node and link failure is considerably minimized. The rarest of the uncommon link failures include link failures between stable intermediate nodes, link failure between the source and the stable intermediate node, and link failure between the destination and the stable intermediate node, however they are all conceivable. In this case, we have many alternatives for dealing with the network problem.

When a connection between two reliable intermediate nodes fails, the detecting node will utilize RQ and RP packets to locate an alternative route to its target. The source of the RQ packet will be advised of the altered route from intermediate node to destination. RQ packets are sent to the source if a new route is not discovered.

After a loss in communication with an intermediate node, a backup route will be used by the source node if it is available. In the event that there is no backup path, the routes will be rediscovered.

The intermediate node will employ RQ and RP packets to explore pathways to the destination from itself and tell the source about the path if the connection between the destination and the stable intermediate node fails.

Multiple Path Discovery (MPD) Algorithm

In the event that there is no pathway between the source and the destination, the source node will send out a barrage of Route request packets (RREQ) to the network in an effort to discover an alternative route to the destination. Thereafter, destination nodes respond to the source node's RREQ by sending back Route reply packets (RREP). The destination node instantly returns the RREP packet to the sender of the initial RREQ packet in order to explore additional pathways. Thus, this is the quickest delayed route to the sender. The destination node uses the proposed MPD algorithm to determine the most disjoint route between any two nodes, or between any two links, that the sender's RREQ packets may take.

Computation of Multiple Path Discovery

Figure 5. Multipath Discovery Algorithm

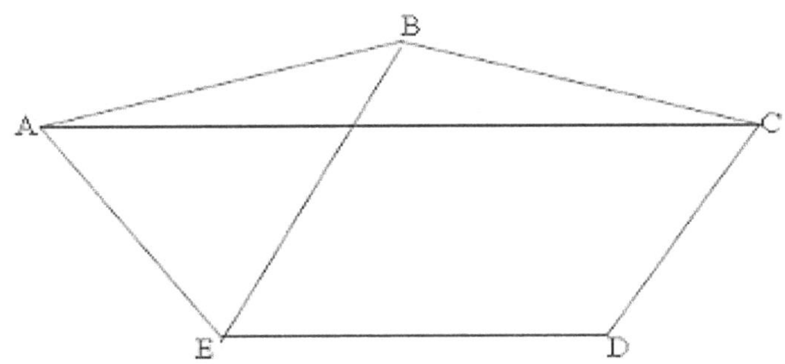

Each node in this diagram has its own unique address. Assume that the source is Node A and the sink is Node D. (Destination). As a result, the destination D keeps track of all potential routes from source A to destination D (ABCD, ABCD, ABED, ABED, ACBED) (D).

Deep Search Algorithm calculates all feasible routes. We may utilize an adjacency list or an adjacency matrix in this technique to discover all of the potential routes. It was decided to use the adjacency list approach in this thesis.

There are six ways to go from point A to point D, as previously indicated. There are six distinct pathways that you may take, each of which is labeled with a number from one to six (P1-ABCD, P2-AED, P3-ACD, P4-AEBCD, P5-ABED and P6-ACBED).

Algorithm to find all Possible Paths from the Source to Destination

Algorithm Depth First Search (n)

Read n number of nodes in the given network.
Read the source and destination node.
Read an array containing the adjacency lists.
Initially mark all the nodes as not visited and visited [ ] is set to zero
Create a queue for DFS.
Mark the current node as visited and enqueue it.

By using the adjacency list, we find all the adjacent nodes of anode.
While queue is not empty dequeue a node from the queue
Get all the adjacent nodes of the dequeued node i.e., source node.
If adjacent node has not been visited, then mark it visited and enqueueit.
If the adjacent node is the last node and it is not a destination node in the adjacency list then remove the last node from the visited [ ]
If this adjacent node is the destination node, then return true or else continue to do Depth First Search.
If step 7 is not satisfied, then return false.
If step 10 returns true, then there is a path from source to destination and print the path (current path indicated as'I').
While queue is not empty then delete all the nodes from queue and assign all these nodes to the path [].
Increment the I value for the next path.

Algorithm to find Multiple Paths

NxN is the number of nodes in the network. Design an adjacency matrix with that size. Any two nodes that have a direct route to each other are marked with a '1' and a '0', respectively.

Read the number of nodes(N)
Read the edges information (0 corresponds to no edge, 1 corresponds to existence of edge)
Select the source and destination nodes.
Find all possible paths from the source to destination node by using Depth First Search algorithm
Calculate path matrix either by using the node disjoint or link disjoint method (Rows corresponds to path and Columns corresponds to nodes).
Find the weight of each path (row sum)
Choose the path with minimum weight as primary path.
Calculate the hamming distance matrix (the hamming distance between two paths = sum (bit-wise Ex-OR of two paths in path matrix)).
Initialize 'sum' array whose size is equals to number of distinct paths with zeros.
Repeat the steps (a) – (d) below until all the paths are ranked (current ranked path indicated with'I').
Add the Path ranked I to sum array
Find the column index corresponding to maximum value from the sum array.
Path corresponding to the column found in the previous step.
(b) Is ranked as I + 1.
I value point to the next value (I = I +1)

Robust Route Selection– Genetic Optimization

The Genetic Algorithm is an evolutionary algorithm with roots in natural selection and biology. Through a process known as optimization, GA determines which node is the most effective at relaying information. Initial population generation, fitness evaluation, selection, crossover, and mutation are the foundation of optimization methods like the Genetic Algorithm. The Genetic Algorithm is a method for reaching the optimal solution. After meeting the termination requirements, the algorithm is complete. Bits are used to refer to each network node. DNA as a method begins with the "Initial Population" points. The starting population is chosen from among the sensor nodes located within the transmission range. Fitness measures a node's effectiveness in extending the network's lifespan. Distance and residual energy are used to determine the nodes' fitness functions. An individual node's chances of staying alive in the network increase with how well suited it is. Once two nodes have been decided upon as parents, the Selection process may begin. Many distinct approaches to making a selection are available. In this study, we employ Tournament Selection, which picks two nodes at random from the existing population. At a certain probability p, the most fit node is chosen, and at a probability of (1-p), the least fit node is chosen.

The process of choosing starts with choosing two nodes to be the parents of the next generation. There are different ways to choose what to do. The Tournament Selection is used in this study. It picks two nodes from the current population at random. When a probability p has already been set, the fit node is chosen, and when the probability is (1-p), the less fit node is chosen.

During crossover or recombination, two nodes swap bits of genetic information with each other. The results of a crossover depend a lot on which nodes are chosen from the population. When a random crossover point is used, the Single Point Crossover method described in this study lets the two parent nodes of a network share data.

Table 1. Single Point Cross Over

	Parent 1	Parent 2
Node	10011|10011001101	11001|01100110101
Offspring	10011|01100110101	11001|10011001101

Table.1 illustrates a single-point crossover. Until the crossover point, the parent nodes' bit sequences are copied to the new offspring nodes in Table.1, and then the second offspring node's bit sequences are transferred to the first parent node's, and so on. In the Mutation process, a random node is chosen from the previous generation's best nodes and a random bit is altered to create new progeny. The Table.2 shows the nodes before and after mutation process in which the eighth bit of the offspring 1 is changed from 1 to 0. Thus, the new offspring is produced after the mutation process

Table 2. Mutation

	Offspring 1	Offspring 2
Before Mutation	1001101100110101	1100110011001101
After Mutation	1001101000110101	1100110011001101

Then, fitness values are computed for new nodes. If the newly produced nodes have a higher health and fitness value than the parent nodes, they are replaced. Examine end-of-life criteria. When a set number of generations pass or the network's energy level drops, the process finishes. The GA method

Figure 6. Flow chart for Fitness based Routing Protocol.

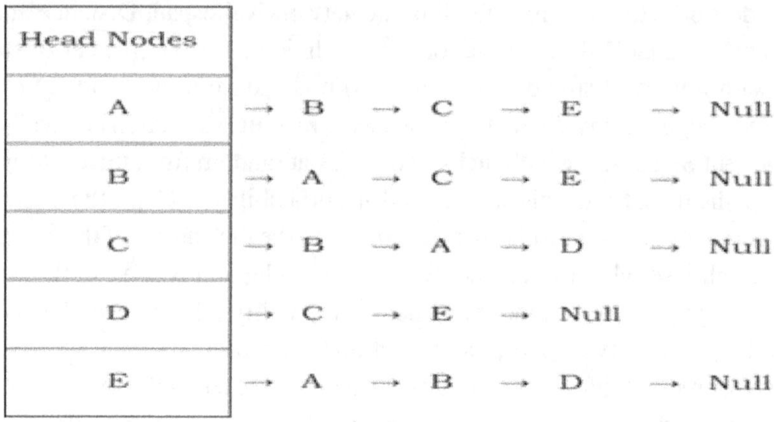

continues seeking for the best match as long as termination requirements are fulfilled. Genetic Algorithm implementation is simple. Electronics, computer science, and sensor networks employ genetic algorithms.

Value assigned by the Fitness-Based Routing Protocol (version 3.3.3.1) The next step is to examine the criteria for ending the project. When either the maximum number of generations has passed or the network's energy has dropped below a predetermined threshold, the algorithm exits. When the GA process reaches the termination criterion, it continues and identifies the optimal node. The implementation of a genetic algorithm is straightforward. Wireless sensor networks are only one of the many places where the versatile Genetic Algorithm has been put to use.

Fitness Based Routing protocol

Using a new fitness-based routing protocol that helps sensor nodes save power might improve the lifespan of existing networks. The fitness value of the node is calculated with the help of the genetic algorithm and then utilized to direct data packets.

Using the node's distance, the fitness function calculates its value. There are four stages to the new protocol: network setup, data collection, fitness computation, and packet routing.

The Fitness-based Routing Protocol flowchart.:

Initializing the hop count throughout the whole network is the first step in the protocol. It receives information from its neighbors within the transmission radius after sensing an incident.

In Genetic Algorithm, packets are routed to the closest node that fits the criteria best depending on the distance between them. The operation will continue as normal until the data packet has reached the base station. To reduce hub and connection disappointment, we have devised a plan to improve the organization's entire availability by extending the hubs' lifespan. Interface dissatisfaction between stable midway hubs, interface disappointment between the source and the stable medium hub, and connection disappointment between objections and the stable moderate hub are all examples of connection disappointments. We'll take care of the organizational problem afterwards. If two stable transitional hubs fail to link, the Hub Identifying Disappointment State will employ RQ and RP bundles to find a stable route.

Fitness Calculation

When a packet has been sent and received, the sending node must decide which node will send the packet to the destination. Using the fitness value, the optimal node is chosen such that the network's lifetime is extended to the greatest extent possible. In this research, we use a Genetic Algorithm to determine which node is the healthiest and then choose it. The fitness value is calculated as a function of the node's distance from the home base. A distance between a set of nodes and a central hub may be calculated with the help of the corresponding formula.

$$dist(i,j) = \sqrt{(x1 - x)^2 + (y1 - y)^2}$$

Fitness = $dist(i,j) + dist(j,bs)$

Where, i, j are parent nodes, bs is the base station.

The new convention has four steps: network initiation, data consolidation, fitness computation, and packet steering. The fitness-based directing convention's stream map. As soon as the hub is set up, a jump check is administered to each of the organization's hubs. After the hub has recognized the event, the transmission sweep gathers data from the adjacent hubs. Genomic algorithms choose the most appropriate hub from a pool of possible hubs based on factors such as distance and direction of package delivery. Until the information package arrives at the base station, this cycle repeats.

RESULTS AND DISCUSSION

Here, NS-2 is used to model the DOS (Dominant Optimization System) strategy. Our model has a channel capacity for mobile hosts of 2 Mbps. IEEE 802.11 provides the foundation for the MAC layer protocol used in wireless local area networks (DCF). Sometimes the network layer is informed when a connection fails.

Mobile node makes 40 movements in a 1000 metre by 1000 metre rectangular space in our experiment during the course of 100 seconds. The speed ranged from 2,4,6,8 and 10 m/s at various points. Constant Bit Rate (CBR) is used to imitate the traffic (CBR).

A network simulator is used to design and build the sensor network, which has N nodes and other characteristics. The random way point model is used to move the mobile nodes throughout each 200-second simulation. As soon as the node arrives at its final destination, it is quickly transferred to a new one.

These speeds are based on a normal distribution and are 5, 10, 15, 20, and 25 m/s for each node. Received packets consume between 200 and 400mW of electricity. The energy consumption model for nodes with ranges less than Rmax may perhaps be found in the NS2 documentation. For different combinations of mobile nodes, the packet delivery rate and end-to-end latency have been figured out.

The simulation is done with the help of the network simulator. This part is all about how to evaluate performance metrics. PDR as a function of the size of the group and the number of nodes at different energy levels. The energy levels of non-pruned nodes have gone down. This has led to a new set of end-to-end multipaths with intermediate nodes that use less energy, which lowers PDR. During packet

Figure 7. Graph of End-to-End Delay

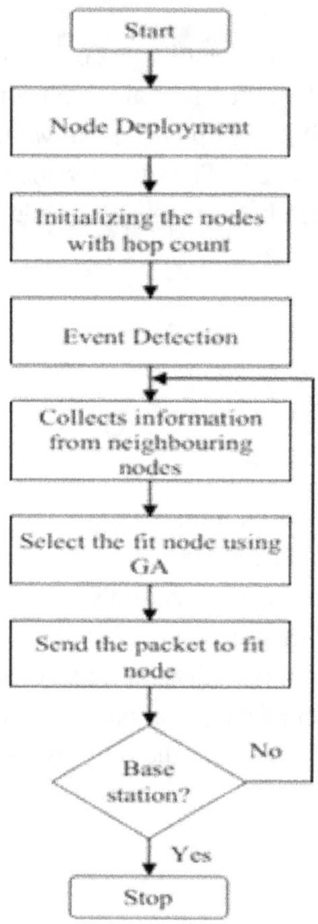

transmission, sometimes a node or connection fails because the PDR value drops because a node's battery power is running low.

Even so, we have a respectable PDR in this scenario since we took into account nodes with more battery life left. Route establishment requires a greater degree of energy (i.e. Our accurate mathematical energy model says that a node's calculated residual energy will be significant even if it uses little energy to send and receive a packet or control information. In Wireless Sensor Networks, the four most important parameters for secure data transmission are as follows:

END TO END DELAY.
PACKET DELIVERYRATIO.
AVERAGETHROUGHPUT.
ENERGYCONSUMPTION.
End-to-End Delay contrasts the suggested low delay ratio with current frameworks.

End-to-End Comparing the projected low delay ratio to current frameworks is called a Delay. It takes less time and a higher speed to transport data more quickly. Extremely minimal delays are expected.

Figure 8. Packet Delivery Ratio

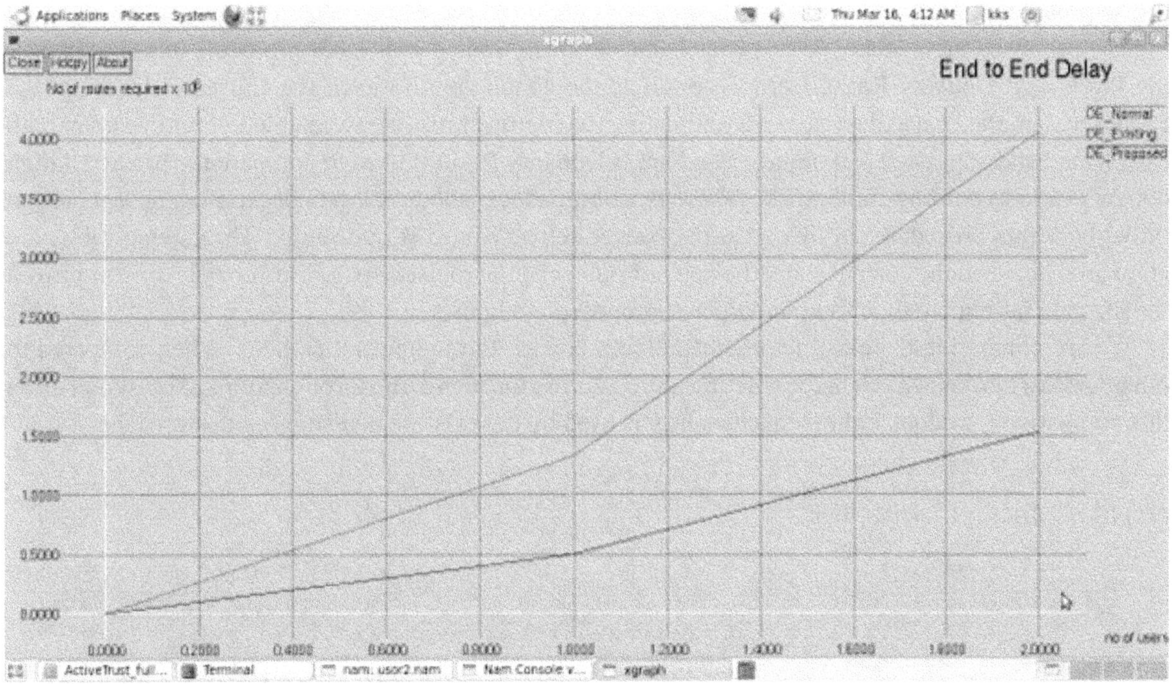

Figure 9. Average Throughput

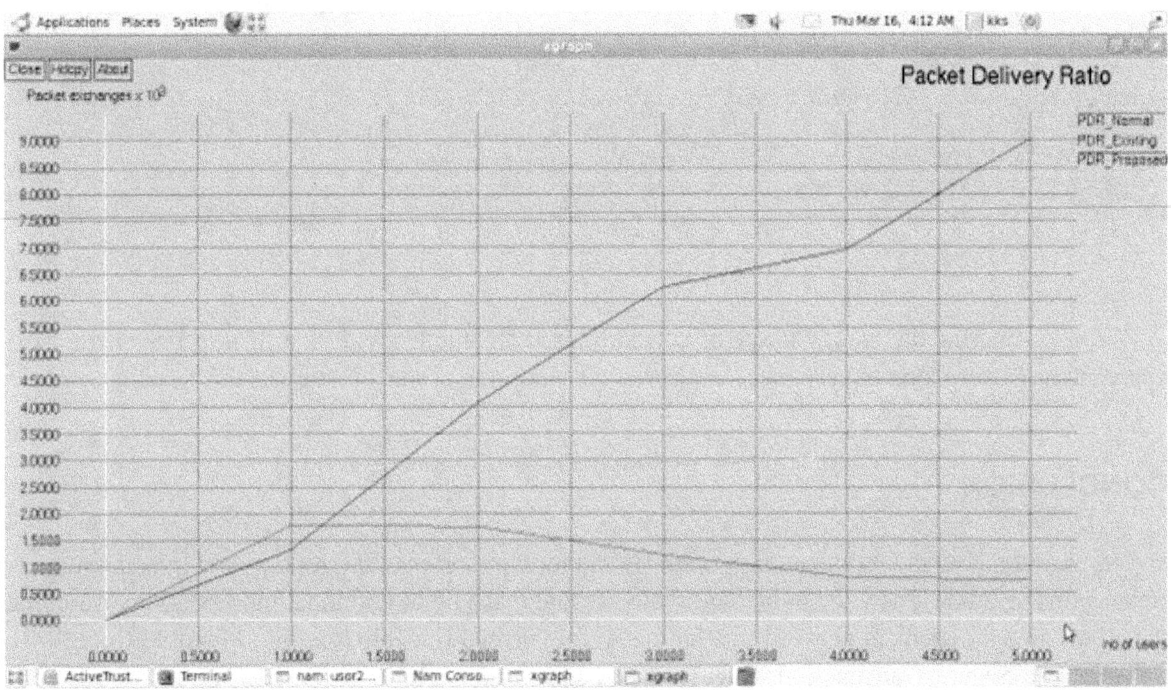

Using the Packet Delivery Ratio, the proposed high packet delivery ratio may be compared to already-in-use protocols.

The proposed high packet delivery ratio may be measured against other protocols in use today using the Packet Delivery Ratio. Here we evaluate the Throughput ratio of the Currently Existing, the Proposed, and the Updated Systems. By adjusting the lifetime ratio, we were able to improve upon both established and suggested techniques. The Packet Delivery Ratio is used to compare the predicted high delivery ratio to existing frameworks. There is a tremendous amount of data packets being sent around throughout this procedure. In this area, the packet delivery ratio is quite high. Throughput ratios are compared between the current state, the desired state, and the revised state. Our modification improved the lifespan ratio more than the standard and proposed methods.

We are comparing the existing, planned, and modified throughput ratios here. When compared to other approaches, ours has a higher lifetime ratio. Across the board, the throughput ratio is excessive. In this comparison, we look at how much energy is used by the existing and suggested approaches.

Figure 10. Energy Consumption

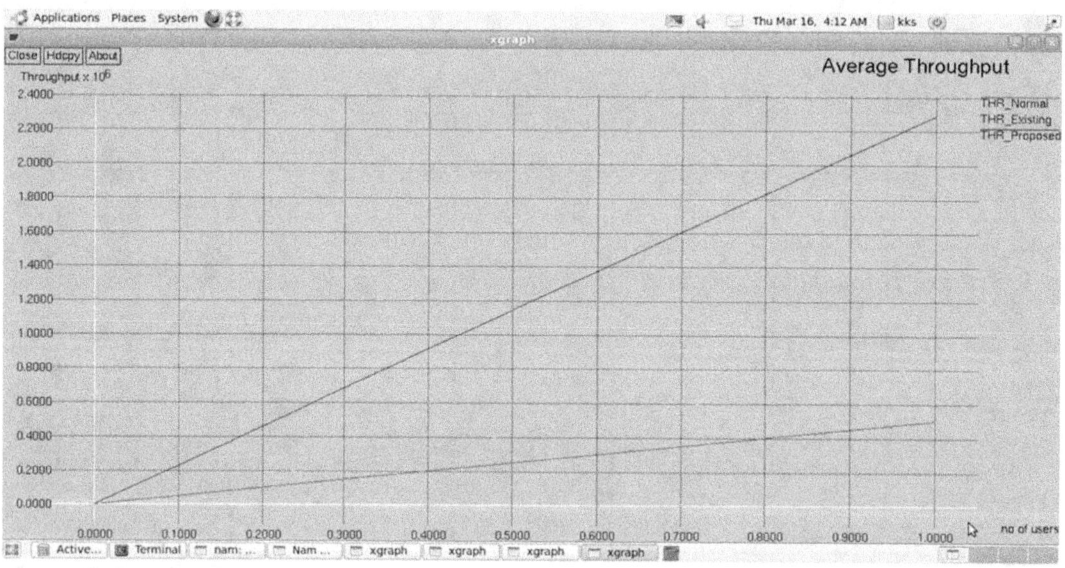

We're comparing the energy usage of currently used techniques with that of new ones here. This approach has a very long battery life. Low power consumption is a key feature of this system.

CONCLUSION

We employed a Hybrid Optimization System (HOS) for wireless network routing and transmission efficiency in this study. There are numerous strategies that make up our dominating strategy to overcome the aforesaid restrictions in networks, including Accomplished Capable Routing; Manifold Coordinate Scheme; and Secure Transmission. Our suggested strategy aims to improve network end-to-end connectivity while reducing the number of node- or link-level failures.

An energy-efficient technique for selecting neighbor nodes forms several pathways from the source to multicast destinations. It finds a stable path between the source and the destination that satisfies the required latency while balancing the load on the node. The proposed protocol outperforms the state-of-the-art in all of these areas: packet delivery ratio, throughput, routing overhead and average end-to-end latency.

Using active detection, a new security and trust routing mechanism has been presented, and it has the following good characteristics: The likelihood of a successful route, security, and scalability are all quite high. By avoiding suspicious nodes and promptly detecting nodal trust, the Trust scheme may fast achieve a near-100 percent success rate while routing data. High efficiency in terms of use of energy. (2). The Trust system builds several detection paths using just residual energy.

It has been proved that our method boosts the chance of effective routing by more than three times, and in certain situations up to ten times. In addition, we've devised a strategy that boosts both network energy efficiency and security. Wireless sensor network security relies heavily on it.

FUTURE WORK

Concurrent scheduling is the next step in our effort to enhance scheduling performance by taking into account numerous networks. In addition, we'll devise distinct time-scheduling approaches for networks with various levels of importance. Nodes and connections may fail at a given rate when the network's size, complexity, and heterogeneity grow fast. This is particularly true for long-term data transmissions. Tolerance for failure has thus grown to be a crucial component of network performance.

The researchers came up with a new security and trust routing scheme based on active detection. It has the following great qualities: (1) High chance of successful routing, security, and ability to grow. The Trust scheme can quickly figure out which nodes can be trusted and then avoid suspicious nodes. This makes it possible to quickly get a nearly 100% chance of successful routing. (2) Good use of energy. The Trust plan uses all of the leftover energy to build several detection routes.

Theoretical analysis and experimentation have shown that our scheme increases the chance of successful routing by more than three times, and in some cases by as much as ten times. Also, our plan makes both the use of energy and the security of the network better. It means a lot for the security of wireless sensor networks.

In order to further enhance scheduling performance, we want to expand our work to include concurrent scheduling, which takes into account numerous networks at once. We'll also come up with a variety of scheduling options for various networks. For long-term data transfers, nodes and connections may fail at a given rate when network complexity, size, and heterogeneity quickly rise. As a result, the ability of a network to tolerate faults has grown in importance. The reliability of the bandwidth scheduling problem in high-performance networks with malfunctioning nodes and connections is a topic of interest to that in the future.

REFERENCES

Bagheri, T. (2012). DFMC: decentralized fault management mechanism for cluster based wireless sensor networks. *Second Int. Conf. on IEEE Digital Information and Communication Technology and it's Applications (DICTAP)*, 67-71. 10.1109/DICTAP.2012.6215406

Balamurugan. (2014). An Energy Efficient Fitness Based Routing Protocol in Wireless Sensor Networks. *ICTACT Journal on Communication Technology*. DOI: doi:10.21917/ijct.2014.0127

Bhandari & Saluja. (2014). Applying Genetic Algorithm in Intrusion Detection System: A Comprehensive Review" Recent Trends in Information, Telecommunication and Computing. *ITC2014*.

Chan, H., & Perrig, A. (2004). ACE: An Emergent Algorithm for Highly Uniform Cluster Formation. *Proc. of 1st European Workshop on Sensor Networks*, 154–171,2004. 10.1007/978-3-540-24606-0_11

Chang, Y.-K., Ting, Y.-W., & Wu, S.-C. (2007). Power-Efficient and Path- Stable Broadcasting Scheme for Wireless Ad Hoc Networks. *International Conference on Advanced Information Networking and Applications Workshops (AINAW'07)*. 10.1109/AINAW.2007.291

Gajjar, Sarkar, & Dasgupta. (2014). Cluster Head Selection Protocol using Fuzzy Logic for Wireless Sensor Networks. *International Journal of Computer Applications, 97*(7).

Gulati & Kumar. (2015). Stable Energy efficient QoS based Congestion and Delay aware Routing (SE-QCDR) Protocol for MANETs. *IEEE ICCSP Conference*.

Hoque, Mukit, & Bikas. (2012). An Implementation of Intrusion Detection System Using Genetic Algorithm. *International Journal of Network Security & Its Applications, 4*(2).

Kim, J. M., Park, S. H., Han, Y. J., & Chung, T. M. (2008). CHEF: Cluster head election mechanism using fuzzy logic in wireless sensor networks. *Proc. of International Conference on Advanced Communication Technology*, 654–659. 10.1109/ICACT.2008.4493846

Kulkarni, R. V., Forster, A., & Venayagamoorthy, G. K. (2011). Computational Intelligence in Wireless Sensor Networks: A Survey. *IEEE Communications Surveys and Tutorials, 13*(1), 68–96. doi:10.1109/SURV.2011.040310.00002

Kumar & Mehfuz. (2017). Energy Efficient Probabilistic broadcasting for mobile ad hoc network. *Journal of the Institution of Engineers (India): Series B, 98*(3), 289–294.

Kumar & Mehfuz. (2019). A PSO based malicious node detection and energy efficient clustering in wireless sensor network. *International Conference on Signal Processing and Integrated Networks (SPIN)*.

Kumar & Sharma. (2012). Distance based Cluster Head Selection Algorithm for Wireless Sensor Network. *International Journal of Computer Applications, 57*(9).

Kumar, S., & Mehfuz, S. (2016, July). Intelligent probabilistic broadcasting in mobile ad hoc network: A PSO approach. *Journal of Reliable Intelligent Environments, 2*(2), 107–115. doi:10.100740860-016-0023-9

Kumar, S., & Mehfuz, S. (2016, June). Efficient Fuzzy Logic Based Probabilistic broadcasting for mobile ad hoc network. *International Journal of Computational Intelligence Systems*, *9*(4), 666–675. doi: 10.1080/18756891.2016.1204115

Mehfuz & Kumar. (2015). Energy Aware Probabilistic Broadcasting for Mobile Ad Hoc Network. *2nd IEEE International Conference on Computing for Sustainable Global Development*, 1028 – 1033.

Mehfuz, S., & Kumar, S. (2014, September). Two-dimensional particle swarm optimization algorithm for load flow analysis. *International Journal of Computational Intelligence Systems*, *7*(6), 1074–1082. doi:10.1080/18756891.2014.963973

Pratapa Reddy & Satyanarayana. (2016). Energy Efficient Stable Multi path Routing in MANET. *International conference on Signal Processing, Communication, Power and Embedded System (SCOPES)*, 1115-1121. 10.1109/SCOPES.2016.7955615

Walikar & Biradar. (2012). *Energy aware multicast routing in mobile ad-hoc networks using NS-2*. Academic Press.

Wazid, Katal, Singh, Goudar, & Singh. (2013). Detection and Prevention Mechanism for Blackhole Attack in Wireless Sensor Network. *International Conference on Communication and Signal Processing*.

Xie, Zhang, Fu, Wen, Han, Zhu, Jiang, & Guo. (2013). The Security Issue of WSNs Based on Cloud Computing. *IEEE Conference on Communications and Network Security*.

APPENDIX

Snapshots

Table 3. Snapshot

```
ns main.tcl
num_nodes is set 42
warning: Please use -channel as shown in tcl/ex/wireless-mitf.tcl
INITIALIZE THE LIST xListHead
Loading connection pattern...
Loading scenario file...
Enter Source Node(0 to 41)
20
Enter Destination Node(0 to 41)
30
```

Figure 11. Node Initialization

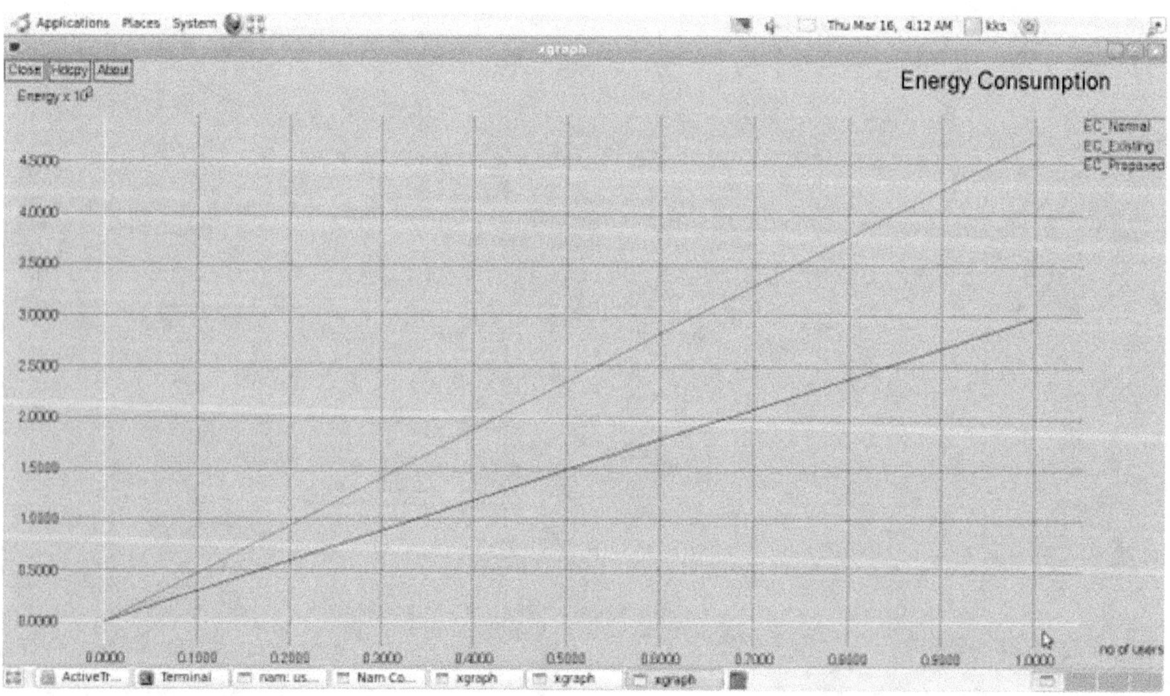

Figure 12. Neighbour Communication

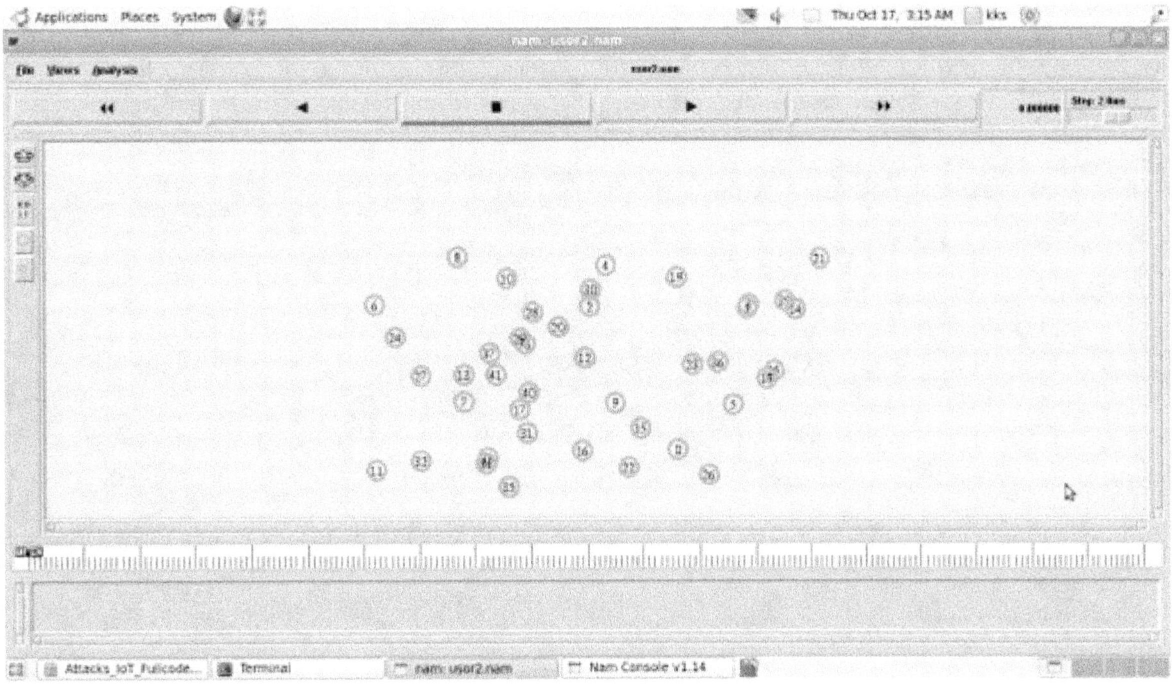

Figure 13. Source, Destination and Route Selection

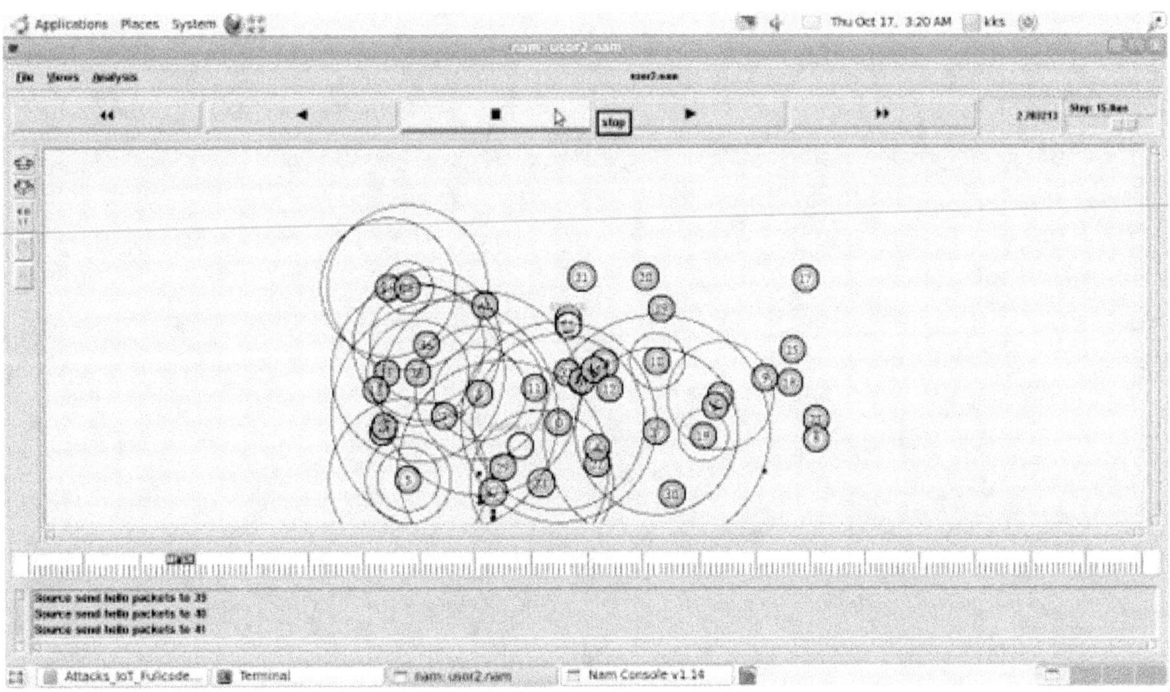

Figure 14. Attacker Detection

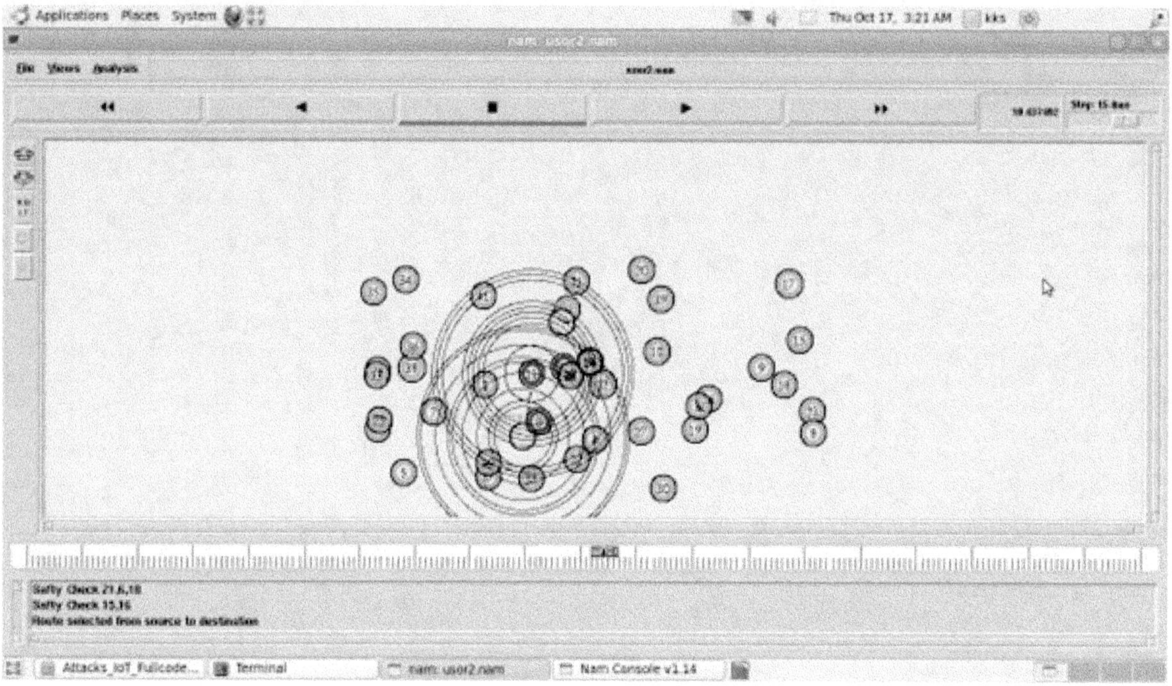

Figure 15. Data Sharing Between Source and Destination

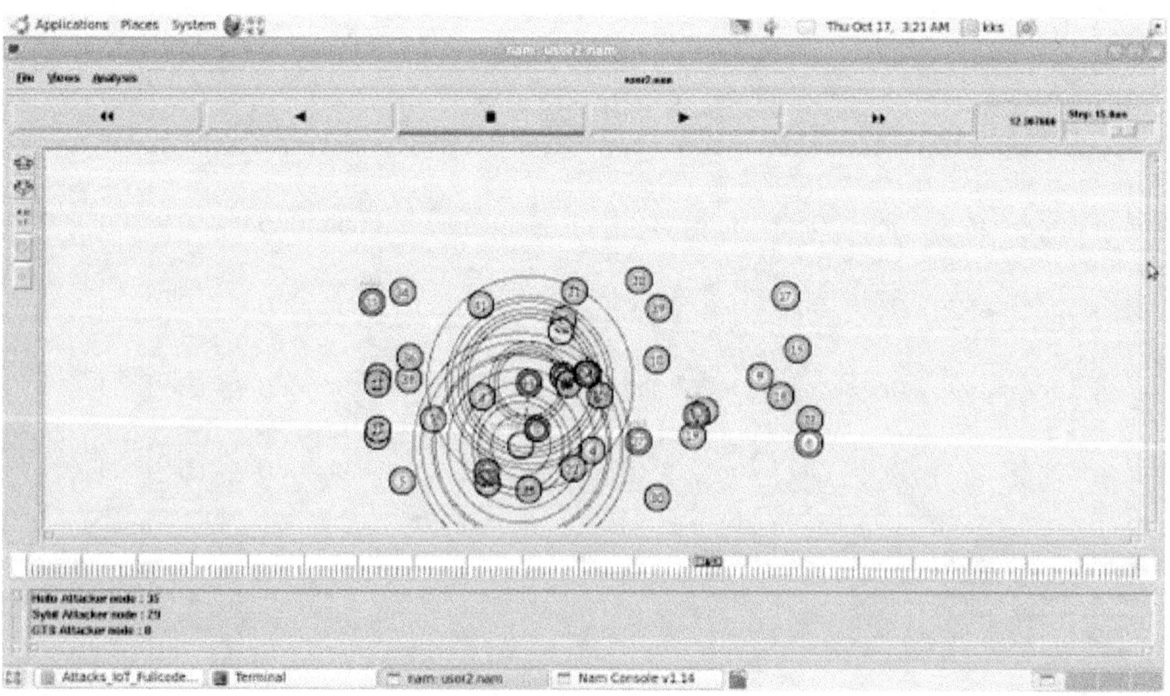

Figure 16.

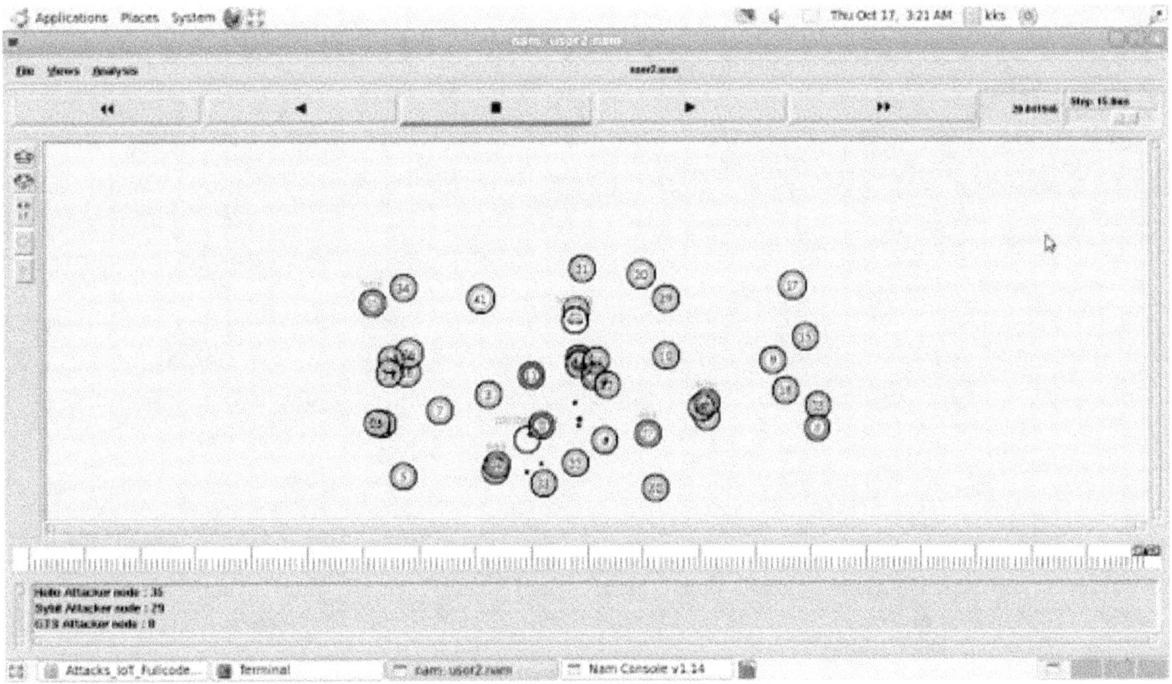

Figure 17.

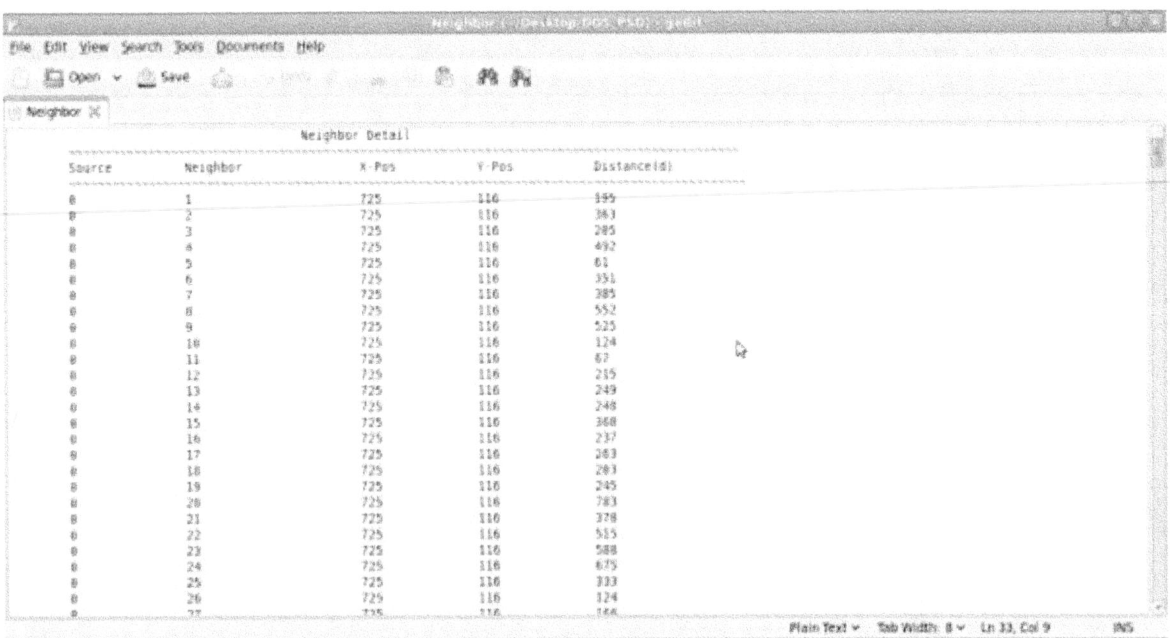

Figure 18.

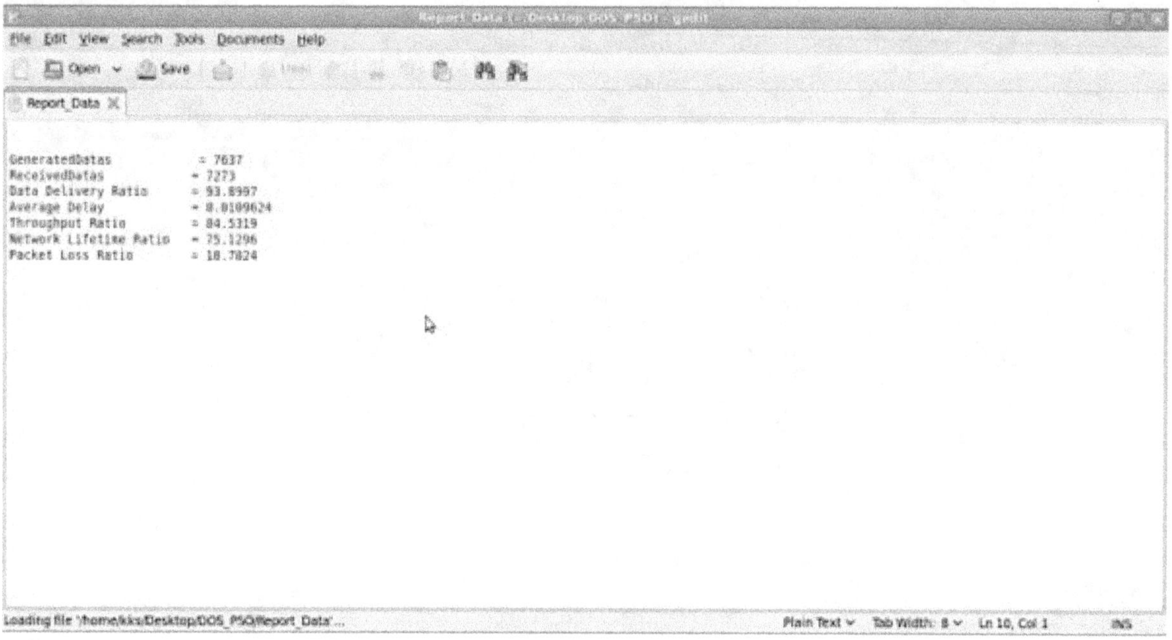
File Edit View Search Tools Documents Help
Open
Save
Report_Data
GeneratedDatas = 7637
ReceivedDatas = 7273
Data Delivery Ratio = 93.8997
Average Delay = 0.0109624
Throughput Ratio = 84.5319
Network Lifetime Ratio = 75.1296
Packet Loss Ratio = 18.7824
Plain Text
Ln 10, Col 1
INS

Figure 19.

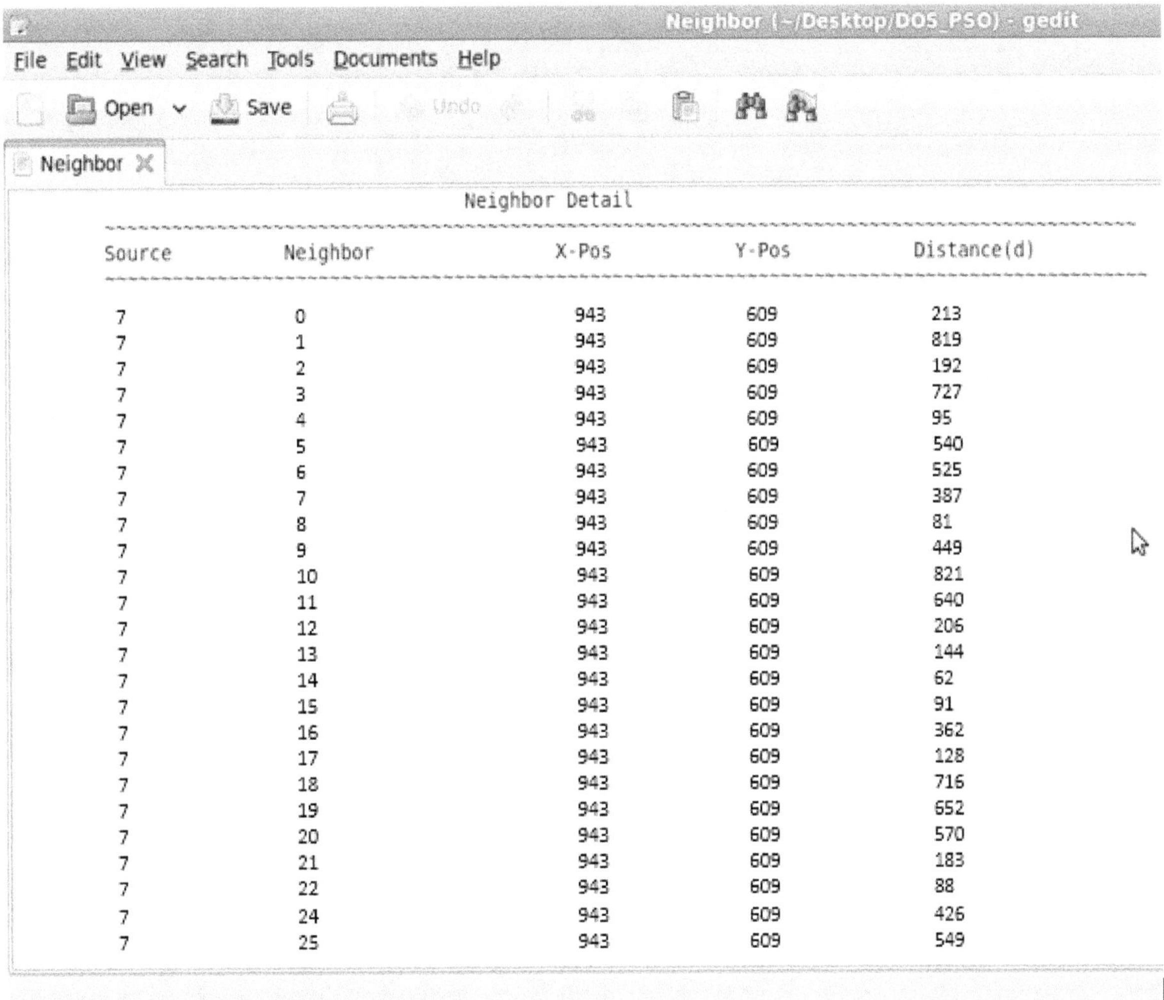

Neighbor Detail

Source	Neighbor	X-Pos	Y-Pos	Distance(d)
7	0	943	609	213
7	1	943	609	819
7	2	943	609	192
7	3	943	609	727
7	4	943	609	95
7	5	943	609	540
7	6	943	609	525
7	7	943	609	387
7	8	943	609	81
7	9	943	609	449
7	10	943	609	821
7	11	943	609	640
7	12	943	609	206
7	13	943	609	144
7	14	943	609	62
7	15	943	609	91
7	16	943	609	362
7	17	943	609	128
7	18	943	609	716
7	19	943	609	652
7	20	943	609	570
7	21	943	609	183
7	22	943	609	88
7	24	943	609	426
7	25	943	609	549

Figure 20.

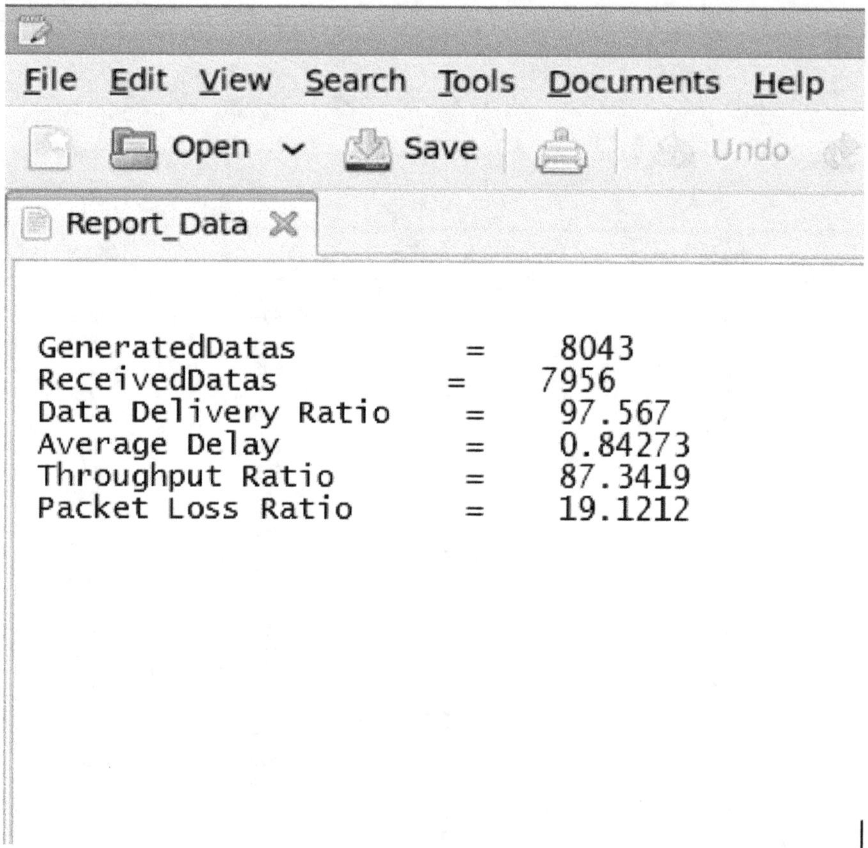
File Edit View Search Tools Documents Help
Open
Save
Undo
Report_Data
GeneratedDatas = 8043
ReceivedDatas = 7956
Data Delivery Ratio = 97.567
Average Delay = 0.84273
Throughput Ratio = 87.3419
Packet Loss Ratio = 19.1212

Figure 21.

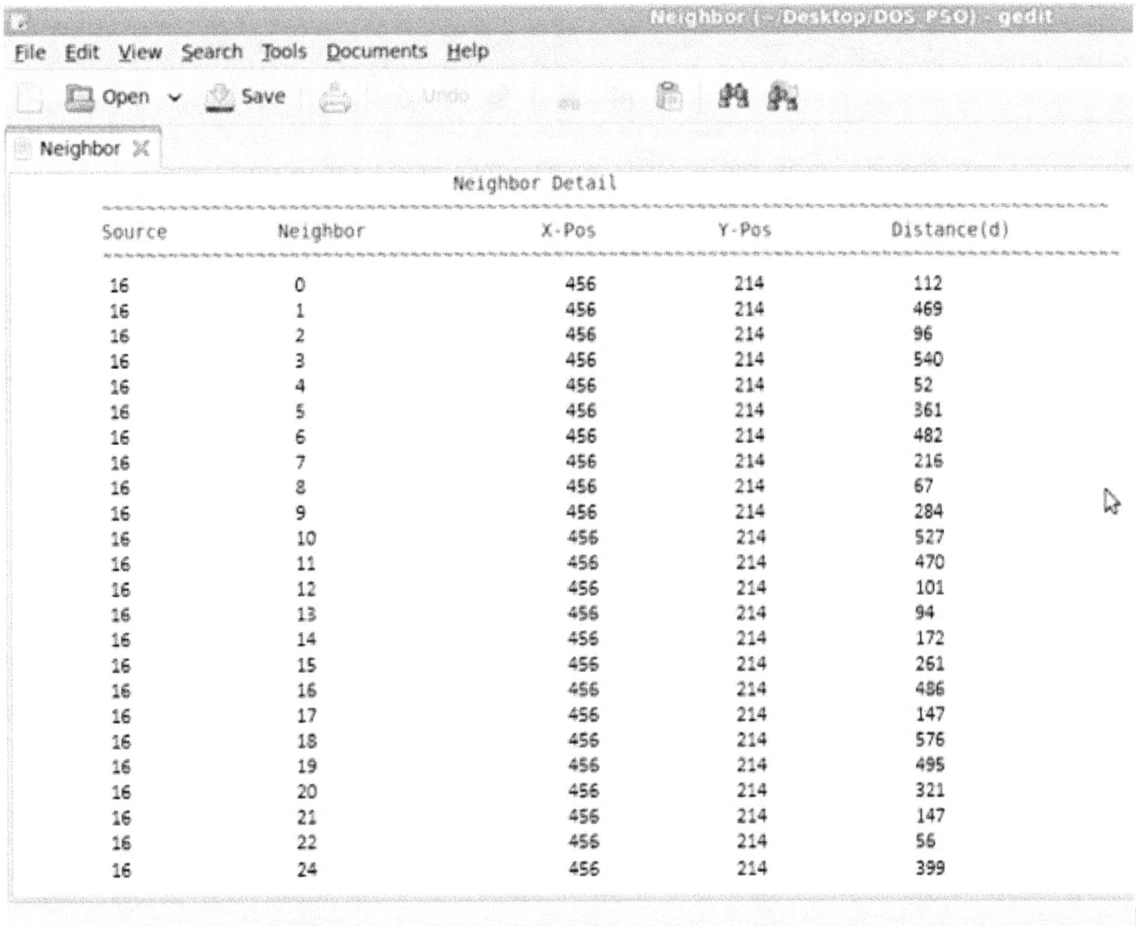

Neighbor Detail

Source	Neighbor	X-Pos	Y-Pos	Distance(d)
16	0	456	214	112
16	1	456	214	469
16	2	456	214	96
16	3	456	214	540
16	4	456	214	52
16	5	456	214	361
16	6	456	214	482
16	7	456	214	216
16	8	456	214	67
16	9	456	214	284
16	10	456	214	527
16	11	456	214	470
16	12	456	214	101
16	13	456	214	94
16	14	456	214	172
16	15	456	214	261
16	16	456	214	486
16	17	456	214	147
16	18	456	214	576
16	19	456	214	495
16	20	456	214	321
16	21	456	214	147
16	22	456	214	56
16	24	456	214	399

Figure 22.

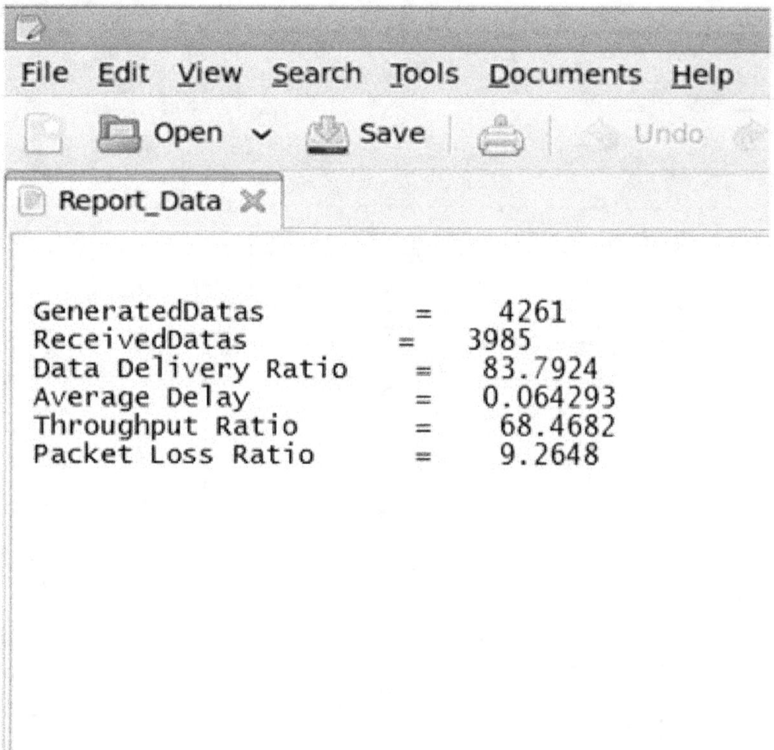
File Edit View Search Tools Documents Help
Open
Save
Undo
Report_Data
GeneratedDatas = 4261
ReceivedDatas = 3985
Data Delivery Ratio = 83.7924
Average Delay = 0.064293
Throughput Ratio = 68.4682
Packet Loss Ratio = 9.2648

Figure 23.

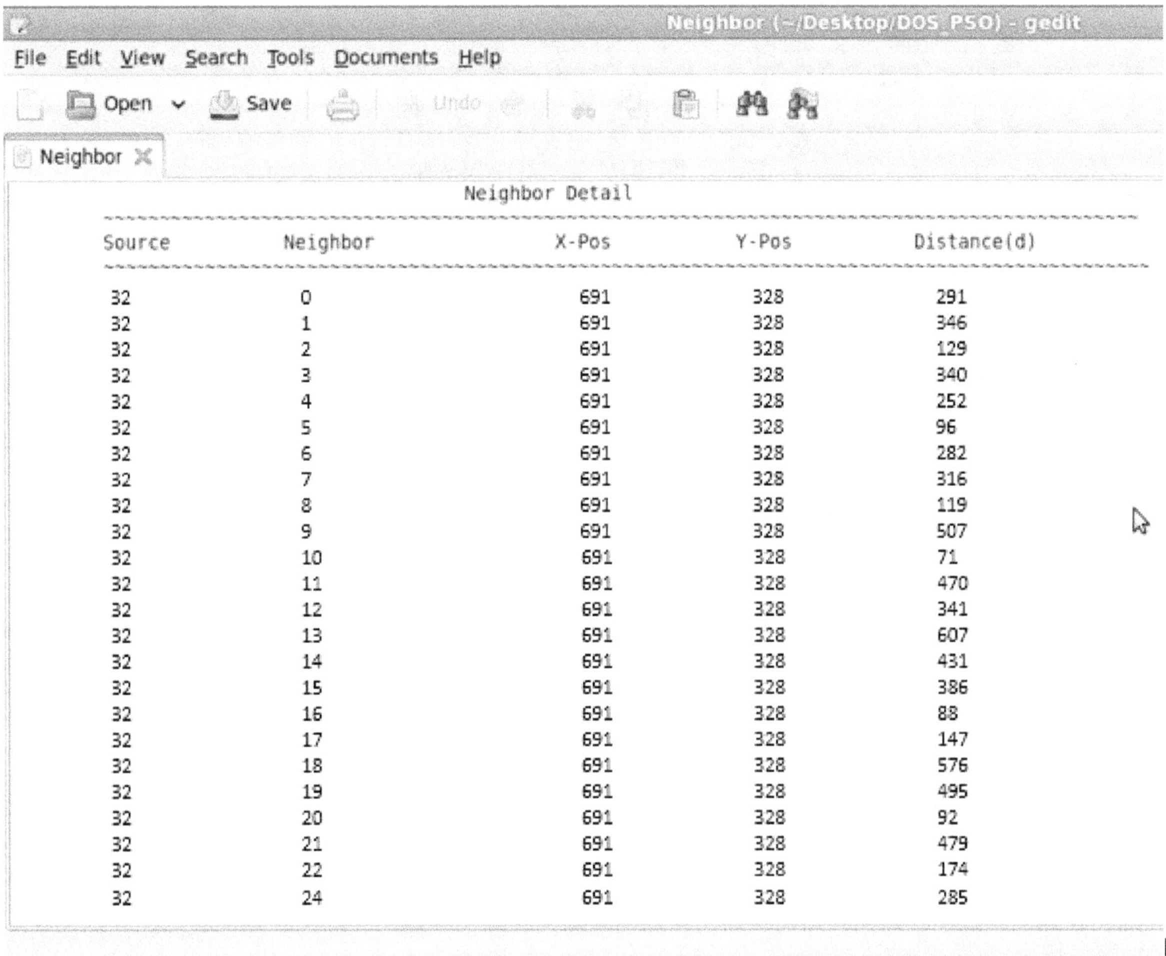

Neighbor Detail

Source	Neighbor	X-Pos	Y-Pos	Distance(d)
32	0	691	328	291
32	1	691	328	346
32	2	691	328	129
32	3	691	328	340
32	4	691	328	252
32	5	691	328	96
32	6	691	328	282
32	7	691	328	316
32	8	691	328	119
32	9	691	328	507
32	10	691	328	71
32	11	691	328	470
32	12	691	328	341
32	13	691	328	607
32	14	691	328	431
32	15	691	328	386
32	16	691	328	88
32	17	691	328	147
32	18	691	328	576
32	19	691	328	495
32	20	691	328	92
32	21	691	328	479
32	22	691	328	174
32	24	691	328	285

Figure 24.

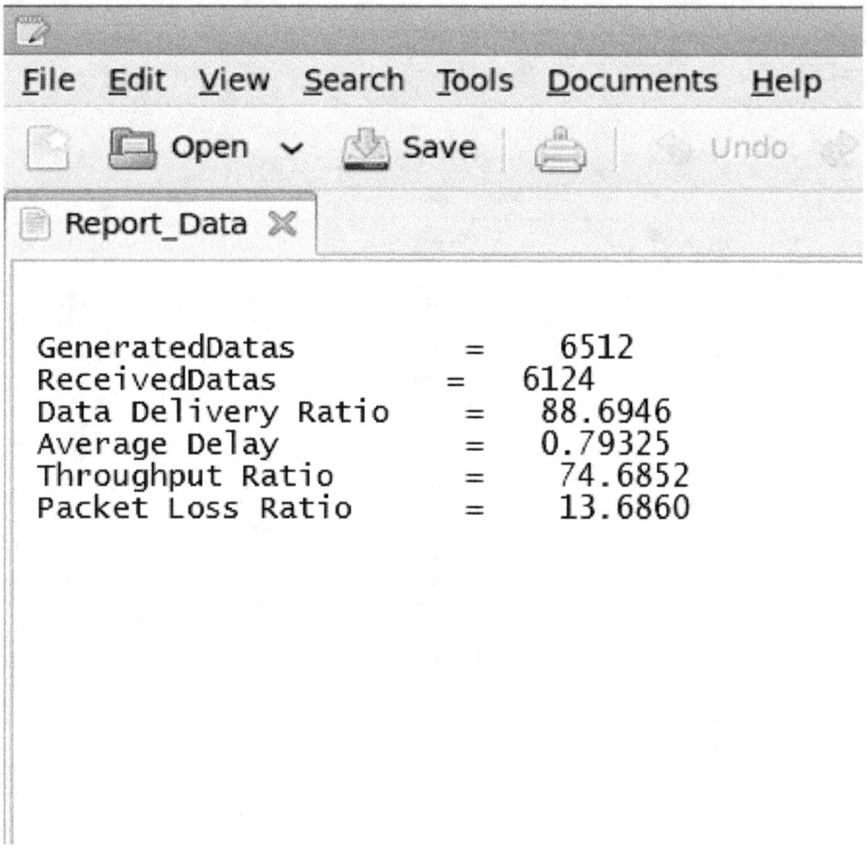
File Edit View Search Tools Documents Help
Open
Save
Undo
Report_Data
GeneratedDatas = 6512
ReceivedDatas = 6124
Data Delivery Ratio = 88.6946
Average Delay = 0.79325
Throughput Ratio = 74.6852
Packet Loss Ratio = 13.6860

Figure 25.

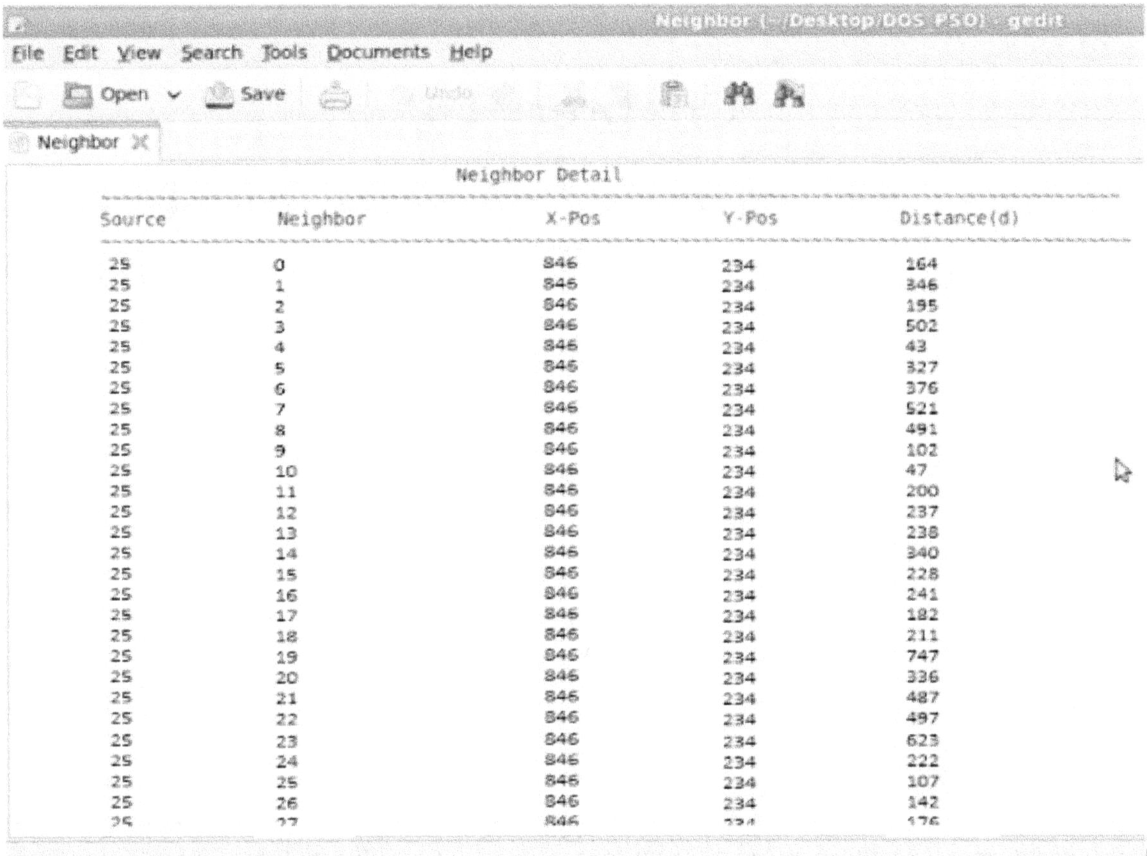

Neighbor Detail

Source	Neighbor	X-Pos	Y-Pos	Distance(d)
25	0	846	234	164
25	1	846	234	346
25	2	846	234	195
25	3	846	234	502
25	4	846	234	43
25	5	846	234	327
25	6	846	234	376
25	7	846	234	521
25	8	846	234	491
25	9	846	234	102
25	10	846	234	47
25	11	846	234	200
25	12	846	234	237
25	13	846	234	238
25	14	846	234	340
25	15	846	234	228
25	16	846	234	241
25	17	846	234	182
25	18	846	234	211
25	19	846	234	747
25	20	846	234	336
25	21	846	234	487
25	22	846	234	497
25	23	846	234	623
25	24	846	234	222
25	25	846	234	107
25	26	846	234	142

Chapter 14
Thermal Management System for Batteries in Electric Vehicles

Saravanan Subramani
Sri Venkateswara College of Engineering, India

Sivaramapandian J.
Sri Venkateswara College of Engineering, India

Ranjith Raj A.
Sri Venkateswara College of Engineering, India

ABSTRACT

The rapid advances in battery packs and improved charging infrastructure have attracted customers to electric vehicles (EVs). Reduced energy dependence on other nations and eco-friendliness are the two major advantages of adopting EVs. Electric vehicles require batteries with a high energy density to provide a large driving range. In addition, long cycle life is important to guarantee a large total driving distance. Currently, lithium-ion batteries are widely used in EVs due to their safety concerns, high stability, and energy density properties. The performance of the batteries depends on their thermal characteristics and thermal management systems. Hence, a thermal management system is necessary to sustain the performance of electric vehicles. In this chapter, the effect of temperature rise and self-discharge of the battery packs are discussed. The techniques followed to manage the thermal characteristics of batteries are described with advantages and disadvantages. The chapter is concluded by a future outlook.

INTRODUCTION

The production of an electric vehicle is increasing day by day since the government banned fossil fuel vehicles. Favorable government subsidies and tax rebates promote the adoption of electric vehicles. The market of automotive sector has countersigned the growth and development of electric vehicles. Tesla, BYD, BMW, Volkswagen and Nissan, etc. manufactures highly innovative electric vehicles (Xia et al., 2017). The compound annual growth rate of electric vehicles is expected to grasp to thirty million units

DOI: 10.4018/978-1-6684-6631-5.ch014

by 2030. Improved charging infrastructure – speed and range per charge attracted consumers towards electric vehicles (Yue et al., 2021). However, Battery Thermal Management (BTM) is a key restraint to be considered. Hence, the successful adaptation of electric vehicles depends only on the BTM. Uniform temperature profile is the essential parameter to be optimized for good battery life. Apart from temperature, the battery management system includes individual cell voltage measurements, battery pack voltage and current monitoring (Tete et al., 2021).

All of the energy required to drive an electric car is stored in the batteries. Primary cells are batteries that can only discharge once, such as triple-A and double-A alkaline batteries. Rechargeable batteries are classified as secondary batteries. Lead-acid, lithium-ion, and nickel-metal hydride batteries are examples of secondary batteries. The energy storage efficiency of batteries should be one of its key characteristics. The energy storing capacity of lithium-ion batteries is more than 90 percent. So, lithium-ion batteries are commonly used in EVs. TThe two most typical forms of lithium-ion batteries are prismatic and cylindrical cells. In EV technology, cylindrical lithium-ion cells are typically preferred since they have the longest cycle life, the highest level of safety, and are less expensive to build (Behi et al., 2020). Due to their shape, they are less effective at volume packing many cells, which is a drawback; however, these shapes are advantageous for cooling the cells. Prismatic cells are used frequently in mobile phones and tablets because they are more volume efficient. In EVs, a battery pack is made up of numerous individual cells. For instance, a Tesla car's 85 kWh battery pack is made up of 7104 individual cells. The cost distribution is determined by the particular battery technology and the various parts (materials and production). Present-day Lithium-ion batteries are less expensive per watt-hour of energy storage due to higher production volume, increased energy density, and decreased cost (Pfrang et al., 2017). As a result, they are currently used extensively in electric mobility. Typically, another constraint for EV technology is the weight of the batteries. Smaller size batteries have a high risk of overheating (Zhang et al., 2014).

The temperature properties and thermal management system affect how well the batteries function. Consequently, an effective thermal management system is required to maintain electric vehicle performance. This chapter discusses the impact of battery pack self-discharge and temperature rise. The methods used to control the thermal properties of batteries are outlined along with benefits and drawbacks.

Comparison of Batteries

The comparison results of lead acid, Li-ion, Ni-MH and Ni-Cd batteries parameters like energy capacity, voltage, operating temperature, life cycle, charging time and discharge percentage are given in table 1. From the comparison, the Li-ion battery is considered effective in this fast-moving world (Hoffmann, 2019). Hence the most of the works selected for the literature are given preference to Li-ion battery. Fast charging time, high cycle stability and high energy density properties of Li-ion battery proves prominent for electric vehicles.

Necessity for Battery Thermal Management

The battery management system, which regulates how the individual cells are charged and discharged and monitors their state, is an essential part of a battery pack. To achieve safe operation, a long cycle life, and ideal performance, this is essential. The thermal behavior of batteries depends on the internal and external physiochemical parameters. The accelerating heat calorimeter (Ziebert et al., 2017), Xenon flash method (Huber & Kuhn, 2015), thermal impedance spectroscopy and insertion of thermocouples

Table 1. **Comparison of batteries** *(Huber & Kuhn, 2015)*

Batteries	Lead acid	Li-ion	Ni-Cd	Ni- MH
Origin	Late 1800	1991	1950	1990
Energy capacity (W/kg)	35 - 40	20 - 40	40 - 60	60 - 80
Cell voltage (V)	2.1	3	1.2	1.2
Operating temperature	-35 to 45	-20 to 60	-40 to 60	-20 to 60
Charging temperature recommended (°C)	-20 to 50	0 to 45	-20 to 60	0 to 45
Fast charging time (h)	8 – 16	1	2-4	<1
Overcharge tolerance	High	Low	Moderate	Low
Discharging temperature recommended (°C)	-20 to 50	-20 to 60	-20 to 65	-20 to 65
Life cycles	200 - 300	200 – 700	500 - 2000	500 - 2000
Charging time (hrs)	8	8	1	1
Self gischarge/month (%)	5	< 10	20	30

are some of the techniques used to quantify the amount of heat generation and dissipation in various operating conditions (Arora, 2018).

Thermal Runaway

The thermal runaway is a rapid self-heating process happening in batteries that do not have any temperature control systems (Wang et al., 2012). Thermal runaway mainly appears during the charging process. It is caused by an exothermic reaction – the liberation of chemical energy. The oxygen evolution on the positive plates during the charging process recombines on negative plates leading exothermic reaction thus increasing the temperature of the battery walls (Feng et al., 2018). The thermal runaway of a lithium-ion battery is shown in fig. 1.

Temperature Effect on Cell Performance and Ageing

Temperature is the key factor influences many attributes of the battery like capacity, ageing, discharge and safety, etc. Thermal management is a significant factor influencing cell performance and life span. A good BTM should have high cooling performance, uniform temperature and a compact weight to size ratio. The effect of high temperature increases the power dissipation. At low temperatures, the ionic conductivity of the electrolytes gets reduced and the movement of ions from the anode to cathode finds difficulty. The effect of the high temperature operating condition of lithium-ion batteries is given in table 2. Performance, discharge capacity, reaction kinetics, and risk of electrolyte freeze are some of the effects of low-temperature battery operation. Ageing of the battery packs depends on two factors namely charging and discharging cycles (Liu et al., 2014).

Figure 1. **Thermal runaway of a Li-ion battery** *(Golubkov et al., 2018)*

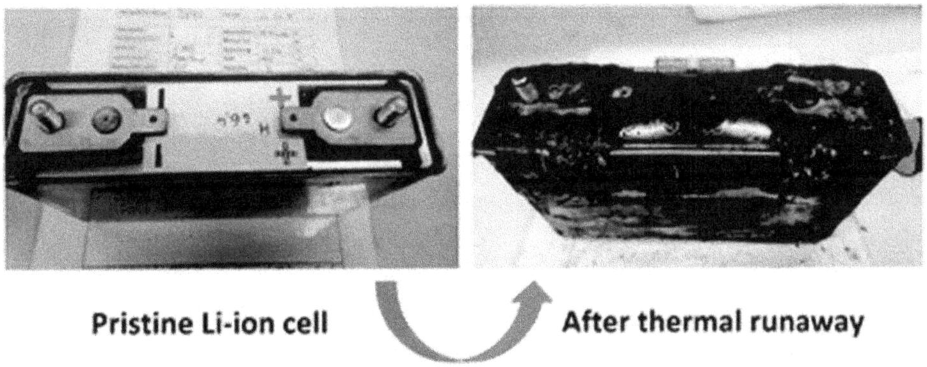

Table 2. Li-ion battery operation temperature

Temperature range	Reaction happening	Result
120°C – 130 °C	Anode and cathode electrodes contact	Internal short circuit and generates more heat
130°C – 150 °C	Breakdown of the cathode	Battery failure
150°C – 180 °C	Cell rupture	Thermal runaway

Temperature Effect on Capacity and Power Grade

The temperature has a strong impact on overall battery performance and life span. Emerging of solid electrolyte interphase, electrolytes oxidation, increasing electrode impedance, cathode degradation phenomena, etc., has a strong impact on capacity and power grade due to the temperature change (Aurbach et al., 2002). Figure 2 represents the temperature effect on the capacity fade of a Li-ion battery. It is inferred from the literature that a 10K temperature rise of the battery doubles the speed of the internal reactions. A high temperature operated Li-ion batteries generates lithium salts which forms a new electrolyte layer causing decomposition of the electrolyte (Methekar et al., 2011).

Self-Discharge

The charging and discharging process of the batteries is always a reversible heat generation process. Batteries lose some percentage of their stored energy due to reversible self – discharge. The self-discharge process is accelerated by the elevated temperature and state of charge. Light weight batteries are more sensible to change in internal temperatures compared to high capacity range batteries (Bandhauer et al., 2011). The internal resistance of the cells depends on the temperature. Hence an optimum temperature should be maintained to keep a safe margin. The monthly self-discharge and capacity fade of the li-ion battery is shown in Fig. 3 (Huber & Kuhn, 2015).

Figure 2. **Temperature effect on capacity fade** *(Huber & Kuhn, 2015)*

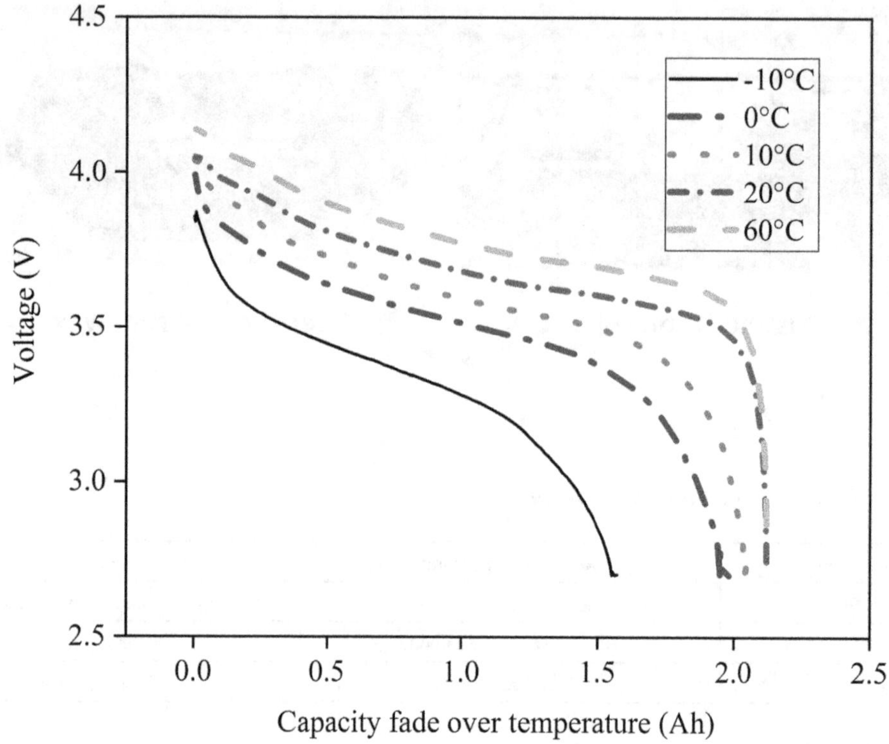

Classification of Thermal Management Techniques in Batteries

Heat transfer between the cells and external surface principally depends on the effect of conduction, convection and radiation (Incropera et al., 2007).

Conduction

Single cells are liked using a metal bus bars to forma battery pack. So cell to cell heat transfer via conduction mode should be considered for effective BTMS. It is inferred that thermal runaway of one cell propagates to whole battery pack. The conductive heat transfer effect is given in the equation (1.1),

$$\dot{Q}_{cond} = \frac{\lambda}{d} A(T_1 - T_2) \tag{1.1}$$

Convection

Most of the heavy weight batteries are cooled by convection mode of heat transfer. a gaseous or liquid are used as coolant depend on the size to weight ratio of the battery pack. the rate of heat transfer depends on the properties of the coolant medium flowrate, velocity and viscosity.

Figure 3. **Self-discharge** *(Huber & Kuhn, 2015)*

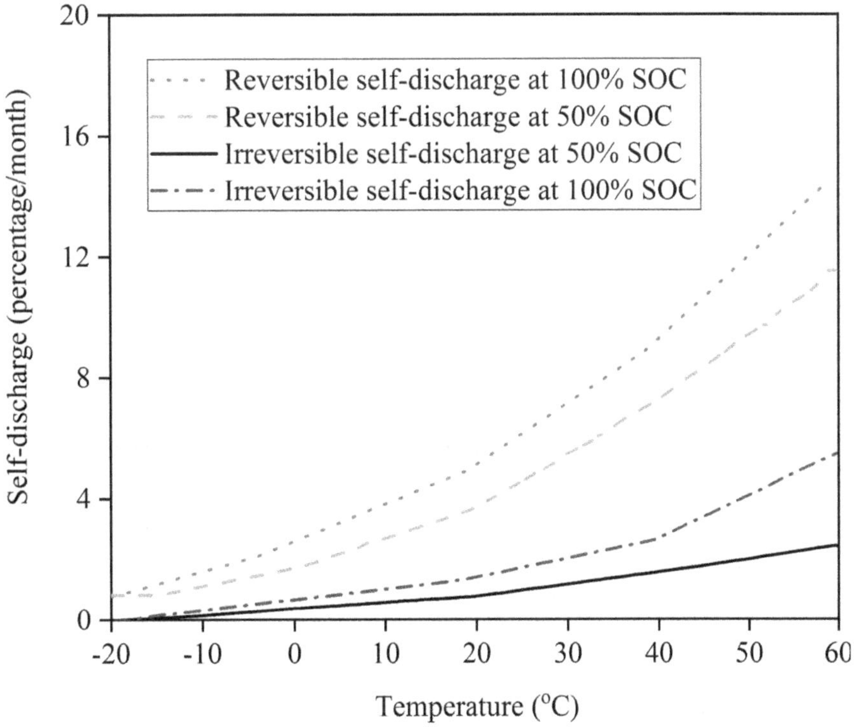

$$\dot{Q}_{conv} = \bar{h}\left(\mathrm{Re}, Nu, \mathrm{Pr}\right) A(T_B - T_\alpha) \tag{1.2}$$

Radiation

Using the Stefan-Boltzmann equation the amount of heat dissipated by the effect of radiation is determined (Equation 1.3) by the product of temperature difference (between the battery and vehicle surface) as well as surface area A and emission coefficient ε.

$$\dot{Q}_{rad} = \varepsilon\sigma A(T_B^4 - T_v^4) \tag{1.3}$$

The total heat flux generated from the battery is given by summation of all the heat sources and sinks.

$$\dot{Q}_{battery} = \dot{Q}_{cond} + \dot{Q}_{conv} + \dot{Q}_{rad} \tag{1.4}$$

Figure 4. Schematic design of natural convection air cooling

Air Cooling

Air cooling uses ambient air as a medium and offers a simple and economic design. Due to the presence of dust particles in the ambient air additional filters are required to prevent the battery packs. The air cooling method is classified into two ways namely natural convection and forced convection (Akinlabi & Solyali, 2020). Cooling fins are used to increase the surface area and thereby reduces the thermal fluctuations (Lu et al., 2016). Fins are commonly used in electronic gadgets and engines. Natural convection air cooling is used earlier in Nissan Leaf and Toyota Prius. Figure 4 displays the simple schematic design of natural convection air cooling. This mode of cooling fails particularly in summer climates. Hence, forced convection cooling is emerged to overcome the issues raised in natural convection. Additional space requirement and background noise created by the fans and blowers are the disadvantages of forced convection. Addition of a fan seems 40% extra weight of cell and the weight outweighs the cooling benefits. Cell spacing, air supply strategy and airflow rate must be considered to optimize the air cooling effect of the battery packs. Figure 5 shows the simple schematic design of forced convection air cooling. So, Tesla insist liquid cooling as safe measure to prevent the battery from thermal runaway. As per the results segregated from the literature proves that air cooling systems requires two to three times more input energy than other systems. Figure 6 depicts the air cooled Nissan Leaf's battery pack outlook.

Liquid Cooling

To improve the cooling effect between the coolant and the cells of the battery packs a liquid media is introduced in the place of air. Liquid coolants performs effectively compared to air since they have higher heat conductivity property. Liquid cooling systems are implemented in Tesla, Checy Volt, Jaguar i-Pace, Ford Focus, BMW i-3 and i-8, etc for maintaining proper temperature and uniformity. However, it has some disadvantages like leakiness in the connectors, type of coolant used (glycol or polyglycol) and swapping of the battery packs. The liquid cooling system is classified into two broad ways namely direct liquid cooling based systems and indirect liquid cooling based systems (Chung & Kim, 2019).

In direct cooling method the battery is submerged into the coolant. Hence the coolant must be low conductivity or zero conductivity liquid. However, this method of cooling in not commercialized anywhere in electric vehicles till to date. Indirect cooling method is the most prominent method followed in market where the coolant is circulated in and around the battery packs through the pipes like the conventional engine cooling system. The type of coolant used in this type should have high heat carrying capacity property and less corrosive in nature. Figure 7 shows the liquid cooled Chevy Bolt EV battery.

Figure 5. Schematic design of forced convection air cooling

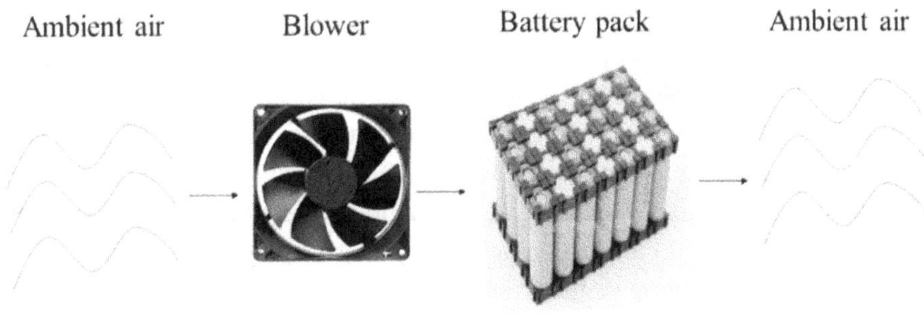

Figure 6. Air cooled Nissan Leaf's battery pack

Figure 7. Liquid cooled Chevy Bolt EV battery

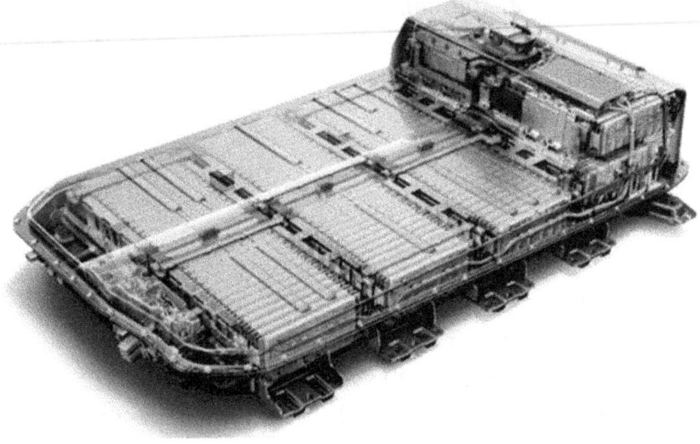

PCM Based Cooling

Phase change material cools the battery pack by changing the phase from solid to liquid. It absorbs large amount of heat energy while changing phase. PCM is used to cool the battery packs however size of the battery is a constraint. Also, volume change occurring in the PCM limits the application. Paraffin's and Lauric acid are commonly used organic PCMs and $LiNO_3$, $CaCL_2$ and water are commonly used inorganic PCMs. and It is used to reduce the internal temperature fluctuations and helps in reducing high temperature peaks. The PCM cannot transfer heat away once collected is the major drawback of this type of cooling. PCM type cooling system is suited only for small scale thermal management system (Ianniciello et al., 2018).

Advance Aspects in Thermal Management For Batteries

a. Minimizing the impact of cold and hot ambient conditions on the battery pack
b. Minimizing the temperature spread between the cells
c. Avoiding cell to cell thermal runaway propagation
d. Minimizing condensation effect of the electrolyte

FUTURE RESEARCH POTENTIAL

Safety is a top priority for EV automobiles, and the safety requirements depend on the battery's composition and usage, such as the charge rate. Researches on crash scenarios like leakage of toxic acids from the batteries should be encouraged. Recycling will be more crucial in the future to recover the materials and elements from lithium-ion batteries since increased battery manufacturing will put pressure on both the cost and the availability of battery components. The battery performance, as well as its safety, is determined by the electrode and electrolyte materials. Therefore, worldwide battery research aims at finding new materials and battery chemistries that improve performance, safety and cost parameters. IIt will take much research to determine which technology can surpass Li-ion batteries' energy density threshold. Although relatively abundant, lithium is only geographically present in a few number of locations worldwide. Additionally, lithium is found in the oceans, it cannot yet be economically extracted from saltwater. Therefore, research on EV batteries takes into account other charge-carrying ions.

CONCLUSION

i. A battery pack must have dual protection – hardware and software
ii. The voltage must be monitored regularly
iii. A safe operating boundary condition is required to avoid thermal runaway. The core temperature differs from the temperature measured I cell surfaces.
iv. Lowering thermal conductivity could effectively limit the temperature rise and thereby thermal runaway.
v. The cycle life of the batteries depends on the operation conditions, including the temperature, the charge rate and the load profile.

vi. New lithium chemistries are being developed to increase the performance of the batteries. Examples of these novel chemistries are lithium-sulfur and lithium-air. These chemistries currently face a number of difficulties, including significant structural changes and sensitivity with the electrolyte.
vii. Adding additives to the liquid cooling system is mandatory to prevent the coolant from anti-freeze and resists corrosion.
viii. The development of solid-state batteries with a solid electrolyte is the focus of study in terms of safety.

REFERENCES

Akinlabi, A. A. H., & Solyali, D. (2020). Configuration, design, and optimization of air-cooled battery thermal management system for electric vehicles: A review. *Renewable & Sustainable Energy Reviews, 125*, 109815. doi:10.1016/j.rser.2020.109815

Arora, S. (2018). Selection of thermal management system for modular battery packs of electric vehicles : A review of existing and emerging technologies. *Journal of Power Sources, 400*, 621–640. doi:10.1016/j.jpowsour.2018.08.020

Aurbach, D., Zinigrad, E., Cohen, Y., & Teller, H. (2002). *A short review of failure mechanisms of lithium metal and lithiated graphite anodes in liquid electrolyte solutions*. Academic Press.

Bandhauer, T. M., Garimella, S., & Fuller, T. F. (2011). A Critical Review of Thermal Issues in Lithium-Ion Batteries. *Journal of the Electrochemical Society, 158*(3), R1. doi:10.1149/1.3515880

Behi, H., Karimi, D., Behi, M., Ghanbarpour, M., Jaguemont, J., Sokkeh, M. A., Gandoman, F. H., Berecibar, M., & Van Mierlo, J. (2020). A new concept of thermal management system in Li-ion battery using air cooling and heat pipe for electric vehicles. *Applied Thermal Engineering, 174*, 115280. doi:10.1016/j.applthermaleng.2020.115280

Chung, Y., & Kim, M. S. (2019). Thermal analysis and pack level design of battery thermal management system with liquid cooling for electric vehicles. *Energy Conversion and Management, 196*, 105–116. doi:10.1016/j.enconman.2019.05.083

Feng, X., Ouyang, M., Liu, X., Lu, L., Xia, Y., & He, X. (2018). Thermal runaway mechanism of lithium ion battery for electric vehicles: A review. *Energy Storage Mater, 10*, 246–267. doi:10.1016/j.ensm.2017.05.013

Golubkov, A. W., Planteu, R., Krohn, P., Rasch, B., Brunnsteiner, B., Thaler, A., & Hacker, V. (2018). Thermal runaway of large automotive Li-ion batteries. *RSC Advances, 8*(70), 40172–40186. doi:10.1039/C8RA06458J PMID:35558218

Hoffmann, D. (2019). Safety tests for li-secondary batteries. *Electrochem. Power Sources Fundam. Syst. Appl.,* 387–453. doi:10.1016/B978-0-444-63777-2.00010-4

Huber, C., & Kuhn, R. (2015). *Thermal management of batteries for electric vehicles*. Elsevier Ltd. doi:10.1016/B978-1-78242-377-5.00013-3

Ianniciello, L., Henry, P., & Achard, P. (2018). Electric vehicles batteries thermal management systems employing phase change materials. *Journal of Power Sources, 378*, 383–403. doi:10.1016/j.jpowsour.2017.12.071

Incropera, F.P., Dewitt, D.P., Bergman, T.L., & Lavine, A. (2007). *Fundamentals of Heat and Mass Transfer.* Academic Press.

Liu, G., Ouyang, M., Lu, L., Li, J., & Han, X. (2014). Analysis of the heat generation of lithium-ion battery during charging and discharging considering different influencing factors. *Journal of Thermal Analysis and Calorimetry, 116*(2), 1001–1010. doi:10.100710973-013-3599-9

Lu, Z., Meng, X. Z., Wei, L. C., Hu, W. Y., Zhang, L. Y., & Jin, L. W. (2016). Thermal Management of Densely-packed EV Battery With Forced Air Cooling Strategies. *Energy Procedia, 88*, 682–688. doi:10.1016/j.egypro.2016.06.098

Methekar, R. N., Northrop, P. W. C., Chen, K., & Braatz, R. D. (2011). Kinetic Monte Carlo Simulation of Surface Heterogeneity in Graphite Anodes for Lithium-Ion Batteries. *Passive Layer Formation, 158*, 363–370. doi:10.1149/1.3548526

Pfrang, A., Kriston, A., Ruiz, V., Lebedeva, N., & Persio, F. (2017). *Safety of Rechargeable Energy Storage Systems with a focus on Li-ion Technology. In Emerg. Nanotechnologies Recharg. Energy Storage Syst.* Elsevier Inc. doi:10.1016/B978-0-323-42977-1/00008-X

Tete, P. R., Gupta, M. M., & Joshi, S. S. (2021). Developments in battery thermal management systems for electric vehicles : A technical review State of Power. *J Energy Storage, 35*, 102255. doi:10.1016/j.est.2021.102255

Wang, Q., Ping, P., Zhao, X., Chu, G., Sun, J., & Chen, C. (2012). Thermal runaway caused fire and explosion of lithium ion battery. *Journal of Power Sources, 208*, 210–224. doi:10.1016/j.jpowsour.2012.02.038

Xia, G., Cao, L., & Bi, G. (2017). Review article A review on battery thermal management in electric vehicle application. *Journal of Power Sources, 367*, 90–105. doi:10.1016/j.jpowsour.2017.09.046

Yue, Q. L., He, C. X., Jiang, H. R., Wu, M. C., & Zhao, T. S. (2021). A hybrid battery thermal management system for electric vehicles under dynamic working conditions. *International Journal of Heat and Mass Transfer, 164*, 120528. doi:10.1016/j.ijheatmasstransfer.2020.120528

Zhang, Z. J., Ramadass, P., & Fang, W. (2014). *18 – Safety of Lithium-Ion Batteries. In Lithium-Ion Batter. Adv. Appl.* Elsevier. doi:10.1016/B978-0-444-59513-3.00018-2

Ziebert, C., Melcher, A., Lei, B., Zhao, W., Rohde, M., & Seifert, H. J. (2017). *Electrochemical – Thermal Characterization and Thermal Modeling for Batteries. In Emerg. Nanotechnologies Recharg. Energy Storage Syst.* Elsevier Inc. doi:10.1016/B978-0-323-42977-1/00006-6

Compilation of References

Abagnale, C., Cardone, M., Iodice, P., Strano, S., Terzo, M., & Vorraro, G. (2015). Derivation and Validation of a Mathematical Model for a Novel Electric Bicycle. Proceedings of the World Congress on *Engineering.*

*Adam, S. A., & Jalil, N. A. (2017). Vertica*l *su*spension seat transmissibility and SEAT values for seated person exposed to whole-body vibration in agricultural tractor preliminary study. Procedia Engineering, 170, 435–442. doi:10.1016/j.proeng.2017.03.0*70*

Affanni, A., Bellini, A., Franceschini, G., Guglielmi, P., & Tassoni, C. (2005). Battery choice and management for new-generation electric vehicles. IEEE Transactions on Industrial Electronics, 52(*5), 1343–1349. doi:10.1109/TIE.*2005.855664

Affanni, Bellini, Franceschini, Guglielmi, & Tassoni. (2005). Battery Choice and Management for New-Generation *Electric Vehicles. IEEE Transac*tions on Industrial Electronics, 52(5), 1343–1349.

Aggarwal, C. C. (2011). An introduction to social network data analytics. In Social network data *analytics (pp. 1–15). Springer. d*oi:10.1007/978-1-4419-8462-3_1

Aguilar-Dominguez, D., Dunbar, A., & Brown, S. (2020). The electricity demand of an EV providing power via vehicle-to-home and its potential impact on the grid with different electricity price tariffs. Energy Re*p., 6, 132–141. doi:10.1016/j.* egyr.2020.03.007

Ahmed, M., Zheng, Y., Amine, A., Fathiannasab, H., & Chen, Z. (2021). The role of artifici*al intelligen*ce *i*n the mass adoption of electric vehicles. Joule, 5(9), 2296–2322. doi:10.1016/j.joule.2021.07.012

Aizpuru, U., Iraola, J., Canales, M., Echeverria, M., & Gil, I. (2013). Passive balancing design for Li-ion battery *packs based on single cell exp*erimental tests for a CCCV charging mode. 2013 International Conference on Clean Electrical Power (ICCEP), 93-98.

Akhavan-Hejazi, H., Mohsenian-Rad, H., & Nejat, A. (2014, September). Dev*eloping a tes*t *da*ta set for electric vehicle applications in smart grid research. In 2014 IEEE 80th Vehicular Technology Conference (VTC2014-Fall) (pp. 1-6). IEEE. 10.1109/VTCFall.2014.6965934

Akinlabi, *A. A. H., &* So*l*yali, D. (2020). Configuration, design, and optimization of air-cooled battery thermal management system for electric vehicles: A review. Renewable & Sustainable Energy Reviews, 125, 109815. *doi:10.1016/j.rser.2020.109815*

Al-Awami, A. T., & Sortomme, E. (2012). Coordinating Vehicle-to-Grid Services With Energy Trading. IEEE Transactions on Smart Grid, 3(1), 453–462. doi:10.1109/TSG.2011.2167992

*Alfaverh, F., Denaï, M., & Sun, Y. (2021). E*lectrical vehicle grid integration for demand response in distribution networks using reinforcement learning. IET Electr. Syst. Transp., 11(4), 348–361. doi:10.1049/els2.12030

Ali, M. S., Wang, *L., Alquhayz, H., Rehman, O. U., & Chen, G.* (2021). Performance Improvement of Three-Phase Boost Power Factor Correction Rectifier Through Combined Parameters Optimization of *Proportional-Integral and Repetitive Controller. IEEE Access*: Practical Innovations, Open Solutions, 9, 58893–58909. doi:10.1109/ACCESS.2021.3073004

Alirezazadeh, A., Rashidinejad, M., Abdollahi, A., Afzali, P., & Bakhshai, A. (202*0). A new flexible model for generation scheduling in* a smart grid. Energy, 191, 191. doi:10.1016/j.energy.2019.116438

Altuntas, S., Dereli, T., & Kusiak, A. (2015). Forecasting t*echnology success based on pate*nt data. Technological Forecasting and Social Change, 96, 202–214. doi:10.1016/j.techfore.2015.03.011

Amin, B. R., Hossain, M. J., Anwar, A., & Zaman, S. (2021). Cyber attacks and faults disc*rimination in intelligent electronic device-based ene*rgy management systems. Electronics (Basel), 10(6), 650. doi:10.3390/electronics10060650

Ananthraj & Ghosh. (2021). Battery Management System in Electric Vehicle. 2021 4t*h Biennial International Conference* on Nascent Technologies in Engineering (ICNTE). doi:10.1109/icnte51185.2021.9487762

Andrea, D. (2010). Battery Management Systems for Large Lithium-Ion Battery Packs. Artech House. https://www.libraryth*ing.com/work/11739168/details/78008584*

Arachchige, B., Perinpanayagam, S., & Jaras, R. (2017). Enhanced prognostic model for lithium ion batteries based on particle ðlter state transition model modi*ðcation. Applied Sciences (Basel, S*wit*z*erland), 7(11), 1172. doi:10.3390/app7111172

Arasaratnam, Tjong, Ahmed, El-Sayed, & Habibi. (2013). Adaptive Temperature Monitoring for Battery Thermal Management. In 2013 *IEEE Transportation Electrific*ation Conference and Expo (Itec). IEEE.

Aristodemou, L., & Tietze, F. (2018). The state-of-the-art on Intellectual Property Analytics (IPA): A literature review on artificial intel*ligence, machine learning and deep l*earning methods for analysing intellectual property (IP) data. World Patent Information, 55, 37–51. doi:10.1016/j.wpi.2018.07.002

Arora, S. (2018). Selection of thermal management system for modular battery packs of electric ve*hicles : A review of existing and emergin*g technologies. Journal of Power Sources, 400, 621–640. doi:10.1016/j.jpowsour.2018.08.020

Arora, U., Lodhi, M. E. A., & Saxena, T. K. (2016). PID parameter tuning using modified BAT Algorithm. Journa*l of Automation and Control Engineering, 4(5), 347–352.* doi:10.18178/joace.4.5.347-352

Asadinejad, A., *& Tomsovic, K. (2017). Optim*al use of incentive and price based demand response to reduce costs and price volatility. Electric Power Systems Research, 144, 215–223. doi:10.1016/j.epsr.2016.12.012

*Auger, F., Hilairet, M., Guerr*ero, J. M., Monmasson, E., Orlowska-Kowalska, T., & Katsura, S. (2013). Industrial Applications of The Kalman Filter: A Review. IEE*E Transactions on Industrial Elec*tronics, 60(12), 5458–5471. doi:10.1109/TIE.2012.2236994

Aurbach, D., Zinigrad, E., Cohen, Y., & Telle*r, H. (2002). A short review of failure mechanisms* of lithium metal and lithiated graphite anodes in liquid electrolyte solutions. Academic Press.

Babu, P. R., R*amprasath, S., & Paranthagan, B. (2012). Modeling an*d Dynamic Simulation of Permanent Magnet Brushless DC Motor (PMBLDCM) Drives. AIM 2012: Mobile Communication and Power Engineering, 556-564.

Bae, S., & Kwasinski, A. (2012). Spatial and Tem*poral Model of Electric Vehicle* Charging Demand. IEEE Transactions on Smart Grid, 3(1), 394–403. doi:10.1109/TSG.2011.2159278

Bagheri, T. (2012). DFMC: decentralized fault management mechanism for cluster based wireless *sensor networks. Second Int. Conf.* on IEEE Digital Information and Communication Technology and it's Applications (DICTAP), 67-71. 10.1109/DICTAP.2012.6215406

Balamurugan. (2014). An Energy *Efficient Fitness Based Routing* Protocol in Wireless Sensor Networks. ICTACT Journal on Communication Technology. Doi:10.21917/ijct.2014.0127

Bandhauer, T. M., Garimella, S., & Fuller, T. F. (2011). *A Critical Review of Thermal Issues* in Lithium-Ion Batteries. Journal of the Electrochemical Society, 158(3), R1. doi:10.1149/1.3515880

Barsukov & Instruments. (n.d.). Battery Cell Balancing: What to Balance and How. Academic Press.

*Bayar, K., & Kh*aneghah, F. S. (2020). Optimal sliding mode control method for active suspension control. IFAC-PapersOnLine, 53(2), 14285–14291. doi:10.1016/j.ifacol.2020.12.1178

Becker, J., Schaeper, C., *& Sauer, D. U. (2012)*. Energy Management System for A Multi-Source Storage System Electric Vehicle. 2012 IEEE Vehicle Power and Propulsion Conference, 407-412. 10.1109/VPPC.2012.6422664

Behi, H., Karimi, D., Behi, M., Ghanbarpour, *M., Jaguemont, J., Sokkeh,* M. A., Gandoman, F. H., Berecibar, M., & Van Mierlo, J. (2020). A new concept of thermal management system in Li-ion battery using air cooling and heat pipe for electric vehicles. App*lied Thermal Engine*ering, 174, 115280. doi:10.1016/j.applthermaleng.2020.115280

Bhandari & Saluja. (2014). Applying Genetic Algorithm in Intrusion Detection System: A Comprehensive Review" Rec*ent Trends in Info*rmat*i*on, Telecommunication and Computing. ITC2014.

Bhowmik, A., Castelli, I. E., Garcia-Lastra, J. M., Jørgensen, P. B., Winther, O., & Vegge, T. (2019). A perspective on inverse design of battery interphases using multi-scale modelling, experim*ents and ge*nerative deep learning. Energy Storage Materials, 21, 446–456. doi:10.1016/j.ensm.2019.06.011

Bibak, B., & Tekiner-Moğulkoç, H. (2021). A comprehensive analysis of vehicle to grid (V2G) s*ystems and sch*olar*l*y literature on the application of such systems. Renew. Energy Focus, 36, 1–20. doi:10.1016/j.ref.2020.10.001

Bolly, V. K., Springer, J., & Dietz, J. *E. (20*1*4,* June). Using Open Source NoSQL technologies in Designing Systems for Delivering Electric Vehicle Data Analytics. In 2014 ASEE Annual Conference & Exposition (pp. 24-1339). 10.18260/1-2--23272

Bolognani, S., *& Zampieri, S. (2013). A dis*tri*but*ed control strategy for reactive power compensation in smart microgrids. IEEE Transactions on Automatic Control, 58(11), 2818–2833. doi:10.1109/TAC.2013.2270317

Boulanger, A. G., Chu, A. C., *Maxx, S., & Waltz, D. L. (*2011). Vehicle electrification: Status and issues. Proceedings of the IEEE, 99(6), 1116–1138. doi:10.1109/JPROC.2011.2112750

Brand, C., Anable, J., & Morton, C. (2019). Lifestyle, efficiency and limits: Modell*ing transport energy and emissions using a socio-technical* appr*o*ach. Energy Effic., 12(1), 187–207. doi:10.100712053-018-9678-9

Burgess-Limer*ick, R. (2016). Measuring and managing workplace whole-body vibration exposures. Acoust*ics Australia, 44(1), 129–135. doi:10.100740857-015-0032-5

C. M., NP., J. (2021). Realization of Cascaded H-Bridge Multilevel Inverter Based Grid Integrated Solar Energy System With Band Stop Generalized Integral Control. IEEE Transactions *on Industry* Appl*ic*ations, 57(1), 764–773. doi:10.1109/TIA.2020.3031546

Cai, H., Jia, X., Chiu, A. S., Hu, X., & Xu, M. (2014). Siting public electric vehicle charging stations in Beijing using big-data informed travel patterns of the taxi fleet. Transportation Research Part D, Transport and Environment, 33, 39–46. doi:10.1016/j.trd.2014.09.003

Cao, J., Schofield, N., & Emadi, A. (2008). Battery balancing methods: A comprehensive review. In IEEE Vehicle Power and Propulsion Conference, VPPC 08 (pp. 1-6). IEEE.

Cao, Y., Song, H., Kaiwartya, O., Zhou, B., Zhuang, Y., Cao, Y., & Zhang, X. (2018). Mobile edge computing for big-data-enabled electric vehicle charging. IEEE Communications Magazine, 56(3), 150–156. doi:10.1109/MCOM.2018.1700210

Cao, Y., Tang, S., Li, C., Zhang, P., Tan, Y., Zhang, Z., & Li, J. (2012). An optimized EV charging model considering TOU price and SOC curve. IEEE Transactions on Smart Grid, 3(1), 388–393. doi:10.1109/TSG.2011.2159630

Chan, H., & Perrig, A. (2004). ACE: An Emergent Algorithm for Highly Uniform Cluster Formation. Proc. of 1st European Workshop on Sensor Networks, 154–171,2004. 10.1007/978-3-540-24606-0_11

Chang, W.-Y. (2013). The State of Charge Estimating Methods for Battery: A review. International Scholarly Research Notices, 2013.

Chang, Y.-K., Ting, Y.-W., & Wu, S.-C. (2007). Power-Efficient and Path- Stable Broadcasting Scheme for Wireless Ad Hoc Networks. International Conference on Advanced Information Networking and Applications Workshops (AINAW'07). 10.1109/AINAW.2007.291

Chen, B. C., Shiu, Y. H., & Hsieh, F. C. (2011). Sliding-mode control for semi-active suspension with actuator dynamics. Vehicle System Dynamics, 49(1-2), 277–290. doi:10.1080/00423111003602376

Chen, H. Y., & Huang, S. J. (2008). A new model-free adaptive sliding controller for active suspension system. International Journal of Systems Science, 39(1), 57–69. doi:10.1080/00207720701669453

Chen, M., Lin, S., Song, W., Lv, J., & Feng, Z. (2020). Electrical and thermal interplay in lithium-ion battery internal short circuit and safety protection. International Journal of Energy Research, 44(8), 6745–6757. doi:10.1002/er.5411

Chen, T. D., Kockelman, K. M., & Khan, M. (2013). Locating electric vehicle charging stations: Parking-based assignment method for Seattle, Washington. Transportation Research Record: Journal of the Transportation Research Board, 2385(1), 28–36. doi:10.3141/2385-04

Chen, X., Shen, W., Vo, T. T., Cao, Z., & Kapoor, A. (2012). An overview of lithium-ion batteries for electric vehicles. 2012 10th International Power & Energy Conference (IPEC), 230–235. doi:10.1109/ASSCC.2012.6523269

Cheng, C. P., Chao, C. H., & Li, T. H. S. (2010). Design of observer-based fuzzy sliding-mode control for an active suspension system with full-car model. IEEE International Conference on Systems, Man and Cybernetics, 1939-1944. 10.1109/ICSMC.2010.5642261

Cheng, Z., Chow, M. Y., Jung, D., & Jeon, J. (2017, June). A big data based deep learning approach for vehicle speed prediction. In 2017 IEEE 26th International Symposium on Industrial Electronics (ISIE) (pp. 389-394). IEEE. 10.1109/ISIE.2017.8001278

Chung, Y., & Kim, M. S. (2019). Thermal analysis and pack level design of battery thermal management system with liquid cooling for electric vehicles. Energy Conversion and Management, 196, 105–116. doi:10.1016/j.enconman.2019.05.083

Clement-Nyns, K., Haesen, E., & Driesen, J. (2011). The Impact of Vehicle-to-grid on the Distribution Grid. Electric Power Systems Research, 81(1), 185–192. doi:10.1016/j.epsr.2010.08.007

Conejo, A., Morales, J., & Baringo, L. (2010). Real-*time demand response model. IEEE Transactions on Sma*rt Grid, 1(3), 236–242. doi:10.1109/TSG.2010.2078843

Cowan, R., & Hultén, S. (1996). Escaping lock-in: The case of the electric vehicle. Technological Forecasting and Social Change, 53(1), 61–79. doi:10.1016/0040-*1625(96)00059-5*

Damiano, A., Gatto, G., Marongiu, I., Porru, M., & Serpi, A. (2014). Vehicle-to-Grid technology: State-of-the-art and future scenarios. J Energy Power Eng, 8(1), 152.

Daowd, M., Omar, N., Van Den Bossche, *P., & Van Mierlo, J. (2011). Passive* and active battery balancing comparison based on MATLAB simulation. In Vehicle Power and Propulsion Conference (VPPC) (pp. 1-7). IEEE. 10.1109/VPPC.2011.6043010

Darus, R., & Sam, Y. M. (2009). Modeling and control activ*e suspension system for a full car mod*el. 5th International Colloquium on Signal Processing & Its Applications, 13-18. 10.1109/CSPA.2009.5069178

De Cauwer, C., Verbeke, W., Coosemans, T., Faid, S., & Van Mierlo, J. (20*17). A Data-Driven Method for E*nergy Consumption Prediction and Energy-Efficient Routing of Electric Vehicles in Real-World Conditions. Energies, 10(5), 608. doi:10.3390/en10050608

Delogu, M., Del Pero, F., & Pierini, *M. (2016). L*ightweight Design Solutions in the Automotive Field: Environmental Modelling Based on Fuel Reduction Value Applied to Diesel Turbocharged Vehicles. Sustainability, 8(11), 1167. doi:10.3390u8111167

de Morais, J. C. S., de Morais, *J. L. S., & Gul*es, R. (2019). Photovoltaic AC Module Based on a Cuk Converter With a Switched-Inductor Structure. IEEE Transactions on Industrial Electronics, 66(5), 3881–3890. doi:10.1109/TIE.2018.2856202

Desbiens, C. (2012). Electric Veh*icle Model for Estimating Distribut*ion Transformer Load for Normal and Cold-Load Pickup Conditions. Rec. IEEE PES Innovative Smart Grid Tech. Conf. (ISGT). 10.1109/ISGT.2012.6175606

Dharmakeerthi, Mithulananthan, & Saha. (2011). Overview of *the impacts* of *p*lug-in electric vehicles on the power grid. In 2011 IEEE PES Innovative Smart Grid Technologies. IEEE.

Diamantoulakis, P. D., Kapinas, V. M., & Karagiannidis, G. K. (2015). Big data analytics for dynamic energy mana*gement in smart grids. Big Data R*esearch, 2(3), 94–101. doi:10.1016/j.bdr.2015.03.003

Diao, W., Xue, N., Bhattacharjee, V., Jiang, J., Karabasoglu, O., & Pecht, M. (2018, January). Active battery cell equalization based on residual available energy maximization. Applied Energy, 210, 690–*698. doi:*10.*1*016/j.apenergy.2017.07.137

Dost & Sourkounis. (2014). Battery management system realisation for electric vehicles. doi:10.1109/MED.2014.6961456

Duan, Caisheng, Zongzheng, Jianfei, Wang, Snyder, & Jiang. (2018). A Solar Power-Assi*sted Battery B*alancing System for Electric Vehicles. IEEE Transactions on Transportation Electrification, 4(2).

Duarte, M. L., de Ara*újo, P. A., Horta, F. C., Del Vecchio, S., & de Carvalho, L. A. (2018). Correlation between weighted acceleration, vibration* dose value and exposure time on whole body vibration comfort levels evaluation. Safety Science, 103, 218–224. doi:10.1016/j.ssci.2017.1*1.008*

*Efe, M. Ö. (2010). Fractional o*rd*er* sliding mode control with reaching law approach. Turkish Journal of Electrical Engineering and Computer Sciences, 18(5), 731–748. doi:10.3906/elk-0906-3

Electric Power Research Institute. (2009). Report to NIS*T on the Smart Grid Interoperabil*ity Standards Roadmap. Author.

Elgendy, N., & Elragal, A. (2014, July). Big data analytics: a literature review paper. In Industrial conference on data mining (pp. 214-227). Springer. doi:10.1007/978-3-319-08976-8_16

Ernst, H. (2003). Patent information for strategic technology management. World Patent Information, 25(3), 233–242. doi:10.1016/S0172-2190(03)00077-2

Errouissi, R., Al-Durra, A., & Muyeen, S. M. (2016). A Robust Continuous-Time MPC of a DC–DC Boost Converter Interfaced With a Grid-Connected Photovoltaic System. In IEEE Journal of Photovoltaics, 6(6), pp. 1619-1629.

ESCTUFF4U. (n.d.). ESCTUFF4U for electronics engineering. ESCTUFF. https://www.ecstuff4u.com/2018/01/advantages-and-disadvantages-of-igbt.html

Eswarawaka, R. (2021). IoT and AI based Pedestrian Detection for Autonomous Vehicles, Indian Patent Publication, 202141056543.

Falabretti, D., & Gulotta, F. (2022). A Nature-Inspired Algorithm to Enable the E-Mobility Participation in the Ancillary Service Market. Energies, 15(9), 3023. doi:10.3390/en15093023

Falcão, E. A. M., Ana, C. R. T., & Sodré, J. R. (2017). Analysis of CO2 emissions and techno-economic feasibility of an electric commercial vehicle. Applied Energy, 193, 297–307. doi:10.1016/j.apenergy.2017.02.050

Fan, Z. (2012). A distributed demand response algorithm and its application to PHEV charging in smart grids. IEEE Transactions on Smart Grid, 3(3), 1280–1290. doi:10.1109/TSG.2012.2185075

Farag, M. (2013). Lithium-Ion Batteries: Modelling and State of Charge estimation [Ph.D. Dissertation].

Feng, X., Ouyang, M., Liu, X., Lu, L., Xia, Y., & He, X. (2018). Thermal runaway mechanism of lithium ion battery for electric vehicles: A review. Energy Storage Mater, 10, 246–267. doi:10.1016/j.ensm.2017.05.013

Freeman, D., & Heacock, D. (1993). Advanced Charger Designs Address Modernbattery Technologies. Proceedings of Wescon 39.

Fujii, H., & Managi, S. (2018). Trends and priority shifts in artificial intelligence technology invention: A global patent analysis. Economic Analysis and Policy, 58, 60–69. doi:10.1016/j.eap.2017.12.006

Gajjar, Sarkar, & Dasgupta. (2014). Cluster Head Selection Protocol using Fuzzy Logic for Wireless Sensor Networks. International Journal of Computer Applications, 97(7).

Gao, Y., Zhu, C., Zhang, X., & Guo, B. (2021). Implementation and evaluation of a practical electrochemical- thermal model of lithium-ion batteries for EV battery management system. Energy, 221, 119688. Advance online publication. doi:10.1016/j.energy.2020.119688

García-Olivares, A., Solé, J., & Osychenko, O. (2018). Transportation in a 100% renewable energy system. Energy Conversion and Management, 158, 266–285. doi:10.1016/j.enconman.2017.12.053

Geels, F. W. (2012). A socio-technical analysis of low-carbon transitions: Introducing the multi-level perspective into transport studies. Journal of Transport Geography, 24, 471–482. doi:10.1016/j.jtrangeo.2012.01.021

Golubkov, A. W., Planteu, R., Krohn, P., Rasch, B., Brunnsteiner, B., Thaler, A., & Hacker, V. (2018). Thermal runaway of large automotive Li-ion batteries. RSC Advances, 8(70), 40172–40186. doi:10.1039/C8RA06458J PMID:35558218

Gong, Q., Mohler, S. M., Marano, V., & Rizzoni, G. (2012). Study of PEV Charging on Residential Distribution Transformer Life. IEEE Transactions on Smart Grid, 3(1), 404–412. doi:10.1109/TSG.2011.2163650

Goransson, L., Karlsson, S., & Johnsson, F. (2010). Integration of Plug-in Hybrid Electric Vehicles in a Regional Wind-thermal Power System. Energy Policy, 38(10), 5482–5492. doi:10.1016/j.enpol.2010.04.001

Gough, R., Dickerson, C., Rowley, P., & Walsh, C. (2017). Vehicle-to-grid feasibili*ty: A techno-economic an*aly*s*is of EV-based energy storage. Applied Energy, 192, 12–23. doi:10.1016/j.apenergy.2017.01.102

Goulden, M., Bedwell, B., Rennick-Egglestone, S., Rodden, T., & Spence, A. (2014). Smartgrids, smar*t users? The role of the* u*ser* in demand side management. Energy Research & Social Science, 2, 21–29. doi:10.1016/j.erss.2014.04.008

Guclu, R., & Gulez, K. (2008). Neural network *control of seat vibrations of a non-linear fu*ll vehicle model using PMSM. Mathematical and Computer Modelling, 47(11-12), 1356–1371. doi:10.1016/j.mcm.2007.08.013

Guille, C., & Gross, G. (2009). A Conceptual Framework for the Vehicle-to-grid (V2G) Imp*lementation. Energy Policy, 37(11), 4379–43*90. doi:10.1016/j.enpol.2009.05.053

Gulati & Kumar. (2015). Stable Energy efficient QoS based Congestion and *Delay aware Routing (SEQCDR) Protocol for MANETs. IEEE ICCSP Conference.*

Guran, I. C., Perişoară, L. A., Florescu, A., & Săcăleanu, D. I. (n.d.). 4-Cell Passive Battery Management System for Automotive Applications. IEEE 27th International Symposium for Design and Technology in Electronic Packaging (SIITME). 10.1109/SIITME53254.2021.9663604

Gyamfi, S., & Krumdieck, S. (2011). Price, environment and security: Exploring multimodal motivation in volu*ntary residential peak demand response. Energy Policy, 39(5), 2993–3004. doi:10.1016/j.enpol.2011.03.012*

Gáspár, P., Szaszi, I., & Bokor, J. (2003). Design of robust controllers for active vehicle suspension using the mixed μ synthesis. Vehicl*e System Dynamics, 40(4), 193–228. doi:10.*1076/vesd.40.2.193.16541

Gómez, J. C., & Morcos, M. M. (2003). Impact of EV Battery Chargers on the Power Quality of Distribution Systems. IEEE Trans*actions on Power Delivery, 18(3), 975–981.* doi:10.1109/TPWRD.2003.813873

Güçlü. (2003). Active control of *seat vibrations of a vehicle model using variou*s suspension alternatives. Turkish Journal of Engineering and Environmental Sciences, 27(6), 361–374.

Habib, S., Kamran, M., *& Rashid, U. (2015).* Impact analysis of vehicle-to-grid technology and charging strategies of electric vehicles on distribution networks review. Journal of Power Sources, 277, 205–214. doi:10.1016/j.*jpowsour.2014.12.020*

*Haq, N., Leksono, E., Iqba*l, M., Sodami, F. N., Kurniadi, D., & Yuliartoet, B. (2014). "Development of Battery Management System for Cell Monitoring And protection. In 2014 International Conference on Electrical Engineering And-computer Science (Iceecs). IEEE. 10.1109/ICEECS.2014.7045246

He, G., Zhang, H., Xu, Y., *& Lu, X. (2017). China's cl*ea*n* power transition: Current status and future prospect. Resources, Conservation and Recycling, 121, 3–10. doi:10.1016/j.resconrec.2016.11.003

Hemavathi, S. (2019). Modeling and Estimation of Lithium Ion Battery State of Charge *Using* Intelligent Techniques. IEEE International Conference on Computing, Power and Communication Technologies (Gucon 2019).

Hemavathi, S. (2020). Li-Ion Battery Health Estimation Based on Battery Internal Impedance Measurement. 1st Nati*onal Conference on Innov*atio*ns in Sustainable Energy and Technology (Iset 2020).

Hemavathi, S. (2020). Modeling of Analog Battery Management System for Single Cell Lithium Ion Battery. Energy Storage, 1-9.

Hemavathi, S. (2020). Overvie*w of cell balancing methods for Li-ion b*attery technology. Energy Storage, e203.

Hemavathi, S. (2021). Modeling and Energy Optimization of Hybrid Energy Storage System. Hybrid Renewab*le Energy Systems. doi:*10.*1*002/9781119555667.ch4

Hoffmann, D. (2019). Safety tests for li-secondary batteries. Electrochem. Power Sources Fundam. Syst. Appl., 387–453. *doi:10.1016/B978-0-444-*63777-2.00010-4

Hoque, M. M., Hannan, M. A., Mohamed, A., & Ayob, A. (2017). Battery charge equalization controller in electric vehicle applications: A review. Renewable and Sustainable Energy Reviews, Els*evier, 75(C), 1363–1385. doi:10.1016/j. rse*r.*20*16.11.126

Hoque, Mukit, & Bikas. (2012). An Implementation of Intrusion Detection System Using Genetic Algorithm. International Journal of Network Security & Its Applications, 4(2).

Hossain Lipu, M. S., Hannan, M. A., *& Tahia, F. (2021). Intelligent algorithms and control stra*tegies for battery management system in electric vehicles: Progress, challenges and future outlook. Journal of Cleaner Production. Advance *online publication. doi:10.1016/j.jclepro.2021.126044*

Huang, S.-C., Tseng, K.-H., Liang, J.-W., Chang, C.-L., & Pecht, M. G. (2017). An online SoC and SOH estimation model for lithium-ion batteries. Energies, 10(4), 512. doi:1*0.3390/en10040512*

*Huang, X., A*charya, A. B., Meng, J., Sui, X., Stroe, D-I., & Teodorescu, R. (2020). Wireless Smart Battery Management System for Electric Veh*icles. 2020 IEEE Energy Conversion Congr*ess *and* Exposition (ECCE), 5620–5625. doi:10.1109/ECCE44975.2020.9236279

Huber, C., & Kuhn, R. (2015). Thermal management of batteries for *electric vehicles. Elsevier Ltd. doi:10.1016/B978-1-78242-377-5.00013-3*

Hung, N. B., & Lim, O. (2019). A simulation and experimental study of dynamic performance and electric consumption of an electric bicycle. Energy *Procedia, 158, 2865–2871. doi:10.1016/j.egypro.*2019.01.937

Hussain, M. M., Beg, M. S., Alam, M. S., & Laskar, S. H. (2020). Big data analytics platforms for electric vehicle integration in transport oriented smart *cities: Computing platforms for platforms for electric ve*hicle integration in smart cities. In Cyber warfare and terrorism: Concepts, methodologies, tools, and applications (pp. 833–854). IGI Global. doi:10.4018/978-1-7*998-2466-4.ch051*

*Iannicie*ll*o*, L., Henry, P., & Achard, P. (2018). Electric vehicles batteries thermal management systems employing phase change materials. Journal of Power Sources, 378, 383–403. doi:10.1016*/j.jpowsour.2017.12.071*

Incropera, F.P., Dewitt, D.P., Bergman, T.L., & Lavine, A. (2007). Fundamentals of Heat and Mass Transfer. Academic Press.

Indragandhi, S. R., & Subramaniyaswamy, V. (2022). Electric Motor Drives and their Applications with Simulation Practices. Ind*ian Patent Publication,202241023774.*

*Isakova, Y. I., Pushka*rev, A. I., & Kholodnaya, G. E. (2011). A differential high-voltage divider. Instruments and Experimental Techniques, 54(2), 183–186. doi:10.1134/S00204*41211020175*

Jahangir, H., Tayarani, H., Ah*ma*dian, A., Golkar, M. A., Miret, J., Tayarani, M., & Gao, H. O. (2019). Charging demand of plug-in electric vehicles: Forecasting travel behavior based on a novel rough artificial neural network approa*ch. Journal of Cleaner Production, 229, 1*029–1044. doi:10.1016/j.jclepro.2019.04.345

Jahromi, A. F., & Zabihollah, A. (2010). Linear quadratic regulator and fuzzy controller application in full-car model of suspension system with mag*netorheological shock absorber. In Proceedings of 2010 IEEE/ASME Internationa*l Con*fer*ence on Mechatronic and Embedded Systems and Applications. IEEE. 10.1109/MESA.2010.5552010

Jander, C. S., & Schneider, R. (2005). Magnetoresistive Sensors for Non-Destructive Evaluation. In Advanced Sensor Technologies for Nondestruc-Tive Evaluation and Structural Health Monitoring (Vol. 5770). International Society for Optics and Photonics.

Jiang, L., & Low, S. (2011). Multi-period optimal energy procurement and demand response in smart grid with uncertain supply. 50th IEEE conference on decision and control and European control conference (CDC-ECC), 4348-53. 10.1109/CDC.2011.6161320

Joo, J., Ahn, S., Yoon, Y., & Choi, J. (2007). Option valuation applied to implementing demand response via critical peak pricing. Proc. IEEE Power Energy Soc. Gen. Meet., 24-28. 10.1109/PES.2007.385559

Jossen, V., Späth, V., Döring, H., & Garche, J. (1999). Reliable Battery Operation—A Challenge for The Battery Management System. Journal of Power Sources, 84(2), 283–286. doi:10.1016/S0378-7753(99)00329-8

Kambatla, K., Kollias, G., Kumar, V., & Grama, A. (2014). Trends in big data analytics. Journal of Parallel and Distributed Computing, 74(7), 2561–2573. doi:10.1016/j.jpdc.2014.01.003

Karvonen, M., Kapoor, R., Uusitalo, A., & Ojanen, V. (2016). Technology competition in the internal combustion engine waste heat recovery: A patent landscape analysis. Journal of Cleaner Production, 112, 3735–3743. doi:10.1016/j.jclepro.2015.06.031

Kathan, D., Aldina, R., Lee, M.P., Medearis, L., Sporborg, P., & Tita, M. (2012). Assessment of demand response and advanced metering. Federal Energy Regulatory Commission Report.

Kavya Santhoshi, B., Mohana Sundaram, K., Padmanaban, S., & Holm-Nielsen, J. B. (2019). K. K., P. Critical Review of PV Grid-Tied Inverters. Energies, 12, 1921. doi:10.3390/en12101921

Kempton, W., & Tomic, J. (2005). Vehicle-to-grid power implementation: From stabilizing the grid to supporting large-scale renewable energy. Journal of Power Sources, 144(1), 280–294. doi:10.1016/j.jpowsour.2004.12.022

Kempton, W., Udo, V., Huber, K., Komara, K., Letendre, S., Baker, S., Brunner, D., & Pearre, N. (2008). A test of vehicle-to-grid (V2G) for energy storage and frequency regulation in the PJM system. Results Ind Univ Res Partnersh, 32.

Kesler, M., Kisacikoglu, M. C., & Tolbert, L. M. (2014). Vehicle-to-grid reactive power operation using plug-in electric vehicle bidirectional offboard charger. IEEE Transactions on Industrial Electronics, 61(12), 6778–6784. doi:10.1109/TIE.2014.2314065

Khan, Ahmad, & Alam. (2019). Fast EV charging station integration with grid ensuring optimal and quality power exchange. Engineering Science and Technology, an International Journal, 22(1), 143-152.

Khan, L., Qamar, S., & Khan, M. U. (2014). Comparative analysis of adaptive NeuroFuzzy control techniques for full car active suspension system. Arabian Journal for Science and Engineering, 39(3), 2045–2069. doi:10.100713369-013-0729-4

Kim, J. M., Park, S. H., Han, Y. J., & Chung, T. M. (2008). CHEF: Cluster head election mechanism using fuzzy logic in wireless sensor networks. Proc. of International Conference on Advanced Communication Technology, 654–659. 10.1109/ICACT.2008.4493846

Kinhekar, N., Padhy, N. P., & Gupta, H. O. (2014). Multi objective demand side management solutions for utilities with peak demand deficit. International Journal of Electrical Power & Energy Systems, 55, 612–619. doi:10.1016/j.ijepes.2013.10.011

Kisacikoglu, M. C., Erden, F., & Erdogan, N. (2017). Distributed control of PEV charging based on energy demand forecast. IEEE Transactions on Industrial Informatics, 14(1), 332–341. doi:10.1109/TII.2017.2705075

Kovalcik, M. (2015). Constructing A Lithium-Ion/Polymer Battery With An Imbed-Ded Protection Circuit Module (PCM). Michigan State University.

Kulkarni, R. V., Forst*er, A., & Venayagamo*orth*y*, G. K. (2011). Computational Intelligence in Wireless Sensor Networks: A Sur*vey. IEEE Communications Surveys and Tutorials, 13(1), 68–96. doi:10.1109/SURV.2011.*040310.00002

Kumar & Mehfuz. (2017). Energy Efficient Probabilistic broadcasting for mobile ad hoc network. Journal of the Institution of *Engineers (India): Series B*, *98*(3), 289–294.

Kumar & Mehfuz. (2019). A PSO based malicious node detection and energy efficient clustering in wireless sensor network. International Confer*ence on Signal Processing and Integrated Networks* (SPIN).

Kumar & Sharma. (2012). Distance based Cluster Head Selection Algorithm for Wireless Sensor Network. International Journal of Computer Applications, 57(9).

Kumar, M. (2020). Esti*mation of* State of Charge and Terminal Voltage of Li-Ion Battery Using Extended Kalman Filter. 2020 6th IEEE International Energy Conference (Energycon).

Kumar, S., & Mehfuz, S. (2016, Ju*ly). Intelligent probabilistic b*roadc*asting in mobile ad hoc network: A PSO approach. Journal of Reliable Intelligent Environments, 2(2), 107–115. doi:10.100740860-016-0023-9

Kumar, S., & Mehfuz, S. (2016, June). E*fficient Fuzzy Logic Bas*ed *P*robabilistic broadcasting for mobile ad hoc network. International Journal of Computational Intelligence Systems, 9(4), 666–675. doi:10.1080/18756891.2016.1204115

Kıvrak, S., Özer, T., *Oğuz, Y., & Erken, E. B. (2020). B*attery management system implementation with the passive control method using MOSFET as a load. Measurement and Control, 53(1-2), 205–213. doi:10.1177/0020294019883401

Lannoye, E., Flynn, D., & O'Malley, *M. (2012).* Ev*a*luation of power system flexibility. IEEE Transactions on Power Systems, 27(2), 922–931. doi:10.1109/TPWRS.2011.2177280

Lassila, J., Haakana, J., Tikka, V., & Part*anen, J. (2012). Methodology to Analyze the Economic Effe*cts of Electric Cars as Energy Storages. IEEE Transactions on Smart Grid, 3(1), 506–516. doi:10.1109/TSG.2011.2168548

LaValle, S., Lesser, *E., Shockley, R., Hopkins, M. S., & Kr*uschwitz, N. (2011). Big data, analytics and the path from insights to value. MIT Sloan Management Review, 52(2), 21–32.

Lawson, B. (n.d.). A software conðgurable battery. I*n Proc. EVS26 Int. B*atter. Hybrid Fuel Cell Electr. Veh. Symp. (vol. 1, pp. 252–263). IEEE.

Lee, C. H., & Wu, C. H. (2015, April). Collecting and mining big data for *electric vehicle systems using battery modeling data. In 2015 12th International Conference* on Information Technology-New Generations (pp. 626-631). IEEE. 10.1109/ITNG.2015.104

Lee, J. (2010). Battery management technology for an electric vehicle. Proceedings of the 16th ACM/IEEE international *symposium on Low power electronics and design - ISLPED '10, 361. doi:10.1145/1840845.*1840923

Lee, J., & Park, G. L. (2016). Electric vehicle charger management system for interoperable charging facilities. Jurnal Teknologi, 78(5-8).

Lee, J., Bagheri, B., & Kao, H. A. (2014, July). Recent advances and trends of cyber-physical systems and big data analytics in industrial informatics. In International proceeding of int co*nference on industrial informatics (INDIN) (pp. 1-6). Academic Press.*

Lee, J., Park, G. L., Cho, Y., Kim, S., & Jung, J. (2015, April). Spatio-temporal analysis of state-of-charge streams for electric vehicles. In Proceedings of the 14th International Conference on Information Processing in Sensor Networks (pp. 368-369). 10.1145/2737095.2742932

Lee, J., Park, G. L., Han, Y., & Yoo, S. (2017, May). Big data analysis for an electric vehicle charging infrastructure using open data and software. In Proceedings of the eighth international conference on future energy systems (pp. 252-253). 10.1145/3077839.3081670

Lee, K. B. A., Ahmed, M., Kang, D. K., & Kim, Y. C. (2020). Deep reinforcement learning based optimal route and charging station selection. Energies, 13(23), 6255. doi:10.3390/en13236255

Lee, M. (2020). An analysis of the effects of artificial intelligence on electric vehicle technology innovation using patent data. World Patent Information, 63, 102002. doi:10.1016/j.wpi.2020.102002

Lehtola, T., & Zahedi, A. (2019). Solar energy and wind power supply supported by storage technology: A review. Sustainable Energy Technologies and Assessments, 35, 25–31. doi:10.1016/j.seta.2019.05.013

Lelie, M., Braun, T., Knips, M., Nordmann, H., Ringbeck, F., Zappen, H., & Sauer, D.U. (2018). Battery Management System Hardware Concepts: An Overview. Applied Sciences, 8(4). doi:10.3390/app8040534

Li, B., Kisacikoglu, M. C., Liu, C., Singh, N., & Erol-Kantarci, M. (2017). Big data analytics for electric vehicle integration in green smart cities. IEEE Communications Magazine, 55(11), 19–25. doi:10.1109/MCOM.2017.1700133

Li, Y., & Li, K. (2019). Incorporating Demand Response of Electric Vehicles in Scheduling of Isolated Microgrids With Renewables Using a Bi-Level Programming Approach. IEEE Access: Practical Innovations, Open Solutions, 7, 116256–116266. doi:10.1109/ACCESS.2019.2936487

Liang, H., Liu, Y., Shen, Y., Li, F., & Man, Y. (2018). A hybrid bat algorithm for economic dispatch with random wind power. IEEE Transactions on Power Systems, 33(5), 5052–5061. doi:10.1109/TPWRS.2018.2812711

Lin, J., Dong, P., Liu, M., Huang, X., & Deng, W. (2020). Research on Demand Response of Electric Vehicle Agents Based on Multi-Layer Machine Learning Algorithm. IEEE Access: Practical Innovations, Open Solutions, 8, 224224–224234. doi:10.1109/ACCESS.2020.3042235

Lin, J.-C. M. (2017). Development of a new battery management system with an independent balance module for electrical motorcycles. Energies, 10(9), 1289. doi:10.3390/en10091289

Liu, G., Guo, L., Liu, C., & Wu, Q. (2018). Evaluation of Different Calibration equations for Ntc Thermistor Applied To High-Precision Temperature Measurement. Measurement, 120, 21–27. doi:10.1016/j.measurement.2018.02.007

Liu, G., Ouyang, M., Lu, L., Li, J., & Han, X. (2014). Analysis of the heat generation of lithium-ion battery during charging and discharging considering different influencing factors. Journal of Thermal Analysis and Calorimetry, 116(2), 1001–1010. doi:10.100710973-013-3599-9

Liu, P.-J., & Chien, L.-H. (2017). A High-Efficiency Integrated Multimode Batterycharger With An Adaptive Supply Voltage Control Scheme. IEEE Transactions on Power Electronics, 33(8), 6869–6876. doi:10.1109/TPEL.2017.2761816

Loiseau, P., Chevrel, P., Yagoubi, M., & Duffal, J. M. (2018). Robust active noise control in a car cabin: Evaluation of achievable performances with a feedback control scheme. Control Engineering Practice, 81, 172–182. doi:10.1016/j.conengprac.2018.09.015

Lokhande, J. S., Daigavhane, P. M., & Sarkar, M. (2020). A Critical Approach Towards a Smarter Battery Management System for Electric Vehicle. 2020 4th International Conference on Trends in Electronics and Informatics (ICOEI), (48184), 232–235. doi:10.1109/ICOEI48184.2020.9142869

Loniza, E., Situmorang, J.A., Kusuma, D.D.A., Cahyadi, A.I., & Wahyunggoro, O. (2016). Passive bal*ancing of battery lithium polymer using shunt resistor circuit method. Adv Sci Technol Soc., 1755, 90011-1-*90011-6.

Lopez, K. L., Gagné, C., & Gardner, M. A. (2018). Demand-Side M*anagement Using Deep Learning for Smart Charging of Electric Vehicles. IEEE Transactions on Smart Grid, 10(3), 2683–2691. doi:10.1109/TSG.2018.2808247*

*L*u, Z., Meng, X. Z., Wei, L. C., Hu, W. Y., Zhang, L. Y., & Jin, L. W. (2016). Thermal Management of Densely-packed EV Battery With Forced Air Cooling Strategies. Energy Procedia, 88, 682–688. d*oi:10.1016/j.egypro.2016.0*6.098

Lua, L., Hana, X., Lia, J., Huab, J., & Ouyanga, M. (2013). A Review on The Key Issues for Lithium-Ion Battery Management In Electri*c Vehicles. Journal of Power Sources, 226, 272–28*8. doi:10.1016/j.jpowsour.2012.10.060

Lv, Z., Qiao, L., Cai, K., & Wang, Q. (2020). Big data analysis technology for electric vehicle networks in smart cities. IEEE Transactions on Intelligent Transportation Systems*, 22(3), 1807–1816. doi:10.1109/TI*TS.2020.3008884

Lü, X., Wu, Y., Lian, J., Zhang, Y., Chen, C., Wang, P., & Meng, L. (2020). Energy management of hybrid electric vehicles: A review of energy optimization of fuel cell hybr*id power sys*tem based on genetic algorithm. Energy Conversion and Management, 205, 112474. doi:10.1016/j.enconman.2020.112474

Malek, Y. N., Najib, M., Bakhouya, M., *& Essaaidi, M. (2021). Multivariate deep learning approach for* electric vehicle speed forecasting. Big Data Mining and Analytics, 4(1), 56–64. doi:10.26599/BDMA.2020.9020027

Mangaiyarkarasi, P., Lakshmi, P., & Sasrika, V*. (2019). Enhancement of vibration based piez*oelectric energy harvester using hybrid optimization techniques. Microsystem Technologies, 25(10), 3791–3800. doi:10.100700542-018-04291-1

Manzetti, S., & Mariasiu, F. (2015). Electric vehicle *battery technologies: From present state to future systems. Renewable* & Sustainable Energy Reviews, 51, 1004–1012. doi:10.1016/j.rser.2015.07.010

Martinez, D. A., Poveda, J. D., & Montenegro, D. (2017). Li-Ion battery management s*ystem based in fuzzy logic for improving electr*ic vehicle autonomy. 2017 IEEE Workshop on Power Electronics and Power Quality Applications (PEPQA), 1–6. doi:10.1109/PEPQA.2017.7981677

Mclean, R., Williamson, C., & Kattan, L. (2021). Simulatio*n Modeling of Urban E-Scoot*er Mobility. International Symposium on Modeling, Analysis, and Simulation of Computer and Telecommunication Systems.

Mech, L. D. (1999). Alpha status, dominance, and division of labor in wolf packs. Canadian Journal of Zoology, 77(8), 1196–1203. doi:10.1139/z99-099

Meelen, T., Doody, B., & Schwanen, T. (2020). Veh*icle-to-grid in the UK fleet market: An a*naly*s*is of upscaling potential in a changing environment. Journal of Cleaner Production, 125203.

Mehfuz & Kumar. (2015). Energy Aware Probabilisti*c Broadcasting for Mobile Ad Hoc Network. 2nd IEEE Internatio*nal Conference on Computing for Sustainable Global Development, 1028 – 1033.

Mehfuz, S., & Kumar, S. (2014, September). Two-dimensional *particle swarm optimization algorithm for load flow analysis. International* Journal of Computational Intelligence Systems, 7(6), 1074–1082. doi:10.1080/18756891.2014.963973

Meiss*ner, E., & Richter, G. (2003). Battery Monitoring* and Electrical Energy Man-Agement: Precondition for Future Vehicle Electric Power Systems. Journal of Power Sources, 116(1-*2), 79–98. doi:10.1016/S0378-7753(02)00713-9*

Meng, Luo, Ricco, Swierczynski, Stroe, & Teodor-Escu. (2018). Overview of Lithium-Ion Battery Modeling Methods for State-Of-Charge Es*timation in Electrical Vehicles. Applied Sci*ences, 8(5), 659.

Mercurio, A., Di Giorgio, A., & Quaresima, A. (2012). Distributed control approach for community energy management systems in presence of storage. *20th Mediterranean conference on control automation (MED),* 1303-8. 10.1109/MED.2012.6265819

Messaoudi, I., & Kamel, N. (2019). A multi-objective bat algorithm for community detection on dynamic social networks. Applied Intelligence, 49(6), 2119–21*36. doi:10.100710489-018-13*86-9

Mestrallet, F., Kerachev, L., Crebier, J.-C., & Collet, A. (2014, *June). Multiphase Interleaved Converter for Lithium Battery Active* Balancing. IEEE Transactions on Power Electronics, 29(6), 2874–2881. doi:10.1109/TPEL.2013.2276152

Methekar, R. N., Northrop, P. W. C., Chen, K., & Braatz*, R. D. (2011). Kinetic Monte Carlo Simulation of Surface Heterogeneity in* Graphite Anodes for Lithium-Ion Batteries. Passive Layer Formation, 158, 363–370. doi:10.1149/1.3548526

Mihet-Popa *& Saponara. (2018). Toward green vehicles digitalization for the next generation of connected and electri*fied transport systems. Energies, 11.

Mihet-Popa, L., & Groza, V. (2015). Battery management system test platform developed for electric vehicle app*lications. 2015 IEEE* 9th International Symposium on Intelligent Signal Processing (WISP) Proceedings, 1–6. doi:10.1109/wisp.2015.7139163

Millner, A. (2010). Modeling lithium ion batt*ery degradation in electric vehicles. IEEE Conf. Innov. Technol. an Effic. R*eliab. Electr. Supply, 349-356. 10.1109/CITRES.2010.5619782

Mishra, M. K., & Karthikeyan, K. (2009). A fast-acting dc-link voltage controller for three-phase DSTATCOM to co*mpensate ac and dc loads. IEEE Transactions on Power Delivery, 24(4), 2291–2299. doi:10.1109/*TPWRD.2009.2027501

Moghe, R., Kreikebaum, F., Hernandez, J. E., Kandula, R. P., & Divan, D. (2011). Mitigating Distribution Transformer Lifetime Degradation Caused by Grid-Enabled Vehicle (GEV) *Charging. Proc. IEEE Energy Conversion Cong. (ECCE), 835-842. 10.1109/ECCE.*2011.6063857

Moore, S., & Schneider, P. (2001). A Review of Cell Equalization Methods for Lithium Ion and Lithium Polymer Battery Systems. In Proceedings of the SAE 2001 World Con*gress. S*AE *M*obilus. 10.4271/2001-01-0959

Munshi, A. A., & Yasser, A. R. M. (2017). Big data framework for analytics in smart grids. Electric Power Systems Research, 151, 369–380. *doi:10.1016/j.epsr.2017.06.*006

Muralidharan, A., Sreelekshmi, R. S., & Nair, M. G. (2020). Cell Modelling for Battery Management System in Electric Vehicles. 2020 Third International Conference on Smart Systems and Inven*tive Technology (IC*SSIT), 558–564. doi:10.1109/ICSSIT48917.2020.9214253

Muro, C., Escobedo, R., Spector, L., & Coppinger, R. P. (2011). Wolf-pack (Canis lupus) hunting stra*tegies emerge from simple rules in co*mputational simulations. Behavioural Processes, 88(3), 192–197. doi:10.1016/j.beproc.2011.09.006 PMID:21963347

Mutyala, M. S. K., Zhao, J., Li, J., Pan, H., Yuan, C., & Li, X. (2014). In-Situ tempera*ture Measurement In Lithium* Io*n* Battery By Transferable Flexible thin Film Thermocouples. Journal of Power Sources, 260, 43–49. doi:10.1016/j.jpowsour.2014.03.004

Nagarkar, M. P., Bhalerao, Y. J., Patil, G. *J. V., & Pa*til, R. N. Z. (2018). GA-based multi-objective optimization of active nonlinear quarter car suspension system-PID and fuzzy logic control. International Journal of Mechanical and Materials Enginee*ring, 13(1), 1–*20. doi:10.118640712-018-0096-8

Nallathambi, A., ... (2022). Autonomous Electric Vehicles Auto Monitor and Control, Indian Patent Publication, 202241015720.

Nallathambi, A., ... (2022a). Design and Development of Advanced Solar *Based Electric Vehicles Charging Power Station*, Indian Patent Publication,202241023774.

Nallathambi, A., ... (2022b). Framework for the protection of the cyber-physical system of electric vehicle charging stations and the el*ectricity grid, Indian Patent Publicati*on, 202241045388.

Nandkeolyar, S., & Ray, P. K. (2022). Identifying households with electrical vehicle for demand response participation. Electric Power Systems Research, 208, 107909. doi:10.1016/j.epsr.2022.107909

National E*nergy Technology Laboratory.* (*20*07). A system view of modern grid. Author.

Negrohol, A., Kaleg, S., & Nugrhol, A. (2017). Passive Balancing Battery Management System using MOSFET Internal Resistance as Balancing Resistor. I*nternational Conference on Sustainable Energy Engineering and Application (ICSEEA)*.

Nikolian, A., de Hoog, J., Fleurbaey, K., Timmermans, J. M., Omar, N., van den Bossche, P., & van Mierlo, J. (2014). Classification of Electric modeling and Characterization methods of Lithium-ion Batteriesfor Vehicle Appli*cations. Proceedings of th*e European Electric Vehicle Congress, 2–5.

Nikoobakht, A., Aghaei, J., Shafie-Khah, M., & Catalao, J. P. S. (2019). Assessing increased flexibility of energy storage *and demand response to acc*ommodate a high penetration of renewable energy sources. IEEE Trans Sustain Energy, 10(2), 659–669. doi:10.1109/TSTE.2018.2843161

Nitta, N., Wu, F., Lee, J. T., & Yushin, G. (2015). Li-Ion Battery Materials: Present and Future. Materials Today, 18(5), 252–264. d*oi:10.1016/j.mattod.2014.10.040*

N*ort*hfield, R. (2016). Greening the smart city. Engineering & Technology, 11(5), 38–41. doi:10.1049/et.2016.0503

Noura, N., Boulon, L., & Jemeï, S. (2020). A Review of Battery State of Health Estimation Methods: *Hybrid Electric* V*eh*icle Challenges. World Electric Vehicle Journal, 11(4), 66. doi:10.3390/wevj11040066

Nunes, P., & Brito, M. C. (2017). Displacing natural gas with electric vehicles f*or grid stabilization. Energy, 141, 87–96. doi:10.1016/*j.*en*ergy.2017.09.064

Omariba, Z. B., Zhang, L., & Sun, D. (2019). Review of Battery Cell Balancing Methodologies for Optimizing Battery Pack Performance in Electric Vehicles. IEEE. doi:10.*1109/ACCESS.2019.2940090*

Ota, Y., Taniguchi, H., Nakajima, T., Liyanage, K. M., Baba, J., & Yokoyama, A. (2012). Autonomous Distributed V2G (Vehicle-to-Grid) Satisfying Scheduled Charging. IEEE Transactions on Smart Grid, 3(1), 559*–564. doi:10.1109/TSG.20*11.2167993

Ovalle, L., Ríos, H., & Ahmed, H. (2022). Robust control for an active suspension system via continuous sliding-mode controllers. Engineering Science and *Technology, an International Journal, 28,* 101026.

Papaefthymiou, G., Grave, K., & Dragoon, K. (2014). Flexibility options in electricity systems. Ecofys. European Copper Institute.

Pazooki, A., Rakheja, S., & Cao, D. (2012). Modeling an*d validation of off-road vehicle ride dynamics. Mechanical Systems and Signal* Processing, 28, 679–695. doi:10.1016/j.ymssp.2011.11.006

Pedro, J. O., Dangor, M., Dahunsi, O. A., & Ali, M. M. (2018). Dynamic neural n*etwork-based feedback linearization control of full-car suspensions using PSO. Applied Soft Computing,* 70, 723–736. doi:10.1016/j.asoc.2018.06.002

Pfenninger, S., Hawkes, A., & Keirstead, *J. (2014). Energy systems* modeling for twenty-first century energy challenges. Renewable & Sustainable Energy Reviews, 33, 74–86. doi:10.1016/j.rser.2014.02.0*03*

Pfrang, A., Kriston, A., Ruiz, V., Lebedeva, N., & Persio, F. (2017). Safety of Rechargeable Energy Storage Systems with a focus on Li-ion Technology. In Emerg. Nanotechnologies Recharg. Energy Storage Syst. Elsevi*er Inc. doi:10.1016/B978-0-323-42977-1/00008-X*

Pilkington, A., Dyerson, R., & Tissier, O. (2002). The electric vehicle: Patent data as indicators of technological development. World Patent Information, 24(1), 5–12. doi:10.1016/S0172-2190(01)00065-5

Pillai, *J. R., & Bak-Jensen, B. (201*1). Integration of Vehicle-to-grid in the Western Danish power system. IEEE Trans. Sustainable Energy, 2(1), 12–19.

Pinto, C., Barreras, J. V., Schaltz, E., & Araujo, R. (2016). Evaluation of Advanced Control for Li-ion Battery Balancing Systems Using Convex Optimization. IEEE Transactions on Sustainable Energy, 7(4).

Plett, G. L. (2004). Extended Kalman Filt*ering for Battery Man*age*m*ent Systems of Lipb-Based Hev Battery Packs: Part 3. State and Parameter Estimation. Journal of Power Sources, 134(2), 277–292. doi:10.1016/j.jpowsour.2004.02.033

Prabakar, R. S., Sujatha, *C., & Narayanan, S. (2013). Response* of *a* quarter car model with optimal magnetorheological damper parameters. Journal of Sound and Vibration, 332(9), 2191–2206. doi:10.1016/j.jsv.2012.08.021

Prassad, G. S., & Mohan, M. K. (2019). A contemporary adaptive air s*uspension using LQR con*tr*ol* for passenger vehicles. ISA Transactions, 93, 244–254. doi:10.1016/j.isatra.2019.02.031 PMID:30837128

Pratapa Reddy & Satyanarayana. (2016). Energy Efficient Stable Multi pa*th Routing i*n MANET. International conference on Signal Processing, Communication, Power and Embedded System (SCOPES), 1115-1121. 10.110*9/SCOPES.2016.7955615*

Qaisar, S. M., Dallet, D., Benjamin, S., Desprez, P., & Yahiaoui, R. (2013). Power efficient Analog To Digital Conversion for The Li-Ion Battery Voltage Monitoring and Measurement. In 2013 IEEE International Instrumentation and Measurement Te*chnology Conference (I2Mtc). IEEE.*

Q*i,* G., Li, X., & Yang, D. (2014). A control strategy for dynamic balancing of lithium iron phosphate battery based on the performance of cell voltage. In Proc. IEEE Conf. Expo Tra*nsp. Electriðc. Asia–Paciðc (ITEC Asia–Pacið*c) (pp. 1–5). IEEE. 10.1109/ITEC-AP.2014.6941212

Quan, O., Chen, J., Zheng, J., & Fang, H. (2018). Optimal Cell-to-Cell Balan*cing Topology Design for Serial*ly C*on*nected Lithium-Ion Battery Packs. IEEE Transactions on Sustainable Energy, 9(1).

Quantmeyer, F., & Liu-Henke, X. (2013). Modeling and identification of lithium-ion bat*teries for electric vehicles. 2013 International Symposium on Electrodynamic and Mechat*ronic Systems (SELM), 19–20. doi:10.1109/SELM.2013.6562959

Qureshi, K. N., Alhudhaif, A., & Jeon, G. (2021). Electric-vehicle energy management and charging scheduling system in sustainable cities and society. *Sustainable Cities a*nd Society, 71, 102990. Advance online publication. doi:10.1016/j.scs.2021.102990

Raghavan, S. S., & Khaligh, A. (2012). Impact of Plug-in Hybrid Electric Vehicle Charging on a Distribution Network in a Smart Grid Environment. Rec. IEEE PES In*novative Smart Grid Tech. Con*f. (ISGT). 10.1109/ISGT.2012.6175632

Rahimi-Eichi, H., & Chow, M. Y. (2014, October). Big-data framework for electric vehicle range estimation. In IECON 2014-40th Annual Conference of the IEEE Industrial Electronics Society (pp. 5628-5634*). IEEE. 10.1109/IECON.2014.7049362*

Rahimi-Eichi, H., Ojha, U., Baronti, F., & Chow, M.-Y. (2013). Battery Manage-Ment System: An Overview of Its Application In The Smart Grid and Electric vehicles. IEEE Industrial Electronics Magazine, 7(2), 4–16. doi:10.1109/MIE.2013.2250351

Rajagopal, K., & Ponnusamy, L. (2014). Biogeography-based optimization of PID tuning parameters for the vibration control of active suspension system. Journal of Control Engineering and Applied Informatics, 16(1), 31–39.

Rajagopalan, S. (2013). Fast charging: An in-depth look at market penetration, charging characteristics, and advanced technologies. In 2013 world electric vehicle symposium and exhibition (EVS27). IEEE.

Ranganathan, P., & Nygard, K. (2011, *October). Smart grid data analytics for decision support. In 2011 IEEE Electrical* Power and Energy Conference. IEEE.

Rao, P. C., Obulesh, Y. P., & Babu, S. (2012). Mathematical modeling and BLDC motor with closed loop speed control using PID controller under various loading conditions. Journal of Engineering and A*pplied Sciences, 7(10).*

Reynoso-Meza, G., Blasco, X., Sanchis, J., & Martínez, M. (2014). Controller tuning using evolutionary multi-objective optimisation:Cu*rrent trends an*d applications. Control Engineering Practice, 28, 58–73. doi:10.1016/j.conengprac.2014.03.003

Richa*rd, B. (2018). State of* Cha*r*ge and Parameter Estimation of Electric Vehicle Batteries. doi:10.25071/10315/35324

Rigas, E. S., Ramchurn, S. D., & Bassiliades, N. (2014). Managing electric vehicles in t*he smart grid using artificial* in*t*elligence: A survey. IEEE Transactions on Intelligent Transportation Systems, 16*(4), 1619–1635. doi:10.1109/TITS.2014.2376873*

Rivera-Barrera, Munoz-Galeano, & Sarmiento-Maldonado. (2017). Soc Estimation for Lithium-Ion Batteries: Review and Future Challenges. Electronics, 6(4).

Rizvi, S. M., Abid, M., Khan, A. Q., Satti, S. G., & Latif, J. (2018). H¥ con*trol of 8 degrees of freedom vehicle active suspension system.* Journal of King Saud University-Engineering Sciences, 30(2), 161–169. doi:10.1016/j.jksues.2016.02.004

Roukerd, S. *P., Abdollahi, A., & Rashidinejad, M. (*20*19*). Probabilistic-possibilistic flexibility-based unit commitment with uncertain negawatt demand response resources considering Z-number method. International Journal of Electrical Power & Energy Systems*, 113, 71–89. doi:10.1*016*/*j.ijepes.2019.05.011

Runkler, T. A. (2020). Data analytics. Springer Fachmedien Wiesbaden. doi:10.1007/978-3-658-29779-4

Russom, P. (2011). Big data analytics. T*DWI Best Practices Report, 19(4), 1-34.*

*S*aber, A. Y., & Venayagamoorthy, G. K. (2011). Plug-in Vehicles and Renewable Energy Sources for Cost and Emissi*on Reductions. IEEE Transactions on Industrial Electronics, 58(4), 1229–1238. doi:10.1109/TIE.2010.2047828*

Sadati, H., Malekzadeh, S., & Ghasemi, M. (2008). Optimal Control of an 8-DOF Vehicle Active Suspension System Using Kalman Observer. Shock and Vibration, 15(5), 493–503. doi:10.1155/2008/725264

Sajfar, M., & Bul*lough, R. (1990). Sealed* Bat*t*eries In Transient Limit-Ing Distribution Networks-Methods of Measuring Their Internal Resistance. In 12th International Conference on Telecommunications Energy. IEEE.

Saji, D., Babu, P. S., & Ilango, *K. (2019). Soc Estimation of Lithium Ion B*attery using Combined Coulomb Counting and Fuzzy Logic Method. In 2019 4th international Conference on Recent Trends on Electronics, Information, Communicati*on & Technology (Rteict). IEE*E. 10.1109/RTEICT46194.2019.9016956

Samadi, P., Mohsenian-Rad, A.-H., Schober, R., Wong, V., & Jatskevich, J. (2010). Optimal real-time pricing algorithm based on utility maximization fo*r smart grid. First IEEE inte*r*nati*o*n*al conference on smart grid communications (SmartGridComm), 415-20. 10.1109/SMARTGRID.2010.5622077

Santhoshi, B. K., Mohanasundaram, K., & Kumar, L. A. (2021). ANN-based dynamic control and energy management of inverter and battery in a grid-tied hybrid renewable power system fed through switched Z-source converter. Electrical Engineering, 103(5), 2285–2301. doi:10.100700202-021-01231-7

Santhoshi, B. K., Sundaram, K. M., Hussain, A., Sanjeevikumar, P., Holm-Nielsen, J. B., & Kaliappan, V. K. (2021). Deep Learning for Fault Diagnostics in Bearings, Insulators, PV Panels, Power Lines, and Electric Vehicle Applications—The State-of-the-Art Approaches. IEEE Access: Practical Innovations, Open Solutions, 9, 41246–41260. doi:10.1109/ACCESS.2021.3064360

Sarker, M. R., Ortega-Vazquez, M. A., & Kirschen, D. S. (2015). Optimal coordination and scheduling of demand response via monetary incentives. IEEE Transactions on Smart Grid, 6(3), 1341–1352. doi:10.1109/TSG.2014.2375067

Sathiyan, S. P. (2018). Effective Design and Analysis of Pedelec e-Bike using MBD Approach. International Journal of Mechanical Engineering and Technology, 9(13), 568–577.

Schwarz, R., Semmler, K., Wenger, M., Lorentz, V. R., & Arz, M. (2015). Sensorless Battery Cell Temperature Estimation Circuit for Enhanced Safety in Battery Systems. In IECON 2015-41St Annual Conference of The IEEE Industrial Electronics Society. IEEE.

Seidi Khorramabadi, S., & Bakhshai, A. (2015). Critic-Based Self-Tuning PI Structure for Active and Reactive Power Control of VSCs in Microgrid Systems. IEEE Transactions on Smart Grid, 6(1), 92–103. doi:10.1109/TSG.2014.2354651

Sekulić, D., Dedović, V., Rusov, S., Šalinić, S., & Obradović, A. (2013). Analysis of vibration effects on the comfort of intercity bus users by oscillatory model with ten degrees of freedom. Applied Mathematical Modelling, 37(18-19), 8629–8644. doi:10.1016/j.apm.2013.03.060

Shahbazitabar, M., & Abdi, H. (2018). A novel priority-based stochastic unit commitment considering renewable energy sources and parking lot cooperation. Energy, 161, 308–324. doi:10.1016/j.energy.2018.07.025

Shankar, R., & Marco, J. (2013). Method for estimating the energy consumption of electric vehicles and plug-in hybrid electric vehicles under real-world driving conditions. IET Intelligent Transport Systems, 7(1), 138–150. doi:10.1049/iet-its.2012.0114

Sharma, S. K., Pare, V., Chouksey, M., & Rawal, B. R. (2016). Numerical studies using full car model for combined primary and cabin suspension. Procedia Technology, 23, 171–178. doi:10.1016/j.protcy.2016.03.014

Shukla & Patel. (2022). Battery Management System by Passive Cell Balancing for Electric vehicle. 2022 2nd International Conference on Power Electronics & IoT Applications in Renewable Energy and Its Control (PARC).

Shurpali, S. (2020, January 24). Role of Edge Computing in Connected and Autonomous Vehicles. https://www.einfochips.com/blog/role-of-edge-computing-in-connected-and-autonomous-vehicles/

Siefert, A., Pankoke, S., & Wölfel, H. P. (2008). Virtual optimisation of car passenger seats: Simulation of static and dynamic effects on drivers' seating comfort. International Journal of Industrial Ergonomics, 38(5-6), 410–424. doi:10.1016/j.ergon.2007.08.016

Sierzchula, W., Bakker, S., Maat, K., & van Wee, B. (2014). The influence of financial incentives and other socio-economic factors on electric vehicle adoption. Energy Policy, 68, 183–194. doi:10.1016/j.enpol.2014.01.043

Sioshansi, R. (2012). OR Forum-Modeling the impacts of electricity tariffs on plug-in hybrid electric vehicle charging, costs, and emissions. Operations Research, 60(3), 506–516. doi:10.1287/opre.1120.1038

Sironval, V. (2020). Evaluating and Predicting the Respiratory Hazard of Particles Used in Li-Ion Batteries, for A Safe and Sustainable Development [Ph.D. Dissertation]. Catholique De Louvain.

Skrúcaný. (*2019). Impact of the electric mobility implementation* on the greenhouse gases production in central European countries. Sustaina*bility, 11.*

Slotine, J. J. E., & Li, W. (1991). Applied nonlinear control. Prentice Hall.

Soh, M., Jang, H., Park, *J., Sohn, Y., & Park, K. (2018*). Development of Preview Active Suspension Control System and Performance Limit Analysis by Trajectory Optimization. International Journal of Aut*omotive Technology,* 19*(*6), 1001–1012. doi:10.100712239-018-0097-x

Soleymani, M., Montazeri-Gh, M., & Amiryan, R. (2012). Adaptive fuzzy controller for vehicle active suspension system based on traffic conditio*ns. Scientia Iranica, 19(3), 443–453. doi:10.1016/j.scient.*2012.03.002

Sovacool, B. K., Noel, L., Axsen, J., & Kempton, W. (2018). The neglected social dimensions to a vehicle-to-grid (V2G) transition: A cri*tical and systematic review. Environmental Research Letters, 13(1), 13001. doi:10.1088/1748-9326/aa9c6d*

Sowmya, R., & Sankaranarayanan, V. (2022). Optimal vehicle-to-grid and grid-to-vehicle scheduling strategy with uncertainty management using improved marine predator algorithm. Computers & Electrical Engineering, 100, 107949. doi:10.1016/j.compeleceng.2022.107949

Sreedh*ar, R., & Karunanithi,* K. *(*2021). Design, simulation analysis of universal battery management system for EV applications. Materials Today: Proceedings. Advance online publication. doi:10.1016/j.matpr.2020.12.136

Sreekanth, T., Lakshminarasamma, N., & Mishra, M. K. (2017). A Single-Stage Grid-Connected High Gain Buck–Boost Inverter With M*aximum Power Point Tracking. IEEE Transactions on* Energy Conversion, 32(1), 330–339. doi:10.1109/TEC.2016.2633365

Sun, J., Liu, Z., & Wu, J. (2018). An optimal driving strategy for the ride co*mfort performance of intelligent vehicles on uneven roads. IFA*C-*P*apersOnLine, 51(31), 984–989. doi:10.1016/j.ifacol.2018.10.052

Supanat, A., Manop, M., & Nisai, F. (2019). On a Design of Adjustable Passive Balancing Circuit Using PWM Technique for Li-*Ion Battery. Research, Invention, and Innovation Congress (RI2C).*

Taha, I. B., & Elattar, E. E. (2018). Optimal reactive power resources sizing for power system operations enhancement based on improved grey wolf optimiser. IET Generatio*n, Transmission & Distribution,* 12(14), 3421–3434. doi:10.1049/iet-gtd.2018.0053

Taiebat, M., & Xu, M. (2019). Synergies of four emerging technologies for accelerated adoption of electric vehicles: Shared mobility, wireless charging, vehicle-to-grid, and vehicle autom*ation. Journal of Cleaner Pro*ductio*n*, 230, 794–797. doi:10.1016/j.jclepro.2019.05.142

Talbi, N. (2019). Design of fuzzy controller rule base using bat algorithm. Energy Procedia, 162, 241–250. doi:10.1016/j.egypro.2019.04.026

T*an, K. M., Ramachan*d*ara*murthy, V. K., & Yong, J. Y. (2016). Integration of electric vehicles in smart grid: A review on vehicle to grid technologies and optimization *techniques. Renewable & Sustainable Energy Reviews, 53, 720–732. doi:10.1016/j.rser.2015.09.012*

Tete, P. R., Gupta, M. M., & Joshi, S. S. (2021). Devel*opments in battery thermal management systems for electric* vehicles : A technical review State of Power. J Energy Storage, 35, 102255. doi:10.1016/j.est.2021.102255

Thejasree, G., & Maniyeri, R. (2019). E-bike System Modeling and Simulation. IEEE International Conference on Intelligent Systems and Green Technology.

Tsai, C. W., Lai, C. F., Chao, H. C., & Vasilakos, A. V. (2015). Big data analytics: A survey. Journal of Big Data, 2(1), 1–32. doi:10.118640537-015-0030-3 PMID:26191487

Tu, Y., Li, C., Cheng, L., & Le, L. (2011). Research on vehicle-to-grid technology. Proceed*ings of the* CDC*IE*M, international conference on computer distributed control and i*ntelligent environmental monitoring, 1013-1016.*

*Turker, H., Bacha, S., Chatroux, D., & Hably, A. (2012). Aging Rate of Low Vo*ltage Transformer for a High Penetration of Plug-in Hybrid Electric Vehicles (PHEVs). Rec. IEEE PES Innovative Smart Grid Tech. Conf. (ISGT). 10.1109/ ISGT.2012.6175651

U.S. Depa*rtment of Energy.* (2006). Benefits of demand response in *electricity markets and r*ecommendations for achieving them. A report to the United States Congress pursuant to Section 1252 of the Energy Policy Act of 2005, U.S. Dept. Energy.

USEV Market. (2016). Compatibility and *Investment. Author.*

*Vaideeswaran, V., Bhuvan*e*s*h, S., & Devasena, M. (2019). Battery Management Systems for Electric Vehicles using Lithium Ion Batteries. 2019 Innovations in Power and Advanced Computing Technologies (i-PACT), 1–9. doi:10.1*109/i-PACT44901.2*019*.*8959965

Vermesan, O., John, R., Pype, P., Daalderop, G., Kriegel, K., Mitic, G., ... Waldhör, S. (2021). Automotive Intelligence Embedded in Electric Connected Autonomous and *Shared Vehicles Technology f*or Sustainable Green Mobility. Frontiers in Future Transportation, 18.

Vezzini, A. (2014). Lithiium-ion battery management. In Lithium-Ion Batteries Advance and Applications. Elsevier.

Viera, Gonzalez, Anton, Campo, F*errero, & Valle-Dor. (2006). Nimh Vs N*icd Batteries Under High Charging Rates. In INTELEC06-Twenty-Eighth International Telecommunications Energy Conference. IEEE.

Vovos, P. N., Kiprakis, A. E., Wallace, A. R., & Harrison, *G. P. (2007). Ce*ntr*a*lized and distributed voltage control: Impact on distributed generation penetration. IEEE Transactions on Power Systems, 22(1), 476–483. doi:10.1109/ TPWRS.2006.888982

Vulligaddala, VVe*rnekar, SAdusumalli, SEle, VBrandl, MM.B., S. (2020,* May). A 7-Cell, Stackable, Li-Ion Monitoring and Active/Passive Balancing IC With In-Built Cell Balancing Switches for Electric and Hybrid Vehicles. IEEE Tran*sactions on Industrial Informatics, 16(5).*

W*al*ikar & Biradar. (2012). Energy aware multicast routing in mobile ad-hoc networks using NS-2. Academic Press.

Wang, C., Gu, W., & Liaw, B. (1998). Micro-Macroscopic Coupled Modeling Ofbatteries and Fuel Cells: I. Model Development. Journal of *the Electrochemical Society, 145(*10), 3407–3417. doi:10.1149/1.1838820

Wang, J., Liu, C., Ton, D., Zhou, Y., Kim, J., & Vyas, A. (2011). Impact of PHEVs o*n Power Systems* w*ith* Demand Response and Wind Power. Energy Policy, 39(7), 4016–4021. doi:10.1016/j.enpol.2011.01.042

Wang, K., Gao, F., Zhu, Y., Liu, H., Qi, C., Yang, K., & Jiao, Q. (2018). Internal resistance and heat gen*eration of soft* pa*ck*age Li4Ti5O12 battery during charge and discharge. Energy, 149, 364–374. doi:10.1016/j.energy.2018.02.052

Wang, K., Y*u, J., Yu, Y., Qian, Y., Zeng, D., Guo, S., Xiang, Y., & Wu, J. (2017). A* survey on energy internet: Architecture, approach, and emerging technologies. IEEE Systems Journa*l, 12(3), 2403–2416. d*oi:10.1109/JSYST.2016.2639820

Wang, L., Cheng, Y., & Zhao, X. (2015). Influence of Connecting Plate Resistanceupon Lifepo4 Battery Performance. Applied Energy, 147, 353–360. doi:10.1016/j.*apenergy.2015.03.016*

*Wang, M., & Abdalla, M. A. A. (2022). Optimal En*ergy Scheduling Based on Jaya Algorithm for Integration of Vehicle-to-Home and Energy Storage System with Photovoltaic Generation in Smart Home. Sensors (Basel), 22(4), 1306. doi:10.339022041306 PMID:35214211

Wang, M., & Xie, N. (2018). Structural optimization design *of magnetic Shock absorber based on* particle swarm optimization*. IOP Conference Series. Materials Science and Engineering, 452(2), 1–9. doi:10.1*088/1757-899X/452/2/022012

Wang, Q., Ping, P., Zhao, X., Chu, G., Sun, J., & Chen, C. (2012). Thermal runaway caused fire and *explosion of lithium ion battery. Journal of Power Sources, 208, 210–224. d*oi:10.1016/j.jpowsour.2012.02.038

Wang, W., Song, Y., Xue, Y., Jin, H., Hou, J., & Zhao, M. (2015). An optimal vibration control strategy for a vehicle's active suspension based on improved cultural algorithm. Applied Soft Com*puting, 28, 167–174. doi:10.1016/j.asoc.2014.11.*047

Wazid, Katal, Singh, Goudar*, & Singh. (2013). Detection and Prevention Mechanism for Blackhole* Attack in Wireless Sensor Network. International Conference on Communication and Signal Processing.

Wei, L., Jie, Lv., Song, W., & Feng, Z. *(2017). Study on passive balancing charac*teristics of serially connected lithium-ion battery string. In IEEE 13th International Conference on Electronic Measurement & Instruments. IEEE. 10.1109/ICEMI.2017.8265862

Wei, W., Minghao, A., Naishi, C., Xianjun, G.*, & Tianjia*o, P. (2016). Multi-Level Feeder Queue Optimization Charging Model of Electric Vehicle and its Implementation of MR Algorithm. International Journal of u-and e-Service, Science and Technology, 9(3), 199–208.

Wu, J., Guo, S., Li, J., & Zeng, D. (2016). Big data meet green challenges: Big data toward green applications. IEEE Systems Journal, 10(3), 888–900. doi:10.1109/JSYST.2016.2550530

Wu, *Y., Liao, X., Chen*, W., & Chen, D. (2011). A Battery Management System for electric vehicle based on Zigbee and CAN. 2011 4th International Congress on Image and Signal Processing, 2517–252*1. doi:10.1109/cisp.2011.6100781*

Xia, G., Cao, L., & Bi, G. (2017). Review article A review on battery thermal management in electric vehicle application. Journal of Power Sources, 367, 90–105. doi:10.1016/j.jpowsour.2017.09.046

Xie, Zhang*, Fu, Wen, Han, Zhu, Jiang*, & Guo. (2013). The Security Issue of WSNs Based on Cloud Computing. IEEE Conference on Communications and Network Security.

Xiong, He, Guo, & Ding. (2011). Modeling for Lithium-Ion Battery used in Electric Vehicles. doi:10.*1016/j.proeng.2011.08.540*

Xiong, R. (2020). Battery Management Algorithm for Electric Vehicles. doi:10.1007/978-981-15-0248-4

Xydas, E., Marmaras, C., & Cipcigan, L. M. (2016). A multi-agent based *scheduling algorithm for adaptive electric vehicles charging.* Applied Energy, 177, 354–365. doi:10.1016/j.apenergy.2016.05.034

Yang, Z., Li, K., Niu, Q., & Xue, Y. (2017). A comprehensive study of economic un*it commitment of power systems integrating various renewable generations a*nd plug-in electric vehicles. Energy Conversion and Management, 132, 460–481. doi:10.1016/j.enconman.2016.11.050

Yao, Y., Chen, H.-Y., Huang, J., & Yang, Y. (2007). Low Voltage and Fastspeed All-Polymeric Optocouplers. *Applied Physics Letters, 90(5), 053509.* d*oi:10.1063/1.2437683*

Yap, K. Y., Beh, C. M., & Sarimuthu, C. R. (2021). Fuzzy logic controller-based synchronverter in grid-connected solar power system with adaptive damping factor. in Chinese Journal of Electrical Engineering, 7(2), pp. 37-49, .

Yeruva, V. (2021, September 27). Council Post: Edge Computing's Applications in Autonomous Driving and Business at Large. Forbes. https://www.forbes.com/sites/forbestechcouncil/2021/09/27/edge-computings-applications-in-autonomous-driving-and-busines s-at-large/?sh=68b9426a96aa

Yoldaş, Y., Önen, A., Muyeen, S. M., Vasilakos, A. V., & Alan, İ. (2017). Enhancing smart grid with microgrids: Challenges and opportunities. Renewable & Sustainable Energy Reviews, 72, 205–214. doi:10.1016/j.rser.2017.01.064

Yong, J. Y., Ramachandaramurthy, V. K., Tan, K. M., & Mithulananthan, N. (2015). A review on the state-of-the-art technologies of electric vehicle, its impacts and prospects. Renewable & Sustainable Energy Reviews, 49, 365–385. doi:10.1016/j.rser.2015.04.130

Yong, J. Y., Ramachandaramurthy, V. K., Tan, K. M., & Mithulananthan, N. (2015). Bi-directional electric vehicle fast charging station with novel reactive power compensation for voltage regulation. International Journal of Electrical Power & Energy Systems, 64, 300–310. doi:10.1016/j.ijepes.2014.07.025

Yu, N., Shah, S., Johnson, R., Sherick, R., Hong, M., & Loparo, K. (2015, February). Big data analytics in power distribution systems. In 2015 IEEE power & energy society innovative smart grid technologies conference (ISGT) (pp. 1-5). IEEE. doi:10.1109/ISGT.2015.7131868

Yue, Q. L., He, C. X., Jiang, H. R., Wu, M. C., & Zhao, T. S. (2021). A hybrid battery thermal management system for electric vehicles under dynamic working conditions. International Journal of Heat and Mass Transfer, 164, 120528. doi:10.1016/j.ijheatmasstransfer.2020.120528

Yunus, K. J., Reza, M., Zelaya-De La Parra, H., & Srivastava, K. (2012). Impacts of Stochastic Residential Plug-In Electric Vehicle Charging on Distribution Grid. Rec. IEEE PES Innovative Smart Grid Tech. Conf. (ISGT). 10.1109/ISGT.2012.6175691

Yusof, M. S., Toha, S. F., Kamisan, N. A., Hashim, N. N. W. N., & Abdullah, M. A. (2016). Battery cell balancing optimisation for battery management system. In International Conference on Mechanical, Automotive and Aerospace Engineering. Semantic Scholar.

Yuvapriya, T. & Lakshmi, P. (2020). Numerical analysis and performance enhancement of active suspension system using bat optimization. International Journal of Dynamics and Control. doi:10.1007/s40435-020-00664-5

Yuvapriya, T., & Lakshmi, P. (2020). Simulation study on ride comfort enhancement for a heavy duty vehicle with fractional order sliding mode controller. SN Appl. Sci., 2(9), 1478. doi:10.100742452-020-03263-3

Yuvapriya, T., Lakshmi, P., & Rajendiran, S. (2018). Vibration suppression in full car active suspension system using fractional order sliding mode controller. Journal of the Brazilian Society of Mechanical Sciences and Engineering, 40(4), 1–1. doi:10.100740430-018-1138-0

Yuvapriya, T., Lakshmi, P., & Rajendiran, S. (2020). Design, Analysis and Experimental Validation of a Fractional Order Terminal Sliding Mode Controller on Active Suspension System. Journal of the Institution of Electronics and Telecommunication Engineers, 1–16.

Yuvapriya, T., Lakshmi, P., & Rajendiran, S. (2020). Vibration control and performance analysis of full car active suspension system using fractional order terminal sliding mode controller. Archives of Control Sciences, 295-324.

Yuvapriya, T., Lakshmi, P., & Rajendiran, S. (2020). Vibration control and performance analysis of full car active suspension system using fractional order terminal sliding mode controller. Archives of Control Sciences, 30(2), 295–324.

Zadeh, M., & Fathi, S. H. (2017). A New Approach for Photovoltaic Arrays Modeling and Maximum Power Point Estimation in Real Operating Conditions. IEEE Transactions on Industrial Electronics, 64(12), 9334–9343. doi:*10.1109/TIE.2017.2711571*

Zehsaz, M., Sadeghi, M. H., Ettefagh, M. M., & Shams, F. (2011). Tractor cabin's passive suspension parameters optimization via experimental and numerical methods. Journal of Terramechanics, 48(6), 439–450. doi:10.1016/j.jterra.2011.09.005

Zerriffi, H., Dowlatabadi, H., *& Farrell, A. (2005). Incorporating stress in electric power systems rel*iability models. Energy Policy, 35(1), 61–75. doi:10.1016/j.enpol.2005.10.007

Zhang, B., & Kezunovic, M. (2016). Impact on power system flexibility by electric vehicle participation in ramp market. *IEEE Transactions on Smart Grid,* 7(3), 1285–1294. doi:10.1109/TSG.2015.2437911

Zhang, C., Li, K., Mcloone, S., & Yang, Z. (2014). Battery modelling methods for electric vehicles - A review. 201*4 European Control Conference (ECC), 2673-2678.* 10.1109/ECC.2014.6862541

Zhang, C., Liu, Y., Qian, Y., & Bao, H. (2020). An optimization framework of electric vehicle (EV) batteries for product eco-design. Procedia CIRP, 90, 366-371. doi:10.1016/j.procir.20*20.01.081*

Zhang, J., & Lee, J. (2011, August). A Review on Prognostics and Health Monitoring Ofli-Ion Battery. Journal of Power Sources, 196(15), 6007–6014. doi:10.1016/j.jpowsour.2011.03.101

Zhang, J., Deng, *Y., Zhang, N.,* Zhang, B., Qi, H., & Zheng, M. (2019). Vibration performance analysis of a mining vehicle with bounce and pitch tuned hydraulically interconnected suspension. Chinese J*ournal of Mechanical Engineering, 32(1), 1–17. doi:10.118610033-019-0315-0*

*Zhang, X., W*u, J., & Kang, G. (2016). SOC estimation of Lithium battery by UKF algorithm based on dynamic parameter model. 2016 13th International Conference on Ubi*quitous Robots and Ambient Intelligenc*e (URAI), 45–950. doi:10.1109/URAI.2016.7734116

Zhang, Z. J., Ramadass, P., & Fang, W. (2014). 18 – Safety of Lithium-Ion Batteries. In Lithium-Ion Batter. Adv. Appl. Elsevier. doi:10.1016/B978-0-444-59513-3.00018-2

Zhao, J., Ku*cuksari, S., Mazhari, E., & Son, Y. J. (2013).* Integrated analysis of high-penetration PV and PHEV with energy storage and demand response. Applied Energy, 112, 35–51. doi:10.1016/j.apen*ergy.2013.05.070*

Zhao, J., *W*ong, P. K., Xie, Z., Ma, X., & Hua, X. (2019). Design and control of an automotive variable hydrauli*c damper using cuckoo search optimized PID method. International Journ*al of Automotive Technology, 20(1), 51–63. doi:10.100712239-019-0005-z

Zheng, L., Zhu, J., Wang, G., Lu, D. D.-C., McLean, P., & He, T. (2017). Model predictive control based balancing strategy for series-connecte*d lithium- ion battery packs. In Proc. 19th Eur.* Conf. Power Electron. Appl. (EPE ECCE Eur) (pp. 1–8). IEEE. 10.23919/EPE17ECCEEurope.2017.8099189

Zheng, Y., Ouyang, M., Lu, L., Li, J., Han, X., & Xu, L. (2014). On-line equalization for lithium-ion battery packs ba*sed on charging cell voltages: Part 1. Equalization based* on remaining charging capacity estimation. Journal of Power Sources, 247, 676–686. doi:10.1016/j.jpowsour.2013.09.030

Zhia, L., Pengb, Z., Zhifua, W., Song, Q., & Yinana, R. (2017). State of Charge Estimation for Li-Ion Battery Based on Extended Kalman Filter. Energy Procedi*a, 105, 3515–3520. doi:10.1016*/j.egypro.2017.03.806

Zhou, J., Zhu, W., Zheng, Y., & Li, C. (2016). Precise equivalent model of small hydro generator cluster and its parameter identification using improved Grey Wolf opt*imiser. IET Generati*on, Transmission & Distribution, 10(9), 2108–2117. doi:10.1049/iet-gtd.2015.1141

Zhou, Z., Zhang, L., Zhu, F., Tu, W., Yuan, Z., & Gou, X. (2020). Dynamic Compensation Pricing Scheme for Demand Resources Based on Deep Re*inforcement Learning. In IEEE/IAS Industria*l *an*d Commercial Power System Asia. I&CPS Asia.

Ziebert, C., Melcher, A., Lei, B., Zhao, W., Rohde, M., & Seifert, H. J. (2017). El*ectrochemical – Thermal Characterization and Thermal Modeling for Batteries. In Emerg. Nanotechnologies Recharg. Energy Storage Syst.* Elsevier Inc. doi:10.1016/B978-0-323-42977-1/00006-6

About the Contributors

S. Angalaeswari is Faculty, School of Electrical Engineering, (VIT), Chennai. She is having 18+years of teaching experience. Got gold medal in M.E degree from Anna University; got best project award for UG level project by "The National council of Engineering – New Delhi". She has published more than 35 international journal and conference papers; she has published a book titled "Electric Circuit Analysis", patent and book chapters. She has organized more than 100 events such as workshops, seminars, faculty development Programs and conferences. she has established MSME certified Center for industrial automation laboratory worth of Rs.35 lakhs in collaboration with CDCE automation PVt. Ltd, Chennai and trained more than 30 students in this course. Done 5 consultancy projects for Daewon India Auto parts for Online Vision Inspection Poke yoke system with unlimited storage, Sekisui Dljm Molding Pvt Ltd for design a drilling Machine for 06M Rear Fender, TVS motors Hosur, Kovai Ortho and Tonglit Pvt. Ltd. Research areas renewable energy sources; integration of distributed sources into the grid; controller design optimization techniques for the radial distributed systems, etc.

* * *

Jegan G. received doctorate degree in the field of microwave Engineering in Sathyabama institute of Science and Technology, India in 2019, received bachelor of Engineering, degree in Electronics and Communication Engineering from Government college of Engineering, Bargur, Madras University, India in 2001. He completed his master degree in Sathyabama Institute of Science and Technology, India in 2006. He has totally 16 years of experience in teaching. He was working as Associate Professor in Electronics and communication department in Sathyabama Institute of Science and Technology, Chennai, India. He has published Research articles in national, international Journals and conferences.

Sivaramapandian J. is an assistant professor in Sri Venkateswara College of Engineering also a research scholar in the Department of mechanical Engineering, Sri Venkateswara College of Engineering, Anna University, Chennai, India. His areas of interests are Electrical Vehicle design, Engineering fracture mechanics, Design of machine elements, Heat Transfer.

Mohana Sundaram K. received B.E. degree in Electrical and Electronics Engineering from University of Madras in 2000, M.Tech degree in High Voltage Engineering from SASTRA University in 2002 and Ph.D. degree from Anna University, India in 2014. His research interests include Intelligent controllers, Power systems, Embedded system and Power electronics. He has completed funded project of worth Rs.30 .79 lakhs sponsored by DST, Government of India. Currently he is working as a Professor in EEE department at KPR Institute of Engineering and Technology, India. He has produced 04 Ph.D candidates under his supervision in Anna University, Chennai. He has published three books and serving as reviewer for IEEE, Springer and Elsevier journals. He is an active member of IE, ISTE and IAENG. He has published around 47 papers in international journals.

SelvaSundar K. completed his M.E. in Control and Instrumentation Engineering and B.E. in Electronics and Instrumentation Engineering from Anna University, Chennai. He is currently working as Research Associate at the Vellore Institute of Technology, Chennai.

Ravindra Kollu is a professor in the Department of Electrical and Electronics Engineering, University College of Engineering Kakinada, JNTU Kakinada, India. His areas of interest include power converters, renewable energy systems, distributed generation, evolutionary algorithms.

Sivakumar N. received his B.E in Mechanical Engineering and M.E in Engineering Design (Mechanical) in the year 2002 and 2004 respectively and PhD from Anna University in the year 2018. He started his career as Lecturer and served as Assistant Professor and Associate professor in various prestigious colleges and universities with a total teaching Experience of more than 17 years as of till date. His research interest is focused on Finite Element Method, Smart Materials, Strength of Materials, Renewable Energy Sources and requiring knowledge ofthe cloud forthe data science. He has contributed as a member in various National and International Committees. He is the active life member of Indian Society of Technical Education, International Association of Computer Science & Information Technology, and International Association of Engineers. He has more than 14 publications in reputed national and international conferences, workshops, and journals. He is associated with a various National and International journals as a reviewer. He has 7 patent publications, One Book Chapter and One Canadian copyright.

Amuthan Nallathambi received his B.E in Electrical & Electronics Engineering and M.E in Applied Electronics (EEE) in the year 1997 and 1999 respectively and PhD from Anna University in the year 2014. He started his career as Lecturer and served as Assistant Professor, Associate professor and Professor and Head of the Department in various prestigious colleges and universities with a total teaching Experience of more than two decades as of till date. His research interest is focused on Power Electronics, Energy Conservation, Auditing, Renewable Energy Sources and requiring knowledge of the cloud for the National Level Integration. He has contributed as a member in various National and International Committees. He is the active life member of Society of EMC Engineers, Solar Energy Society of India, Artificial Intelligence Community (AIC), International Association of Computer Science & Information Technology, and International Association of Engineers. He has more than 31 publications in reputed national and international conferences, workshops, and journals. He is associated with a various National and International journals as a reviewer. He is presently guiding 6 Ph.D students under VTU. He has 8 patent publications, 2 Book Chapters, 2 Canadian copyright and 2 textbooks. He has received Alibaba Cloud Most Valuable Professional (MVP) award for the year 2021&2022.

Prasanthi Nunna is working as Associate Professor in the EI Dept of SoE in UPES and has 18 years of teaching experience. Her areas of interest include Antenna Design & RF circuits, Metamaterials, IMPLANTABLES, low power VLSI circuits and EVs. Under her guidance, 1 PhD has been successfully completed. She owns two patents in women supportive models, has 21 research publications. Apart from these, conducted workshops on antenna design in HFSS, guided many student projects. She has raised labs in the EI Departments where she worked. She has worked as core member in accreditation team for the university.

Kavipriya P. obtained her Bachelor's in Engineering (BE) in the year 2000 from AVC College of Engineering, Bharathidasan University. She received her Master's degree in Applied Electronics, first class with distinction, from Sathyabama University, Chennai in 2005 and also she obtained her doctorate in the field of wireless networks and communication in the year of 2015. Her current area of interest is Wireless networks, Wireless communication, Wireless Sensor Networks and Circuit Analysis. She joined as a teaching faculty in the year 2001 and presently working as an Associate Professor in Sathyabama Institute of Science and Technology, Chennai, Tamil Nadu, India. She had published more research articles in reputed Journals and Conference Proceedings. She taught a wide variety of courses for both Under Graduate and Post Graduate levels over a period of twenty years.

Lakshmi P. received her B.E. from Government College of Technology, Coimbatore, Masters and Ph.D. degrees from College of Engineering, Guindy, Anna University, Chennai. Presently, she is working as a Professor, EEE Department, College of Engineering, Guindy, Chennai, Tamil Nadu. She published papers in many national and international conferences and national and international journals. Her areas of interest are intelligent controllers, process control and power system stability.

Ranjith Raj is an assistant professor in Sri Venkateswara College of Engineering also a research scholar in in the Department Production Engineering, National Institute of Technology, Tiruchirappalli, India. His areas of interests are 3D modelling and manufacturing, additive manufacturing, manufacturing and industrial engineering, Electrical Vehicle design.

B. V. A. N. S. S. Prabhakar Rao has obtained Bachelor of Science in Electronics, Master of Computer Applications (MCA), M. Tech. Computer Science & Technology from Andhra University, Visakhapatnam and Ph.D from Jawaharlal Nehru Technological University Kakinada. He published several papers in both national and inter-nation journals in the area of Software Engineering, Machine Learning, Soft Computing, Agriculture, Health Care, and Engineering Economics. He attended and presented several papers at both national and international conferences. He volunteered as reviewer for many books, international conferences & journals. He is a Life Member in CSTA, IACSIT, IAENG, ISCA and ISTE.

Hemavathi S., B.E., ME., PhD Scholar (AcSIR), is serving as a Scientist at the Central Electro Chemical Research Institute (CSIR-CECRI) in Chennai, and she is also an Assistant Professor in the Academy of Scientific and Innovative Research. Her qualifications include a Bachelor of Electrical and Electronic Engineering from Anna University, Trichy, and a Master of Power Systems from Annamalai University, Chidambaram. She has over two decades of experience in academics and research & innovation with the Institute of Electrical Engineering, Karpaga Vinayaga College of Engineering and Technology, Chennai. She was an Electrical Engineer in Her previous employment includes Santriya Technologies & Yashika Industries, both in Chennai. e of Power System Division, Central Power Research Institute, Bangalore, India. She has more than 20 articles in Scopus/SCI-indexed journals/ conferences to her credit and peer-reviewed reputed journals. She has delivered many guest lectures and keynote speeches on various occasions. She has authored 3 book chapters in Wiley, and Springer publications. In her research, she has focused on Sustainable, Renewable Energy, EV, Solar, and Power supplies.

Lakshmi S. is presently working as an Associate Professor in Sathyabama Institute of Science and Technology, Chennai, Tamil Nadu, India. She is keenly interested in Mobile Ad hoc Networks, Wireless Networks, Wireless Sensor Networks, Wireless Communication, VLSI Design, Micro Electronics and Network Security areas. She has published many publications to her credit in reputed International Journals and International Conference Proceedings. She taught a wide variety of courses in both Under Graduate and Post Graduate levels over a period of twenty years.

Shyam Sihare was a Ph.D. candidate. In 2003, he completed a Master's degree in computer science at Nagpur University in Nagpur, India, and finished his M. Phil. in 2005 at Madurai Kamraj University in Madurai, India. In India, he passed the GSLET (Gujarat) and MS-SET (Maharashtra) Professor Eligibility Tests in 2011 and 2018, respectively. In 2011, he earned his MCA from IGNOU in New Delhi, India. He is now employed with Dr. APJ Abdul Kalam Govt. College in Silvassa, Dadra & Nagar Haveli (UT), India, as an assistant professor of computer science and applications. Quantum computing, quantum algorithms, quantum cryptography, and classical computer algorithms are some of his areas of interest.

Rabindra Kumar Singh is working as Associate Professor in School of Computer science and Engineering at Vellore Institute of Technology, Chennai Campus since 2013. He has 20 years of Teaching Experience and 4 years of Industry experience. He has completed M.E. in Computer Science and Engineering from Anna University, Chennai in 2014. Ph.D from Vellore Institute of Technology, Chennai Campus. He is interested in teaching the subject like Operating System, Computer Network, Data Mining, Software Project Management, Distributed Computing, Python Programming, R Tools, etc. His area of research is Machine Learning, Data Mining, and Bioinformatics, etc.

Saravanan Subramani is an assistant professor in Sri Venkateswara College of Engineering also a research scholar in in the VIT University – Vellore, India. His areas of interests are IC Engines, combustion and emission, electric vehicles.

Yuvapriya T. received her B.E. degree in Electronics and Instrumentation Engineering from Thangavelu Engineering College, Chennai, M.E. Control and Instrumentation Engineering from College of Engineering, Guindy, Anna University, Chennai and Ph.D. in EEE department, College of Engineering, Guindy, Chennai, Tamil Nadu. She is working as Assistant Professor Senior Grade 1 in the School of Electrical Engineering, Vellore Institute of Technology, Vellore. She has published several articles in international conferences and journals. Her areas of interest are vibration control, control systems and optimization techniques. Corresponding author.

Chandrika V. S. received her B.E. degree in EEE from Madras University in 2001, M.E. in PED from Anna University, Chennai in 2005 and Ph.D in Electrical Engineering from Anna University, Chennai in May 2015. Started her career as lecturer in EEE in 2001 and is currently working as Professor in EEE at KPR Institute of Engineering and Technology, Coimbatore. She has a total experience of 19 years. She has published 18 papers in reputed national and international journals, 24 papers in national and international conferences and owns three patents. Currently working in the area of Intelligent Drives for Electric Vehicles. She is a Life member of ISTE.

Shanmukha Naga Raju Vonteddu is working as an Assistant Professor on a contract basis in the Department of Electrical & Electronics Engineering at Jawaharlal Nehru Technological University Kakinada, Andhra Pradesh, has about 09 years of teaching experience. He is currently pursuing his Ph.D. in Renewable Energy (Part-Time) at the University of Petroleum and Energy Studies, Deharadun, Uttarakhand, India. He received his B.E degree in Electrical and Electronics Engineering from Anna University, Chennai and M.Tech. degree in Power Systems (High Voltage) Engineering with distinction from Jawaharlal Nehru Technological University Kakinada. His research interests are Electric Vehicles, Intelligent Transportation Systems, Energy Storage, Renewable Energy, Artificial Intelligence Techniques, IoT. He is a member of IEEE; IEEE Intelligent Transportation Systems Society; IEEE Power and Energy Society; IEEE Vehicular Technology Society.

Index

www.ingramcontent.com/pod-product-compliance
Lightning Source LLC
LaVergne TN
LVHW062135300926
849161LV00010B/2537
9781668466315